STRATHCLYDE UNIVERSITY LIBRARY

30125 00079174 8

KU-345-675

£12·75

APPLIED GEOCHEMISTRY

APPLIED GEOCHEMISTRY

FREDERIC R. SIEGEL

The George Washington University

A WILEY-INTERSCIENCE PUBLICATION

John Wiley & Sons

New York • London • Sydney • Toronto

Copyright © 1974 by John Wiley & Sons, Inc.

All rights reserved. Published simultaneously in Canada.

No part of this book may be reproduced by any means, nor
transmitted, nor translated into a machine language without the written permission of the publisher.

Library of Congress Cataloging in Publication Data:

Siegel Frederic R.
Applied Geochemistry.

"A Wiley-Interscience publication."
Bibliography: p.
1. Geochemistry.   2. Geochemical prospecting.

I. Title.
QE515.S49   622'.1   74-13486
ISBN 0-471-79095-8

Printed in the United States of America

10 9 8 7 6 5 4 3 2 1

D

622·1

SIE

DEDICATION

This book is the result of education, environment, and experience, and because of this cannot be dedicated to a single person. The education is due to my parents, Eva and Louis Siegel, who gave me the opportunity to study, to my professors at Harvard University and The University of Kansas, and to my colleagues and students with whom I have always had the opportunity to exchange ideas. The environment is due to my wife, Felisa M. Puszkin de Siegel, and to my daughters Gabriela and Galia Siegel; a happy home and an intelligent environment is a home where one may work well; those who know my family know what environment is. Experience comes from where I worked, The Instituto Miguel Lillo, Universidad Nacional de Tucuman, Tucuman, Argentina, The State Geological Survey of Kansas, and The George Washington University.

To all those mentioned and to those not mentioned because they are so numerous, I dedicate this manuscript.

ANDERSONIAN LIBRARY
★
WITHDRAWN FROM LIBRARY STOCK
★
UNIVERSITY OF STRATHCLYDE

PREFACE

Water, food, fibers, construction materials, energy resources, natural resources for industry, and capacity for the disposal of wastes are all necessary to man's survival on the earth. Applied geochemistry is concerned with these needs. It can assist man to survive and, hopefully, to increase his understanding and appreciation of the finite limits and resources of our solid planet and its liquid and gaseous spheres of life control.

More specifically, applied geochemistry is concerned with prospecting for natural resources (metals and nonmetals), including fossil fuels (petroleum and natural gas) and sources of nuclear fuel. The exploration geochemist uses the principles and concepts of both geology and chemistry to locate mineral deposits within large areas. In addition, the extension of geochemistry into floral and faunal studies (biogeochemistry) and into gaseous or atmospheric analysis further aids the exploration geochemist in his search. Finally, the geochemist can use statistical analysis to provide additional decision-making data for the selection of target zones, and for drilling out and/or outlining the limits of a prospect.

As an adjunct to using geochemical data for indicating mineralized areas, the geochemist is able to contribute to other fields of research, such as health and pollution control. He is aided by his experience, as well as compiled geological-geochemical data, and his understanding of factual ideas as well as conceptual projections. Conversely, the applied geochemist can utilize data generated by health, pollution, and agricultural programs in the search for economic mineral deposits.

Absolute age dating of geological and other materials is an established yet developing branch of geochemistry. Several innovative techniques, now considered to be only theoretical even when apparently good dates have been obtained with them, are described in the final section of the book. Although they may seem too abstract to many geochronologists to warrant such coverage, it should be remembered that the same criticism was applied to the continental drift concept of Wegener. Nevertheless, we see now that Wegener's concept has gained acceptance and is now viewed as one of the most exciting ideas enveloping the earth sciences. Older theories with modern approaches, mechanisms, and proofs may be used in mineral prospecting. Although not treated in this book, continental drift and sea-floor spreading and the comprehension of lithosphere plate tectonics are being evaluated by geochemists to discover possible metalliferous and hydrocarbon provinces. Once designated, these areas can be explored geochemically to find exploitable deposits within the major provinces.

No single directive or set of rules can be used in the mineral exploration phase of applied geochemistry. Indeed, the geochemist may play only a follow-up role in locating ore deposits. In Finland, for example, follow-up work on boulders (in glacial terrain) found by the public which had been made boulder-conscious by state geologists-geochemists, reportedly resulted in the discovery of 10 mines. Each study area provides its own unique set of geologic and other conditions. Hence each applied geochemistry project, whether in geochemical prospecting, health, or pollution, must be evaluated on the basis of this uniqueness using, whenever possible, the experience acquired in studies of similar and dissimilar geoenvironments. Thus a knowledge of the varying conditions that may influence geochemical distributions, and an appreciation of the limitations of interpretations imposed by the data available are fundamental. This book presents some of the ideas developed during many years and reported by geologists and geochemists in various parts of the world. Such experience is of extraordinary value in applied geochemistry projects.

FREDERIC R. SIEGEL

Washington, D.C.
August 1974

CONTENTS

APPLIED GEOCHEMISTRY

1

INTRODUCTION

Classically, geochemistry is concerned with the physical-chemical principles that influence and/or control the fractionation, migration, deposition, and distribution of the chemical elements (and their isotopes) in the various rock classes that comprise the Earth's crust. Laboratory studies on the physical-chemical relations of rocks under simulated subcrustal high temperatures and pressures, together with data from meteorite investigations, earthquake waves and artificially induced seismic waves, spectra of celestial bodies, and astronomical observations and measurements, allow the geoscientist to indirectly study the interior of the Earth and propose reasonable interpretations on the physical nature and the chemistry of the interior and on the chemical processes that have been or are active there.

The cosmos has been of immense interest to geologists who have been constantly involved in controversies concerning the origin of the universe and the solar system and the genesis of the chemical elements that compose our galaxy. Research on atomic energy and nuclear atomic physics has provided us with both the theory and instrumentation to work effectively on these problems. When the Soviet Union put Sputnik I into Earth orbit in November, 1957, a new era began in which the stimuli towards studies on the chemistry of the solar system were greatly amplified and intensified. Less than 10 years after the Russian feat, the United States (September, 1967) soft-landed the Surveyor V satellite with its alpha-scattering experiment on the moon and provided Earth scientists with the first direct measurements on the chemical composition of the lunar surface material. One month later, the Soviet satellite Venus IV alighted on the surface of Venus and gave world scientists important information on the composition of atmosphere of Venus and on the temperature changes through its atmosphere to the planetary surface. Other descents and soft-landings have occurred since then as well as the notable Mars and Mercury overfly missions, but nothing captured and maintained the attention and admiration of human beings so much as the first landing of a manned vehicle on the lunar surface, the descent onto that surface of two North American astronauts, and their subsequent return to Earth with 30 kg of moon rock and valuable personal observations concerning the lunar surface. The seismometer that these two men left behind continued generating important information on the movements within the moon well after their return to Earth. The measurements now being made on the lunar rock samples from the first Apollo mission and later Apollo and Luna missions and the analyses being made in more than 140 specially prepared world-wide laboratories are allowing scientists to better their astrochemical and geochemical (or selenochemical) concepts and develop others.

The application of geochemical principles and data to the resolution of immediate Earth problems, such as the finding of natural resources and the conservation of health and the environment, has been extremely important during the past decade. This has been especially true in prospecting for deposits of metals, nonmetals, and hydrocarbons, for both the localization of potential economic deposits and the extension of known deposits. Derry (1971) has briefly considered geochemistry as a link between ore genesis studies and exploration. He wrote that whether processes of formation or disintegration and decomposition of ore bodies close to the Earth's surface or at a depth of 20 km in the Earth's crust are considered, the action process is mobilization or remobilization of the elements. Thus, the more learned about the migration and concentration of metals and nonmetals, the more effectively can the exploration geologist use this knowledge in the search for hidden element concentrations or accumulations close to or at depth below the Earth's surface. Mitcham noted that for many years preceding his 1967 report, increasing importance in applied geology (and geochemistry) had been given to laboratory geology at the expense of field work that the laboratory techniques were to serve; further, he believes, as do many of the workers and teachers today, that in preparation for exploration geochemistry or economic geology more emphasis must be put on reasoning and the logical development of hypotheses than on methodology developed in the laboratory. Mineral exploration can be carried out effectively when good maps are available and exploration parameters are well defined (based on critical geological observations, a knowledge of the region being investigated, and an understanding of the probable ore type and structure being targeted and its origin). Mitcham tended towards increased quantification of mineral exploration data, but realized via experience that quantification is not an end to itself and must be used only as an adjunct to other geological knowledge and as such could better the geoscientists chances of recognizing the possible presence of a hidden ore deposit.

The need to bring together specialists working in geochemical prospecting in all parts of the world and in many and varied environments was thought to be essential so that one could learn from another, experiences could be shared, and communication could be established. This need was satisfied by having International Geochemical Exploration Symposia beginning in Ottawa, Canada, in

1966 and being held every other year thereafter. Results of the symposia have been outstanding; in 1970 The Association of Exploration Geochemists was chartered, and in 1972 the Association-sponsored *Journal of Exploration Geochemistry* commenced publication with the aim of covering all aspects of the application of geochemistry to the search for mineral deposits, including hydrocarbons. The Association also sponsored the publication of an *Exploration Geochemistry Bibliography* (under the direction of H. E. Hawkes covering January, 1965 to December, 1971 and will sponsor annual additions to the bibliography. In 1966 the first symposium on Trace Elements in Environmental Health was held at the University of Missouri and symposia on the same topic have continued annually and the proceedings published. The response to these meetings was so enthusiastic that by 1968 The Society for Environmental Geochemistry and Health was formed and is now publishing a newsletter that keeps Society members aware of new findings and advances in the field.

The present book has been written with the aim of providing the student of geochemistry with a compendium of rather recent data on applied geochemistry that the author feels are most important to consider in one's self-preparation for carrying forth a geochemistry program whose results can be confidently used by all geoscientists working on a given topic. The book may be considered as an adjunct to the few texts that treat the field of applied geochemistry, most notably, *Geochemistry in Mineral Exploration* (Hawkes and Webb, 1962) and *Principles of Geochemical Prospecting* (Ginzburg, 1960).

In order to discuss geochemical (as much for applied geochemistry as for academic geochemistry), astrochemical, and now selenochemical processes, it is necessary to understand the nature of matter beyond the fact that it is composed of solids, liquids, and gases; that is, we must have a good understanding of the elementary particles that comprise matter. The so-called elementary particles are generally considered as the proton, the neutron, and the electron. Such a designation is not "elementary" for all scientific disciplines. As Weiskopf (1967) emphasized, scientists identify with elementary particles by means of energy data. For example, chemists work with atoms and their energies of 1 to 10 eV, atomic physicists work with atomic nuclei and electrons (1 to 100 eV), nuclear physicists work with nucleons, electrons, and neutrinos (10^5 to 10^7 eV), and high-energy physicists work with nucleons in excited states, light and heavy electrons, neutrinos, mesons, quarks, and so on, with energies in the range of 10^7 to 10^{10} eV.

As geochemists, we are very concerned with nuclear processes because of our need to understand and measure geologic time. However, more important to the geochemist may be the understanding of reaction mechanisms that are ultimately dictated by the number and disposition of electrons around an atomic nucleus, an understanding attributed to the reader of this text. In this way, we achieve the capability of explaining why different elements are found together in various rock types, predicting which elements should be found together in an area of mineralization and evaluating the significance of analyses showing different element associations in geochemical samples.

REFERENCES

Derry, D. R., 1971, Geochemistry – the link between ore genesis and exploration, in *Geochemical Exploration* Special Vol. No. 11, (Boyle, tech. ed.), Canadian Institute of Mining and Metallurgy, pp. 1–4.

Ginzburg, I. I., 1960, *Principles of Geochemical Prospecting,* Pergamon, New York, 311 pp. (translated from the original Russian, 1957, edition).

Hawkes, H. E., and Webb, J. S., 1962, *Geochemistry in Mineral Exploration,* Harper & Row, New York, 415 pp.

Hawkes, H. E. (compiler), 1971, *Exploration Geochemistry Bibliography,* Period January 1965 to December 1971, Special Vol. No. 1, The Association of Exploration Geochemists, Toronto, 118 pp.

Mitcham, T. W., 1967, Emphasis, *Econ. Geol.,* **62,** 421–428.

Weiskopf, V., 1967, Third Pasteur Lecture, Georgetown University, Washington, D. C., September 28, 1967.

2

THE SIGNIFICANCE OF THE GEOCHEMICAL PROSPECTING ANALYSIS WITH COMMENTS ON EQUIPMENT FOR ANALYSIS

A geochemical prospecting program can be divided into various phases each one of which is based on others in the idealized sequence given below:

1. Knowledge of the regional and/or local geology, personal experience in the study zone, and existent bibliography.
2. Geologic observations in the field (with the aid of geologic-topographic maps and air photos where they are available.
3. Pilot study of an area (or areas) representative of the zone of interest in order to determine the best sample (or samples) to be used in the zone.
4. Collection and field description of the geochemical samples and geologic-geomorphic description of each sampling site.
5. Qualitative-semiquantitative analyses in the field (especially where access is difficult).
6. Quantitative analyses in the laboratory.
7. Statistical determination of regional and local background values and of the so-called anomaly.
8. Presentation of the results.
9. Interpretation of the data.
10. Evaluation of the prospect.

The significance of the geochemical analysis depends on many factors, some of which are objective and controllable statistically and others which are subjective and controllable only by personal experience, effective proper planning of the prospecting program, and communication in all the project stages between the geologist and geochemist in the laboratory.

In this Chapter, the influence of the field sampling and laboratory analyses on the significance of and confidence in the results will be considered. All phases of the field–laboratory operation must be methodically regulated by constancy because this constancy is the factor of utmost importance in exploration geochemistry. Because of this, the director of a project in geochemical prospecting should have experience in both the field work and the laboratory work in order to maintain a strict control over the methodology employed; the more people in the program who have such experience, the better the program will be carried to completion.

Before considering the field sampling and the laboratory geochemical analysis, one must define briefly what the geochemical sample represents areally, a topic that will be developed more completely in another chapter. The sample of an active stream sediment provides us with a representation of the rocks (upstream) in the drainage basin. Thus, the type of geochemical sample used in the first stages of a regional study is taken from the principle rivers and the affluent streams; in this way it can be determined if zones of economic-mineralogic interest exist in areas of hundreds of square meters to hundreds and thousands of square kilometers. Once a zone of possible anomalies is localized, the sampling of stream sediments can be intensified or detailed by shortening the distance between samples until a zone of maximum interest is localized in a relatively large area. Generally, this step is followed by a geochemical sampling of the residual soil of the area carried out on a well-defined (by the probable size and shape of the target) grid or pattern when it is possible, because the geochemical value that comes from a residual soil represents only the area of the actual sample, that is, some hundreds of square centimeters. For some time Russian exploration geochemists have been using the flora of target areas to localize mineralization. This is because the flora, for example, trees, provide a geochemical sample of the entire subsurface reached by root system, and instead of having a sample representation of some hundreds of square centimeters (soil), one may have a subsurface representation of several square meters. It is understood then that the different kinds of geochemical samples, three of which have been mentioned, provide geochemical data representative of very distinct areas.

In a regional study for which active stream sediments are selected as the geochemical sample (often used in general regional reconnaissances), one can make an analysis on the total sample or on a size fraction of the sample. In this latter situation the size fractions most frequently used are −80 or −100 mesh, but in each study the fraction selected is that which will provide the most significant contrast between the background values and the values of the samples.

It must be noted and emphasized that the designation −80 mesh or −100 mesh size fraction is often not qualified by the type sieve being used, this being in the opinion of many a serious omission that must be corrected in future publications. In every case the sample used in a project must be equal for all the analyses in order to have data that are worthy of comparison. The same constancy must be followed with any class of sample chosen for a study; in soils, not only must the size fraction to be used for the analysis be selected, but also the soil horizon (OA, A_1, A_2, B, or C) from which the sample will be taken; of course, a channel sample should be taken, or if a spot sample is used, it should come from the same part of the horizon elected for the sampling. For water samples, the analysis can be made on the total sample or on the filtered sample, but in this case the type of container that is used to transport the water must be considered as well as the time lapse before making the analysis. It is known that samples of marine waters or lake waters can exchange ions with those of glass containers (with more exchange per unit of time) and thus alter the result of the analysis; with polyethylene containers the cations may adhere to the polyethylene and analyses on the liquid can be imperfect. Botanogeochemical samples may be leaves, twigs, bark, wood, stems, roots, or needles. Leaves may be compared with leaves, or twigs with twigs of the same species, but leaves cannot be compared with twigs nor with similar samples from distinct species; nor can new twigs and twigs from a previous growth period be compared. Thus it should be clear that the field geoscientists must receive very precise instructions from the director of a program before initiating their field work.

Also to be considered is the fact that there are almost as many methods to prepare a sample and analyze it as there are different types of samples. One cannot send samples to a geochemical laboratory and simply ask for an analysis of a given element or assemblage of elements. The sample preparation procedure to be used, the extraction technique, the analytical method, and, of course, the type of sample (silicate, carbonate, sulfide, etc.) must be indicated. For example, for a black sand from the Balneario del Norte of Rio Negro, Argentina, a geochemical analysis of the magnetic and nonmagnetic fractions of the complete sample via emission spectrography was requested; if an analysis for zinc alone had been sought, the resulting report would have shown 1000 ppm of Zn in the magnetic fraction and < 200 ppm of Zn in the nonmagnetic fraction. But because a general emission spectrograph analysis was required, the resulting report showed interesting data for various elements (see Table 2-1). Other analytical techniques could result in different values due to the sensitivity limits and the accuracy and precision of each instrumental or wet chemical technique.

Table 2-2 shows clearly the large variations that can result from different methods of preparation and extraction of the metal being sought (in this case nickel) from the sample although the instrumental analytical technique was the same for all the geochemical samples (atomic absorption). Table 2-2 is self-

Table 2-1   Geochemical Analysis of a Sample of Black Sand from the Balneario del Norte of Rio Negro, Argentina, by Emission Spectrography (author's data courtesy of Geochemical Surveys, Inc.)

Element	Magnetic Fraction (ppm)	Nonmagnetic Fraction (ppm)
Ba	50	75
B	50	20
Au	<10	<10
Pb	15	35
Bi	<10	<10
Ga	100	20
Mn	1,500	1,000
Cr	100	150
Ni	200	150
V	750	75
Be	<1	<1
Ti	>10,000	>10,000
Mo	10	2
Sn	<10	<10
Y	100	100
Cd	<50	<50
Cu	75	75
Zr	50	500
Ag	2	2
La	350	150
Zn	1,000	<200
Sc	15	75
Co	75	50
Sr	50	50
As	<200	<200
Sb	<100	<100
W	<50	<50
Fe	100,000	200,000

explanatory and it should be very obvious from the table that the trituration or grinding of the total sample to the −100 mesh fraction resulted in a dilution of the fraction with nickeliferous magnetite and the minimum values for nickel were obtained; on the other hand, the selection of the natural −100 mesh fraction resulted in a great concentration of nickeliferous magnetite with respect to the other sample components and the maximum value for nickel was obtained. This type of test can be made in all geochemical laboratories to demonstrate to the field geochemist the importance of his designating clearly for the laboratory geochemist the preparation and treatment a sample should

Table 2-2 Comparison of Six Analyses of the same Stream Sediment for Nickel Using the Atomic Absorption Technique on Aliquots Prepared or Extracted in Different Ways (from Bondar, 1969)

	Preparation		Chemical	Content of
	Ground	Fraction	Attack	Nickel (ppm)
Laboratory A	No	− 80 mesh	70% $HClO_4$	20
Laboratory B	No	− 80 mesh	1:3 HNO_3	60
Laboratory C	No	− 80 mesh	HNO_3−HCl mixed acid	150
Laboratory D	No	− 80 mesh	1:1 HCl	320
Laboratory E	Yes	−100 mesh	1:1 HCl	14
Laboratory F	No	−100 mesh	1:1 HCl	1120

Sample description:

Stream sediment with 0.5% magnetite

Magnetite contains an average of 0.28% Ni

80% of the magnetite is in the −100 mesh size fraction

99% of the sample falls in the +100 mesh size fraction and
    96% is in the +80 mesh size fraction

Source. Reprinted from *Colorado School of Mines Quarterly*, **64**, No. 1, by permission of the Colorado School of Mines. Copyright © 1969 by the Colorado School of Mines.

receive in order to obtain results that are good (accurate and precise) instead of results with differences of up to 8000% as shown in Table 2-2.

To avoid problems in the analytical chemistry phase of a program, the director of a geochemical prospecting project should establish an open communication with the director of the geochemical laboratory. The analytical geochemist must study the sample type or the range of sample types with which he is going to deal. In this way he or she can select the best preparation and extraction technique with respect to the instrumentation or wet chemical method to be used so that the sensitivity limits are sufficient to detect the minimum background values, and indicate the quantity (by weight and volume) of sample necessary for proper laboratory work. Thereafter, the analyst will provide results that are the most accurate and precise possible (and the interferences at a minimum) within the limits of the analytical technique followed. In addition, the laboratory geochemist has the experience that will permit changes in the routine analytical method in order to treat different problems that may arise since a quantitative analysis flow sheet cannot always be used as a cookbook. Also, after a consideration of the problem to be undertaken, the analyst–geochemist can ascertain the analytical technique that will result in the maximum efficiency and best data for the minimum cost and time per analysis, a factor of significant importance in all economic geochemistry projects.

The daily routine and work in a geochemical analytical laboratory can be very boring, and because of this, it is desirable to present the analysts with a problem and explain the entire project to them (instead of merely sending in groups of samples to be analyzed). This is important because a knowledge of the project and of the elemental concentration values that may be indicative of anomalies, for example, will often result in a more rapid communication to the field team and assist in daily or weekly planning (or modifications) in the prospecting program. The communication between colleagues working on a project will put emphasis on the phases of special and potential interest, and thus the decisions on the continuation or termination of the geochemical prospecting in an area or a change in emphasis of the project can be made with a degree of confidence that might "approach" chemical precision.

Obviously, it is of great importance to realize that when the geologic conditions of a study zone change and there is a change in characteristics of the samples, this fact must be communicated immediately to the geochemical laboratory. Each mineral assemblage or individual mineral has its own minerochemical properties and may have to be considered differently in the laboratory with respect to the preparation, chemical attack, analysis, and potential interferences. If laboratory conditions change or if analysts change, this fact must be communicated to the geologist–geochemist in the field in order that he might recognize that a difference in values from one group of samples to another may be the result of changes in the laboratory conditions, technique, or personnel. This is especially necessary in areas where a minimum contrast exists between background values and possible anomalous values; in such a situation, the accuracy and precision must be the best possible.

Of the errors that tend to hide subtle anomalies which can be very important in exploration geochemistry, the two already considered can be quantified using statistical methods: (1) the error due to the sampling itself and (2) the error due to the chemical analysis.

Galle (1964) demonstrated very well the large errors that can come about due to two methods used to collect a sample and to two distinct sample preparations before the chemical analysis. He collected a channel-type sample and a spot-type sample from each of five limestone outcrops in Kansas, and after the analyses found a real average percent deviation for major and minor components of 0.01 to $\sim 3.5\%$. The largest relative errors were in the minor components. Also, Galle analyzed unmixed (unhomogenized) aliquots of the samples in triplicate and found great differences between the values of unmixed triplicate samples from each outcrop and, of course, the same differences when he compared the analysis of each nonmixed sample with its corresponding mixed channel or spot sample. Figure 1 (Galle, 1964) illustrates his results graphically. Quantitative analytical chemistry texts give as the permissible deviation limit that can exist between duplicate or triplicate samples ±0.2%, a figure generally not achieved in the cited study. Parker (1967) and Canney (1969) wrote that the analytical values for the rock standards G-1 and W-1 varied significantly between laboratories using the same analytical techniques, especially with respect to the elements present in quantities of parts per million.

Garrett (1969) proposed a simple method for evaluating the usefulness of geochemical samples from an area and determined if subtle anomalies are due to the errors already mentioned or if they are really errors. In order to make this type of evaluation, he wrote that at first one must do duplicate sampling a few feet (60 to 90 cm) from the principal sample and make analyses corresponding to more than 30 samples or 10% of all the samples from the reconnaissance area. With the resulting data one can calculate the quantities to be used to determine the variance ratio. The combined variance due to the sampling and the analysis is calculated with the following equation:

$$\sigma_{SA}{}^2 = \frac{1}{N} \sum_{i=1}^{N} \frac{(X_{1i} - X_{2i})^2}{2}$$

in which $\sigma_{SA}{}^2$ = the combined variance due to the sampling and the analysis
X_{1i} = the logarithm of the value of the principal sample
X_{2i} = the logarithm of the value of the duplicate sample
N = the number of duplicate samples
The complete variation of the data is calculated by

$$\sigma_D{}^2 = \frac{1}{N-1} \sum_{i-1}^{N} (X_{1i} - \overline{X}_1)^2 = \frac{1}{N} \left[\sum_{i-1}^{N} X_{1i}^2 - \frac{(\sum_{i-1} X_{1i})^2}{N} \right]$$

in which $\sigma_D{}^2$ = the variance of the data
$\overline{X}_1$ = the median value of the logarithms of the N duplications
Now one can use the two variances calculated by the above equations to make an analysis of variance to determine if $\sigma_{SA}{}^2$ is significantly less than $\sigma_D{}^2$, by means of the variance ratio F:

$$F = \frac{\sigma_D{}^2}{\sigma_{SA}{}^2}$$

In general, in order that the sampling and analysis errors be considered minimum at the 95% confidence level, this ratio should be greater than 4.0 (for 30 duplications). Logically the F value will be less when the number of duplications is increased (3.8 for 120 duplications).

Garrett (1969) used this same technique of duplication analysis to determine the analytical precision with 10% of the samples (but between 30 and 100) selected to represent the range of values and reanalyzed; then the precision at the 95% confidence level is given by the equation

$$\text{Precision} = \frac{1.98 \, \sigma_A}{\overline{X}} \times 100\%$$

in which $\overline{X}$ = the median value of the logarithm values of the duplicate analyses.

Michie (1973) found that the test used by Garrett (1969) to investigate the magnitude of the sampling and analytical errors is incorrect. He indicated that there was a serious problem in the choice of the correct degrees of freedom for the analysis of variance and pointed out that if the statistically correct degrees of freedom are used, the value for F of 1.74 for 30 pairs would be obtained instead of the F value of 4.0 given by Garrett. Michie states further that error variance can be partitioned into analytical and site sampling components by a method similar to that presented by Howarth and Lowenstein (1971). Garrett (1973) accepted the critique advanced by Michie (1973), but emphasized that he used a conceptual approach in assessing sampling and analytical errors in geochemical prospecting. This approach and method, Garrett believes, could then be used by unskilled field sampling teams in regional reconnaissance projects where well-designed hierarchical sampling plans are unlikely to be used, so that the proposed method might be a worthwhile field compromise.

Both the care in determining the representativeness of samples and proper selection of subsamples, as well as the precision and accuracy (as well as limitations) of the chemical analyses of these samples, are basic to the end aim of all geochemical programs whether they be academic or applied in character, that is, the logical interpretation of the meaning of the accumulated data in terms of the project being carried out. Such interpretation derives not only from an evaluation of the numerical facts elaborated during a program, but also from experience obtained during the course of one's own work and that of one's colleagues who are gracious enough to share their experiences via publications and lectures.

As just noted, the significance of the geo-(chemical) analysis depends very much on the communication that exists between the geologist or geochemist who does the sampling (and observing) in the field and the chemist or geochemist who does the analyses in the laboratory. If good communication is established during the planning of a project, the element analyses will be very meaningful and the accuracy and precision of the results will depend on the analyst and the method or methods used in the laboratory.

Generally all geochemical laboratories have capabilities for wet chemical and colorimetric analyses and with these capabilities can work well with very few extraordinary expenditures. In addition, colorimetric methods can be adapted for use in the field, sacrificing, of course, the accuracy and precision of the results, in order to find anomalies with respect to the average (or other limit) of chemical values established in the field itself. Colorimetry kits for use outside the laboratory are being used with marked success.

However, the instrumentation branch of chemistry and geochemistry has evolved tremendously during the past 15 to 20 years, and now many laboratories have sophisticated colorimetry, flame photometry, chromatography, fluorimetry, emission spectrography, and atomic absorption spectrometry available. The two most recent and most important instrumentation advances for geochemical prospecting are, in this writer's opinion, atomic absorption spectrometry and direct-reading emission spectrometry. Methods of analysis based on neutron activation should be mentioned as well as analysis with the electron microprobe because these allow chemical determinations of very small amounts of material or of small areas (to a few square microns for the better probes); unfortunately, these techniques are not routinely available in most laboratories and, in general, will not provide more significant data for geochemical prospecting programs (especially within the limits of time and economics) than the other techniques cited.

Because of its simplicity of operation, its sensitivity (parts per billion to parts per million), the accuracy of data, especially for the metallic elements that require separation into aliquots and sometimes complicated preparations in colorimetric analysis, the rapidity once the samples have been prepared for analyses, and the multielemental capacity with the same preparation, atomic absorption equipment (which has been used as a functional method for about a decade) is the most useful and the most utilized today in geochemical laboratories. Figure 2-1 shows schematically the principles of atomic absorption spectrometry, but for two of the better expositions on the subject the reader is referred to the work of Kahn (1966; 1972) or one of the several books that have been published in recent years on atomic absorption techniques. Many investigations are being carried out on techniques related to identifying and eliminating interferences and to sample preparation. With respect to the latter, Medlin, Suhr, and Bodkin (1969) published their results on the analysis of silicates after preparation of the sample using a fusion with anhydrous lithium metaborate, a method that gave very good results when comparisons of data with published rock standards data were made (Table 2-3). In addition, it should be noted that the United States Geological Survey has already mounted an atomic absorption unit in a truck that tows its own generating unit and has thus mobilized the geochemical laboratory for field operations; these operations have been most successful. The United States Geological Survey has its

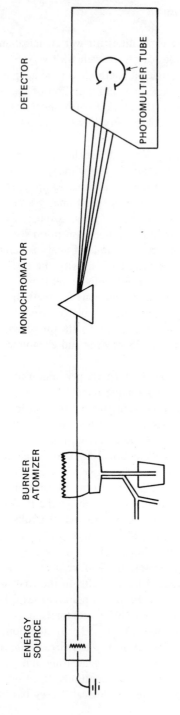

ENERGY
SOURCE

BURNER
ATOMIZER

MONOCHROMATOR

DETECTOR

PHOTOMULTIER TUBE

The energy source is a
hollow cathode or vapor
discharge lamp with the
cathode made of the
element being determined.
Because the emission beam
is characteristic only of
the metal being analyzed,
it is absorbed only by
the atoms of that metal
and, according to theory,
other metals do not
cause interference.

The sample is aspirated
into a vaporizing burner
and a part of the light
that passes through the
flame is absorbed by the
element being determined.

The light that remains
passes to a monochromator
which, by prism or grating,
isolates the analytical
wavelength for the metal
being determined.

The quantity of the
characteristic emission
wavelength that is
absorbed in the flame
is measured by means of
a photomultiplier tube.
The amount of absorption
corresponds directly to
the concentration of the
element being determined.

Figure 2-1. Atomic absorption spectrometry. For more than 100 years it has been known that atoms of some elements are excited when they are vaporized and aspirated into a flame. When these atoms return to their minimum energy state, they emit a characteristic radiation that can be measured. However, the majority of the elements are not easily excited in a flame and the major part of their atoms remain in the minimum energy state. These unexcited atoms can absorb the energy of a beam of light of the same characteristic wavelength. This principle is the basis for atomic absorption spectrometry. (Adapted from photo W-5-D of the U. S. Geological Survey.)

Table 2-3 Comparison of Cu and Zn Values in Standard Silicates Prepared by Fusion with Anhydrous Lithium Metaborate with Values Obtained from the Same Silicates Prepared Using other Techniques (modified from Medlin, Suhr, and Bodkin, 1969)

	This Study Atomic Absorption	Other Studies	Atomic Absorption with Corrections[a]	
Sample – Cu				
Granite, G-1	14	11		
Diabase, W-1	106	110	117.9	113.7
Peridotite, PCC-1	5	10		
Basalt, BCR-1	30	27	19.8	19.3
Granodiorite, GSP-1	42	45	33.8	31.6
Andesite, AGV-1	52	69	61.5	61.0
Granite, G-2	22	11	11.8	11.4
Granite, GH	12	12		
Granite, GR	420	370		
Basalt, BR	70	70		
Tonalite, T-1	50	47		
Syenite, S-1	20	24		
Sample – Zn				
Granite, G-1	54	45		
Diabase, W-1	88	82	97.6	90.3
Peridotite, PCC-1	60			
Basalt, BCR-1	130		147.5	141.7
Granodiorite, GSP-1	106		112.6	108.9
Andesite, AGV-1	118		100.6	95.8
Granite, G-2	88		92.7	89.0
Granite, GH	88	60, 97		
Granite, GR	64	56		
Basalt, BR	150	150		
Tonalite, T-1	164	190		
Syenite, S-1	216	193		

For references to the other studies, see Table III of the study by Medlin, Suhr, and Bodkin (1969).
[a] From Fletcher (1970), left-hand column is uncorrected value, right-hand column is corrected value.

own publication describing standard operating procedures in atomic absorption analyses (Ward et al., 1969).

Where it is possible, the emission spectrograph is used in the first stage of geochemical analysis because with a single arcing or burn of a sample one can obtain a film or photographic plate representation of the spectrum that can be

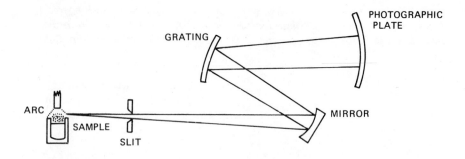

When a sample of material is correctly excited by an electric arc or a spark, each element in the sample emits light of a characteristic wavelength. The light enters the spectrograph via a very narrow opening and falls on a diffraction grating. The grating separates the reflected light of each wavelength by a different angle. The dispersed light is focused and registered on a photographic film or photographic plate in form of lines of the spectrum.

The comparator—densitometer is used to measure the intensity, or darkness, of the spectrum lines registered on the spectrograph photographic plate. Using standard films or plates for each element, the spectrograph markings are changed to per cent concentration. For the visual estimate, a special screen permits a comparison of the spectrum of the sample with the spectra of standards containing known element concentrations.

Figure 2-2. Emission spectrography. Emission spectrography is a method that allows the determination of major, minor, and trace elements in many materials. Approximately 70 elements can be determined in rocks and other geological materials. (Adapted from photo G-24-D of the U. S. Geological Survey.)

used to identify and/or quantify a majority of elements. Most often, geochemical prospecting can be carried out effectively with a knowledge of the presence of less than the 15 to 30 elements routinely analyzed with the emission spectrograph. However, this analytical technique is comparatively slow because after the preparation and the burn of a sample, development of the film or photographic plate under controlled conditions is required; this is followed by a measurement of the spectrum lines with a comparator–densitometer (microphotometer) and a comparison of the spectral line densities of a sample with standards in order to determine the quantities of elements present. Figure 2-2 shows graphically the principles related to the emission spectrograph. Although the experienced spectrographer can provide an accurate and precise analysis, such an analysis takes time, in some cases more and in some cases less than that required for other analytical techniques. In order to improve upon the effectiveness and efficiency of the emission spectrograph method, Cruft and Giles (1967) described a modification of the spectrographic method in which films or photographic plates were not used, but rather calibrated photomultiplier tubes were set in their place on the focal curve of the equipment so as to receive and store electronic signals from specific parts of the spectrum during the burn of a sample; when the burn terminates, the stored signals are emitted, in sequence, to an electronic system that measures the intensities of the spectral lines and presents the quantitative results by means of a typewriter hookup (Figure 2-3), or the results can be transferred directly onto computer cards or magnetic tape for use with computers. This equipment has been given the designation of direct-reading emission spectrometer. It is more rapid than the emission spectrograph per se, more accurate, and more precise, and it does not require a specialist to run it as is the case with the spectrograph. A complete analysis for up to 30 elements takes about 6 minutes. The cost of this equipment is more than double that of atomic absorption equipment, but if cost per analysis per day is considered, it might be found worthwhile to invest in the direct-reading emission spectrometer (if there are technical personnel readily available to maintain and repair it). Although the United States Geological Survey has made the top-quality emission spectrograph mobile, it has not yet done so with the direct-reader unit; however, it would be physically easier to mount the direct reader in a truck or laboratory trailer, but this would require a very large investment. A direct-reading emission spectrometer with capacity for 15 elements can be installed in a permanent laboratory for approximately $25,000, and for 30 elements for approximately $35,000, but to mobilize the unit would require about twice the investment.

In his consideration of the modern geochemical laboratory, Debnam (1969) made some comparisons of the efficiency in analyses per unit time via colorimetry, atomic absorption, emission spectrography, and direct-reading emission spectrometry. He estimated that with the direct-reading emission spectrometer four persons could analyze 200 samples for 15 elements each in 8 hours, for a total of 3000 analyses. To do the same with emission spectrography would require seven persons, of whom two should be trained technicians, plus complementary equipment (dark room, microphotometer–comparator). An atomic absorption laboratory would need eight persons with two atomic absorption units to do the same. And finally, a colorimetry laboratory would require more than 20 persons to make the 3000 analyses daily. It is obvious then not only that the direct-reading emission spectrometer is more efficient than the other methods, but also that it would serve best in centers for regional geo-

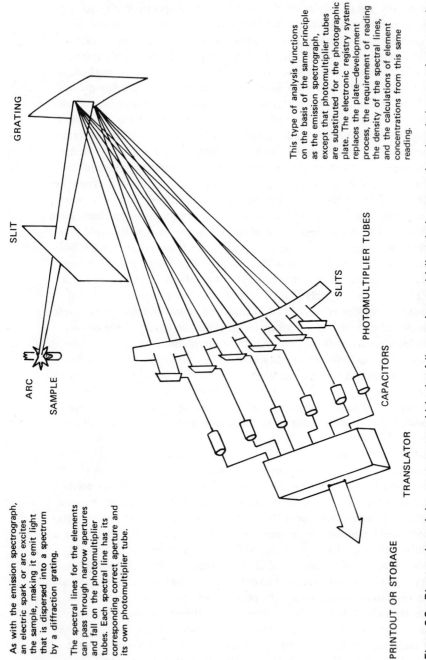

As with the emission spectrograph, an electric spark or arc excites the sample, making it emit light that is dispersed into a spectrum by a diffraction grating.

The spectral lines for the elements can pass through narrow apertures and fall on the photomultiplier tubes. Each spectral line has its corresponding correct aperture and its own photomultiplier tube.

This type of analysis functions on the basis of the same principle as the emission spectrograph, except that photomultiplier tubes are substituted for the photographic plate. The electronic registry system replaces the plate–development process, the requirement of reading the density of the spectral lines, and the calculations of element concentrations from this same reading.

GRATING

SLIT

ARC

SAMPLE

PHOTOMULTIPLIER TUBES

SLITS

CAPACITORS

TRANSLATOR

PRINTOUT OR STORAGE

Figure 2-3. Direct-reader emission spectrometry. Light that falls on a photomultiplier tube is converted to electrical current that accumulates as a charge on a capacitor. This charge or voltage is the measure of the concentration of an element in the sample. Some systems have an indicator dial for each element to supply a direct concentration readout. Others use a translator and typewriters to present and record the results directly in percentage concentration. (Adapted from photo G-26-D of the U. S. Geological Survey.)

18

chemical exploration programs where hundreds of samples come in each week. Local programs can generally do their jobs well with atomic absorption or colorimetry. Whatever the program, it is imprudent to compare the results obtained with one method with those obtained by another method because each class of equipment has its own characteristics with respect to sensitivity, accuracy, and precision.

The geochemical laboratories that are involved in the search for hydrocarbons must be equipped as are organic chemistry laboratories, especially with a gas chromatograph with which one can identify and determine quantities present of, for example, ethane, propane, isobutane, *n*-butane, isopentane and *n*-pentane in soils, rocks, soil and rock gases, drilling mud, and local emanation of gas. Also such laboratories use the mass spectrometer for studying isotopes related to rocks of a petroleum reservoir for studying the hydrocarbons themselves to try to localize the source bed(s) of the hydrocarbons from a determined zone or indicate the direction of migration of oil and natural gas. Recently Thode and Rees (1970) reported that a study of the measurements of the natural variation of the principal isotopes of sulfur, ^{32}S and ^{34}S, in petroleum from the Middle East showed that it is not necessary to postulate four different epochs for the formation of petroleum in this area, but that the actual distribution is the result as much of extensive vertical migration as of horizontal migration from a common source. Actually, in the world today there are probably approximately 200 to 250 scientists working on projects related to the geochemistry of organic materials and the search for petroleum and natural gas using indirect geochemical methods.

There are some private (commercial) laboratories that are offering their services for neutron activation analysis. This method generally offers a much greater sensitivity than other techniques available and, in truth, this is a sensitivity that is most often not necessary in geochemical prospecting. However, Hoyte, Martinez, and Senftle (1967) considered the potential utility of activation analysis techniques in exploration geochemistry and constructed an equipment unit comprised of a positive ion accelerator and detector [with a 3-in. NaI(Tl) crystal] to be mounted in a Jeep and taken to the field for use in the search for Ag. The equipment gave good field results (Figure 2-4) and Senftle (1970) experimented with other similar methods (based on ^{252}Cf as a neutron source and a drifted germanium–lithium crystal as a detector) to be used in field geochemical prospecting for several elements; for example, Ni was detected beneath 55 in. of soil in some recent tests. It is very probable that such a method might be used in the near future for the *in situ* analysis of marine sediments and rocks (Senftle, et al., 1969, 1970, 1971; Siegel, 1971). Recently Plant and Coleman (1973) described the use of neutron activation analysis for Au in placers to assess the economic potential of alluvium and fluvio-glacial deposits of the Strath of Kildonan, Sutherland, England; they achieved a detec-

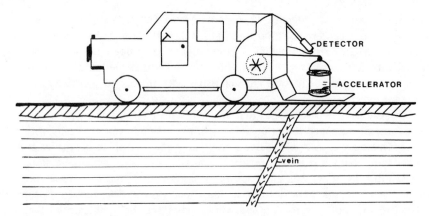

Figure 2-4. Mobile equipment for activation analysis from a jeep or truck. (Modified from Hoyte, Martinez, and Senftle, 1967.)

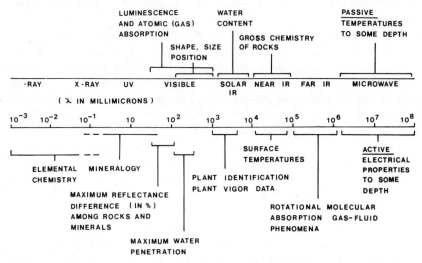

Figure 2-5. Types of information and/or properties of materials that can be obtained from observations of various parts of the electromagnetic spectrum. (From NASA, 1968.)

tion limit of 0.0005 ppm with the reaction ^{197}Au (n, γ) ^{198}Au using the area beneath the 0.41-MeV ^{198}Au peak on the resulting gamma spectrum. However they have not yet mobilized the equipment for *in situ* work.

It should be noted that the electromagnetic spectrum is being studied for obtaining data from satellites that can be used in terrestrial geochemistry. Of course this method is neither portable, nor is it standard laboratory procedure, but it may be able to provide very useful information in the future if initial results are considered indicative. NASA (1968) published a graphic representation (Figure 2-5) showing the types of information related to properties of materials that can be interpreted from space or aerial photographs, or from other measurements of reflectance or absorbance by materials of the energies in the far regions of the spectrum. As NASA (1968) noted, the electromagnetic waves are equal in their character but different in their wavelengths, their frequencies, and their energies. Because the energies are different, the interactions of materials are different; the absorption, reflectance, dispersion, and transmission of light by material is characteristic of the light and the material. This fact allows that such interactions be used for practical applications. Lyon and Lee (1970) provided an excellent review of remote sensing in the exploration of mineral deposits (Tables 2-4 and 2-5). They noted that "remote sensing" is a new term for the old practice of exploration by means of aircraft and emphasize that the depth of penetration of the majority of the sensors is slight, but that the method offers an opportunity for rapid reconnaissance of large areas in the search for diagnostic surficial phenomena deriving from mineral deposits hidden at depth.

Up to now we have discussed permanent laboratory installations or sophisticated mobile equipment for making geochemical analyses directly in the field, but how many times are roads or trails actually available for trucks or trailers that house the mobile laboratories? In the mountains, in the jungle, and in areas of strong relief (topography) one cannot count on these large mobile laboratories. It is obvious, however, that a mobile laboratory can be set up at a main camp relatively near the exploration areas if the geochemists want to do accurate analyses in the field in order to plan program changes on the basis of these analyses (and field observations). But if a geologist–geochemist can carry 3 to 4 kg extra with him, he can do field colorimetric analyses with a kit, as has been described by Bloom (1955) and modified in studies performed by the United States Geological Survey, the Geological Survey of Canada (see references), and other organizations or individuals. With this type of truly portable equipment, preliminary guide analyses can be made *in situ,* to establish a background (and threshold) value for a zone, localize possible anomalies, and make more detailed collections of samples immediately for subsequent quantitative analyses in the central laboratory. Of equal or greater importance are the additional and more intensive geologic observations that should be made at and around the sites of possible anomalies indicated on the basis of the results given by the portable equipment.

Other portable equipment has been adapted for or made for geochemical exploration. In 1970 Durst described the capacity of specific-ion electrodes to be used in the field with portable pH–mV meters. There are commercial electrodes available for measuring at least 18 specific ions (Cd^{2+}, Ca^{2+}, Cu^{2+}, H^+, Pb^{2+}, K^+, Ag^{2+}, Na^+, Br^-, Cl^-, CN^-, F^-, BF_4^-, I^-, NO_3^-, ClO_4^-, S^{2-}, and

Table 2-4 Status of Electromagnetic Remote Sensors from Viewpoint of Economic Geology (from Lyon and Lee, 1970)

Sensor (in Priority)	Application in Economic Geology[a]	Operating Mode		Rationale
		Spacecraft	Aircraft	
1. Photography (0.4–1.0 μ)	Structural geology (lineaments, faults, folds, etc.)	Best		Affords essential synoptic view and orthogonal presentation
2. Radar	Structural geology (lineaments, faults, folds, etc.)		Best	It is assumed that resolutions from aircraft will be finer than those from spacecraft; swath widths for both are comparable
3. Infrared imaging	Convective mass transfer at surface (volcanoes, geothermal areas)		Best	Areas are small, localized, and/or need repeated monitoring
	Conductive heat transfer to surface			
	Landslides, oxidizing orebodies		Best	Areas are small and localized near lineaments or intersections of lineaments

	Outcrop maps (soil/rock inter-faces, physical composition)	Best	Diurnal rate of change of ΔT is essential, not achievable except from geosynchronous orbit
Spectroradiometric	Rock and soil composition (chemical-mineralogical)	Best	Spectrometer field of view is fixed by available energy and smallest ground resolution is essential; spacecraft smear is higher than with aircraft
4. Ultraviolet	Rock-type discrimination	Best	Increasing altitude cuts off useful lower wavelengths
5. Microwave	Ground penetration ability	?	Field of view wider than acceptable except when used in aircraft. Needs research and development

Source. Reproduced from *Economic Geology*, **65**, 786 (1970).

[a] Economic geology defined here includes engineering geology, but excludes hydrology and oceanography.

Table 2-5 Definition of Target Phenomena and Remote Sensor Selection (from Lyon and Lee, 1970)

Target	Target Surface Phenomenon	Photography				Low Sun Angle (LSAP)	Multiband Scanner (Vis & IR)	Infrared Spectro-meter	Side-Looking Airborne Radar (SLAR)	Infrared Scanner
		Black and White	Color	Color Infrared (CIR)	Multi-band					
1. General geology										
a. Rock distribution	Tone or color differences	B	A	C	D	—	E	—	—	—
	Vegetation differences	—	—	A	B	—	C	—	—	—
b. Rock type	Color	—	A	B	C	—	D	—	—	—
	Thermal property differences	—	—	—	—	C	—	—	—	A
	Surface texture	B	—	—	—	C	—	—	A	—
	Chemistry, mineralogy	—	—	—	—	—	—	A	—	—
c. Structure	Surface relief, lineaments	—	—	—	—	A	—	—	B	C
2. Ore deposits	Color of alteration zone	—	A	B	C	—	D	—	—	—
	Texture of alteration zone	—	—	A	B	B	C	—	A	—
	Vegetation differences	—	—	A	B	A	C	—	B	—
	Lineaments	—	—	—	—	A	—	—	—	C
	Thermal anomalies	—	—	—	—	—	—	—	—	A
	Gravel and placer deposits	D	—	—	—	B	—	—	A	C

							As 1a and 1b above			B	C
3. Hydrocarbons	Source rock and reservoir rock distribution										
	Structural trap	D	–	–	–	–	–	–	–	B	C
4. Groundwater	Topography and drainage	–	–	–	–	A	–	–	–	B	C
	Vegetation differences	–	D	A	B	–	C	–	–	–	–
	Soil moisture	B	–	–	–	–	–	–	–	–	A
	Springs	–	–	–	–	–	–	–	–	–	A
	Permafrost indicators				see 6 below						
5. Geothermal energy	Convective and advective heat transfer	–	–	–	–	–	–	–	–	–	A
	Vegetation anomalies	–	–	A	B	–	C	–	–	–	–
	Discoloration	–	A	B	C	–	D	–	–	–	–
6. Permafrost	Polygonal ground	A	B	–	–	D	–	–	–	E	C
	Patterned Rocks	A	B	–	–	C	–	–	–	D	–
	Thermokarst structures	A	B	–	B	C	–	–	–	D	E
	Beaded (button) drainage	A	D	B	–	–	–	–	–	–	C
	Pingos (hydrolaccoliths)	A	B	–	–	C	–	–	–	D	E
	Vegetation differences	–	–	A	B	–	C	–	–	–	–

Source. Reprinted from *Economic Geology*, **65**, 798 (1970).
[a]Letters indicate approximate order of probable use. With varying conditions these priorities will of course change. Availability and cost are considered, so that first priority sensor may not be the best, but simply easier to obtain and use.

SCN⁻) in aqueous media or in soils dispersed in distilled water. Recently three papers were presented in which this technique was applied; total and cold-extractable fluoride were determined in soils and stream sediments with an ion-sensitive electrode (Pluger and Friedrich, 1973), water extractable chloride (plus F^-, Cu^{2+}, and S^{2-}) was analyzed in powdered rocks by use of selective ion electrodes (Van Loon, Kesler and Moore, 1973), and copper was analyzed in natural waters with a copper-sensitive electrode (Friedrich, Pluger, Hilmer, and Abu-Abed, 1973). Although there may be interferences involved in such determinations, research is continuing, and this method seems to have excellent future prospects for geochemical exploration application because of its specific and quantitative nature with respect to many elements of geochemical interest. It may be that measurements at depth in diamond-drill-hole groundwaters can also be made. Bolviken et al. (1973) described an instrument developed for *in situ* measurements of pH and Eh of groundwater in 46-mm diameter diamond-drill holes that can also be used for self-potential (SP) determinations; in one development area (Joma, Nord-Trondelag County, Norway), a Cu- and Zn-containing massive pyrite deposit in metamorphic volcanics with 5 to 10% $CaCO_3$, tests showed that the instrument can be used to 350 meters depth (35 atm) and that SP readings clearly indicate both massive and impregnation types of pyritic ore.

REFERENCES

Bloom, H., 1955, A field method for the determination of ammonium citrate-soluble heavy metals in soils and alluvium, *Econ. Geol.,* **50,** 533–541.

Bolviken, B., Logn, Ø., Breen, A., and Uddu, O., 1973, Instrument for *in situ* measurements of pH, Eh and self-potential in diamond drill holes, in *Geochemical Exploration 1972* (Jones, ed.), Institute of Mining and Metallurgy, London, pp. 415–420.

Bondar, W. F., 1969, Some principles of geochemical analysis, *Quart. Colo. School Mines,* **64,** 19–22.

Canney, F. C., 1969, What is a geochemical analysis (Roundtable discussion) – Analyst's Viewpoint, *Quart. Colo. School Mines,* **64,** 9–11.

Chamberlain, J. A., 1964, A field method for determining uranium in natural waters, Field and laboratory methods used by the Geological Survey of Canads in Geochemical Surveys, Paper 64-14, 7 pp.

Cruft, E. F., and Giles, D. L., 1967, Direct reading emission spectrometry as a geochemical tool, *Econ. Geol.,* **62,** 406–411.

Debnam, A. H., 1965, Determination of hydrocarbons in soils by gas chromatography, Field and laboratory methods used by the Geological Survey of Canada in Geochemical Surveys, Paper 64-15, 17 pp.

Debnam, A. H., 1969, The modern geochemical laboratory, *Quart. Colo. School Mines,* **64,** 217–221.

Durst, R. A., 1970, Pick an ion, any ion, *Ind. Res.*, 36–39.

Fletcher, K., 1970, Some applications of background correction to trace metal analysis of geochemical samples by atomic absorption spectrometry, *Econ. Geol.*, **65**, 588–589.

Friedrich, G. H., Pluger, W. L., Hilmer, E. F., and Abu-Abed, I., 1973, Flameless atomic absorption and ion-sensitive electrodes as analytical tools in copper exploration, in *Geochemical Exploration 1972* (Jones, ed.), Institute of Mining and Metallurgy, London, pp. 435–443.

Galle, O. K., 1964, Comparison of chemical analyses based upon two sampling procedures and two sample preparation methods, *Trans. Kansas Acad. Sci.*, **67**, 100–110.

Garrett, R. G., 1969, The determination of sampling and analytical errors in exploration geochemistry, *Econ. Geol.*, **64**, 568–569.

Garrett, R. G., 1973, The determination of sampling and analytical errors in exploration geochemistry – a reply, *Econ. Geol.*, **68**, 282–283.

Gilbert, M. A., 1963, Laboratory methods for determining copper, zinc, and lead, Field and laboratory methods used by the Geological Survey of Canada in Geochemical Surveys, Paper 59-3, 21 pp.

Holman, R. H. C., 1963, A method for determining readily soluble copper in soil and alluvium, Field and laboratory methods used by the Geological Survey of Canada in Geochemical Surveys, Paper 63-7, 5 pp.

Howarth, R. J., and Lowenstein, P. L., 1971, Sampling variability of stream sediments in broad-scale regional geochemical reconnaissance, *Trans. Inst. Mining Met. Sect. B*, **80**, B363–372.

Hoyte, A. F., Martinez, P., and Senftle, F. E., 1967, Neutron activation method for silver exploration, *Trans. AIME*, **238**, 1–8.

Jardine, M. A., 1963, A laboratory method for determining antimony in soils and rocks, Field and laboratory methods used by the Geological Survey of Canada in Geochemical Surveys, Paper 63-29, 11 pp.

Kahn, H. L., 1966, Instrumentation for atomic absorption, *J. Chem. Educ.*, **43**, 15 pp.

Kahn, H. L., 1972, The use of atomic absorption in analysis of the environment, *Ann. N.Y. Acad. Sci.*, **199**, 145–161.

Lavergne, P. J., 1965, Preparation of geological materials for chemical and spectrographic analysis, Field and laboratory methods used by the Geological Survey of Canada in Geochemical Surveys, Paper 65-18, 23 pp.

Lynch, J. J., and Mihailov, G., 1963, Method for determining arsenic, Field and laboratory methods used by the Geological Survey of Canada in Geochemical Surveys, Paper 63-8, 12 pp.

Lyon, R. J. P., and Lee, K., 1970, Remote sensing in exploration for mineral deposits, *Econ. Geol.* **65**, 785–800.

Medlin, J. H., Suhr, N. H., and Bodkin, J. B., 1969, Atomic absorption analysis of silicates employing $LiBO_2$ fusion, *At. Absorption Newsletter*, **8**, 25–29.

Michie, U. McL., 1973, The determination of sampling and analytical errors in exploration geochemistry, *Econ. Geol.*, **68**, 281–282.

NASA, 1968, *Application of Biogeochemistry to Mineral Prospecting – A Survey*, National Aeronautics and Space Administration, Special Publication-5056, Washington, D.C., 134 pp.

Parker, R. L., 1967, Composition of the Earth's Crust, *U. S. Geol. Surv. Profess. Papers*, **440-D**, 17 pp.

Plant, J., and Coleman, R. G., 1973, Application of neutron activation analysis to the evaluation of placer gold concentrations, in *Geochemical Exploration 1972* (Jones, ed.), Institute of Mining and Metallurgy, London, pp. 373-381.

Pluger, W. L., and Friedrich, G. H., 1973, Determination of total and cold-extractable fluoride in soils and stream sediments with an ion-sensitive fluoride electrode, in *Geochemical Exploration 1972* (Jones, ed.), Institute of Mining and Metallurgy, London, pp. 421-427.

Senftle, F. E., 1970, Mineral exploration by nuclear techniques, *Mining Congr. J.,* **56,** 21-28.

Senftle, F. E., Duffey, D., and Wiggins, P. F., 1969, Mineral exploration of the ocean floor by *in situ* neutron absorption using a californium-252 (^{252}Cf) source, *Marine Technol. Soc. J.,* **3,** 9-16.

Senftle, F. E., Wiggins, P. F., Duffey, D., and Philbin, P., 1971, Nickel exploration by neutron capture gamma ray, *Econ. Geol.,* **66,** 583-590.

Siegel, F. R., 1971, Marine geochemical prospecting – present and future, in *Proc. 3rd Intern. Symp. Geochem. Exploration,* Special Paper 11, pp. 251-257.

Smith, A. H., 1964, Cold extractable "heavy metal" in soil and alluvium, Field and laboratory methods used by the Geological Survey of Canada in Geochemical Surveys, Paper 63-49, 9 pp.

Smith, A. H., 1967, Tin in soils and stream sediments, Field and laboratory methods used by the Geological Survey of Canada in Geochemical Surveys, Paper 67-50, 7 pp.

Stanton, R. E., 1966, *Rapid Methods of Trace Analysis for Geochemical Applications,* Arnold, London, 103 pp.

Thode, H. G., and Rees, C. E., 1970, Sulphur isotope geochemistry and Middle East oil studies, *Endeavor,* xxx, 24-28.

Van Loon, J. C., Kesler, S. E., and Moore, C. M., 1973, Analysis of water-extractable chloride in rocks by use of a selective ion electrode, in *Geochemical Exploration 1972* (Jones, ed.), Institute of Mining and Metallurgy, London, 429-434.

Ward, F. N., Lakin, H. W., Canney, F. C., and others, 1963, Analytical methods used in geochemical exploration by the U. S. Geological Survey, *U. S. Geol. Surv. Bull.,* **1152,** 100 pp.

Ward, F. N., Nakagawa, H. M., Harms, T. F., and Van Sickle, G. H., 1969, Atomic absorption methods of analysis useful in geochemical exploration, *U. S. Geol. Surv. Bull.,* **1289,** 45 pp.

3

THE GEOCHEMICAL DISTRIBUTION
OF THE ELEMENTS DURING THE
FORMATION OF THE EARTH AND
THE CRYSTALLIZATION OF MAGMAS

The distribution and dispersion of the chemical elements that comprise the Earth is a topic with several subtopics. For example, a primary distribution took place while the planet was being developed in the form as we know it now, that is, consisting of a nucleus, a mantle, and a crust. Second, there is a chemical distribution when a magma is solidifying to form igneous rocks and a dispersion of elements in the rock enclosing the magma by hydrothermal alteration. Third, there is a distribution and dispersion of the chemical elements when any class of rock is being weathered. Fourth, sediments are formed as a result of an intense elemental segregation and one factor that may strongly influence the sedimentary chemical distribution is the biological effect. Finally, metamorphic action causes another redistribution of the elements until the point of anatexis is attained with the formation of magmas from preexisting rocks.

The physical–chemical factors that control the cited distributions are many, among them are pressure, temperature, element concentration, presence or absence of volatiles, pH, Eh, solubilities, and the biochemical effects. However, all the factors mentioned depend totally on the tendency of a chemical system (a rock) to attain and maintain chemical equilibrium under the ambient conditions. This tendency itself is influenced by the ability of component elements in the system to form the strongest chemical bonds so that the system has the minimum of free energy. In the ultimate instance, this ability depends on the extranuclear atomic structures of the elements or, more precisely, on the number and disposition of the electrons around the atomic nucleus.

As for the Earth in its totality, we already know with an excellent degree of accuracy various facts about its physical properties such as the mass, the volume (and from these, the density), its structure, and the distribution of mass and volume by general structural unit. Such data come from studies in physical astronomy (the constant of the precession of the equinoxes), geophysics (the acceleration of gravity at the Earth's surface and the gravitational constant), seismology (the propagation of natural and artificially-induced sound waves through the Earth and in rocks placed under high-pressure and high-temperature conditions in the laboratory), and solid-state physics (the changes in elastic constants of rocks with increases of pressure and temperature). As a complement to these data are those on the probable composition (chemical) of the materials that comprise our planet (from analyses of naturally occurring rocks and those produced in the laboratory, under probable conditions of temperature, pressure, and chemical environment of the distinct parts of the Earth, from meteorites and residual slags and alloys formed in smelting operations and in the laboratory, and from celestial spectrographic measurements and lunar material analyses). An evaluation of these data with chemical equilibrium and physical-chemical calculations, has allowed hypotheses to be proposed on the chemical composition and the physical-crystallographic state of the interior of the Earth, which, to a certain degree, have become accepted as facts, as, for example, the hypothesis that the core is composed mainly of metallic iron with a notable nickel content. Figure 3-1 shows the generalized structure of the Earth graphically and gives some physical and chemical properties that change with distance from the surface. Figure 3-2 presents graphs in some detail showing how seismic waves, from which we derive our concept of the structure of the planet, act with depth, and their relations with variations of pressure and density with depth.

When the general chemical compositions of the whole Earth, the crust, the mantle, and the core are compared (see Figure 3-1), it is immediately apparent that there was an initial distribution or differentiation of the chemical elements during the Earth's lithic development. For example, iron is present in greatest quantity in the core, magnesium is abundant in the mantle, and the elements potassium, sodium, calcium, and aluminum (and others) comprise important percentages of the crust. The minor and trace elements should be present in the phases of the Earth that include the major elements with which they are most related by their physical–chemical properties that influence the chemical bonds they can form. This primary distribution is explained on one hand with the concept of electronegativity (a measure of the tendency of an element to form covalent bonds), which states that the elements less electronegative (or more electropositive) than iron could form stronger bonds with a greater ionic component when the silicates of the mantle and crust were solidifying, and thus, the iron, perhaps as metallic particles, was displaced from the mantle and gravitated towards the center of the planetary mass where it was concentrated in the core. Following this line of reasoning, it may be predicted that other elements more electronegative (or less electropositive) than iron should

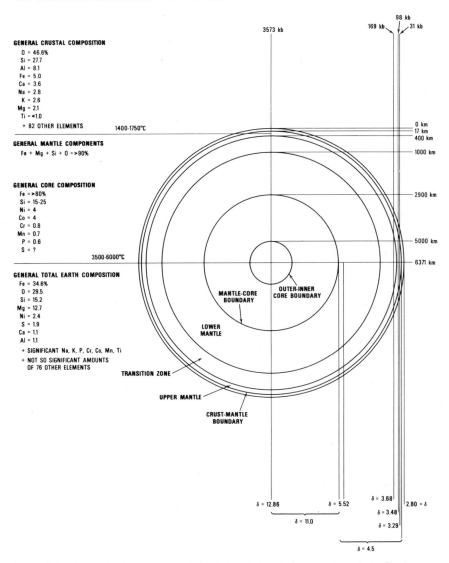

Figure 3-1. Generalized structure and physical and chemical parameters of the Earth.

also be found in the core. On the other hand, the ionization potential or the energy that must be expended to liberate an electron completely from its atom has been considered in the search for an explanation for the primary distribution of the chemical elements in the Earth. The elements for which less energy is required to liberate an electron tend to form ionic bonds and are those that bonded with anions during the solidification of the planet with the result-

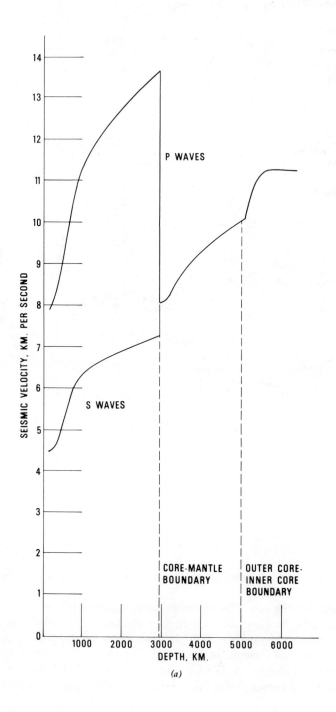

(a)

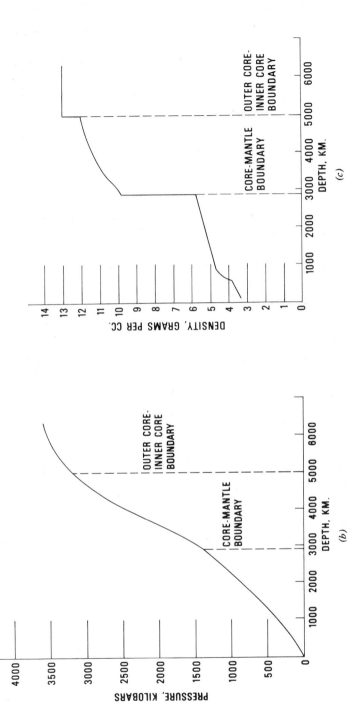

Figure 3-2. (a) Change of seismic velocity with depth in the Earth; (b) change of pressure with depth in the Earth; (c) changes of density with depth in the Earth. (Plotted from data of Clark and Ringwood 1964.)

33

ing displacement of iron, which, because of its density, moved towards the
center of the mass. Of course, electronegativity is related to the ionization
potential; Mulliken (1934) showed that the relation could be described by

$$X = \frac{I + A}{2} \approx \frac{I}{2}$$

in which X = electronegativity

$\quad\quad I$ = ionization potential of an atom with charge I

$\quad\quad A$ = affinity for electrons, and $A \ll I$

Goldschmidt (1937) reviewed data on the distribution of the chemical elements
in natural and artificially prepared materials, recognized the elemental associations according to the class of material, and formulated his concept of the affinities of the chemical elements for the fundamental phases that comprise the
Earth (see Table 3-1). Thus, the siderophile elements are those that are not
combined when they are in the metallic state and are rich in free electrons;
the chalcophile elements are those that tend to form covalent bonds and very
often combine with sulfur in sulfide minerals; and the lithophile elements are
those that tend to form ionic bonds, commonly in silicate minerals. The elements classified as atmophile tend to exist in the form of molecules of gas or
of simple gaseous compounds, and those that comprise organisms or are necessary for the life processes of the organisms can be classified as biophile elements. From Table 3-1 it can be seen that the chemical elements can be incorporated into more than one phase; however, if an element has two or more
possibilities for entering into a crystal lattice during the formation of minerals
and rocks, it preferentially enters that phase in which the strongest bonds are
formed.

As we know from an understanding of the extranuclear structure of atoms,
the manner in which an element acts in a physical–chemical environment depends on its capability to form the strongest bond possible, and this is related
directly to the number of electrons and their configurations around the atomic
nucleus. Because of this, the chemical elements in the classification based on
affinity for the fundamental phases that compose the Earth (see Table 3-1) are
seen to be very regularly presented in groups or subgroups, the extranuclear
relations of which are well defined. For example, in the case of the siderophile
elements Fe, Ni, and Co; Ru, Rh, and Pd; Re, Os, Ir, and Pt; there are d-orbi-
tals being filled with electrons below an outer orbital; thus, the elements of
each subgroup cited have crystallographic–chemical behaviors very much related and all three subgroups have notably similar tendencies as regards physical-
chemical reactions. In the same way, some chalcophile elements should be
cited that can be grouped or subgrouped by their electron configurations and
the corresponding tendencies to form polar or covalent bonds such as S, Se,

Table 3-1 Affinities of the Chemical Elements for the Principal Phases that Comprise the Earth. (compiled from several sources but based on the original presentation of Goldschmidt, 1937)

Siderophile Phase	Chalcophile Phase	Lithophile Phase	Atmophile Phase	Biophile Phase
Fe,Ni,Co	((O)),S,Se,Te	O,(S),(P),(H),(C)	(H),C,N,O	C,H,O,N,P
Ru,Rh,Pd	Fe,Cr,(Ni),(Co)	Si,Ti,Zr,Hf,Th	F,Cl,Br,I	S,Cl,I
Re,Os,Ir,Pt,Au	Cu,Zn,Cd,Pb	Li,Na,K,Rb,Cs	Ar,He,Ne,Kr,Xe	(Ca),(Mg)
Ge,Sn,Sb	Sn,Ge,Mo,(Os)	F,Cl,Br,I		(K),(Na)
(Pb),C,(As),P	As,Sb,Bi	B,Al,(Ga),Sc		(V),(Mn)
Mo,W	Ag,(Au),Hg	Y,RE(La-Lu)		(Fe),(Cu)
(Nb),Ta	Ru,(Pt),(Rh)	Be,Mg,Ca,Sr,Ba		
Se,Te	Ga,In,Tl,(Pd)	(Fe),V,Cr,Mn		
Cu,Ga		Nb,Ta,W,U		
		(Tl),(Ge),(Zn)		
		(N)		

and Te; Cu, Zn, and Cd; As, Sb, and Bi; Ag and Hg. Finally, one can consider
various lithophile elements that are subgrouped into elements that have their
own subaffinities as is the case for Li, Na, K, Rb, and Cs; Be, Mg, Ca, Sr, and
Ba; Ti, Zr, Hf, and Th; B, Al, and Ga; and the rare earths, due to their similar
extranuclear electron structure, but all of which preferentially form the ionic
bond. This tendency for grouping of two or three chemical elements (or
more) is a factor of utmost importance in geochemical prospecting and economic geology. In the case of the search for ore by means of geochemistry,
the geochemist very often inititates a program by analyzing his samples for a
pathfinder element instead of the principal element or elements for reasons
that we will discuss in the following pages. It is worth noting some examples
of pathfinders here, however, such as Mo in the search for copper porphyries,
As in the search for Au-bearing veins, Se in the search for epigenetic sulfide
deposits, and Hg in the search for complex ores of Pb–Zn–Ag. With respect to
economic geology, it is well known that there are sometimes deposits that cannot be exploited economically for a principal element, but when one considers
the recovery of secondary products, the deposits may become economically exploitable; these secondary products may not be elements that have the same
fundamental affinity for a phase as the principal element.

We can consider, then, the physical–chemical properties of the elements that
are directly related to their capabilities to form bonds and that are influenced
by the number and disposition of their electrons. Of course, one thinks immediately of the energy with which the electrons are held by the nucleus (we
have already mentioned the ionization potential) or, in other words, the availability of the electrons to enter into combination with other electrons and
establish bonds of different classes (we have noted electronegativity) in a given
physical–chemical environment. But in a sense these may be considered as
secondary properties. Very early in the evolution of geochemistry or chemical
geology, the ionic radius and the electron charge were recognized as two basic
factors necessary not only to understand the element affinity for a natural terrestrial phase, but also to explain substitutions of one element for another in
natural materials.

The relations between the size of an ion and its charge, on the one hand,
and the ability of the ion to coordinate correctly and maintain the electrical
balance in a mineral structure, on the other hand, are obvious. Geometrically
the ratio between the ionic radius of a cation and that of an anion with which
it is bonding determines in great part the coordination (or the number of anions that can group around the cation) that a cation being combined with an
anion will have. The ratios with values between 0.15 and 0.22 permit a coordination of 3, between 0.22 and 0.41 a coordination of 4, between 0.41 and
0.73 a coordination of 6, between 0.73 and 1.00 a coordination of 8, and
greater than 1.00 a coordination of 12; these coordinations correspond, respec-

Table 3-2 Chemical Composition (Major Elements) of the Lithosphere

Oxide	Igneous Rocks[a]	pC of Finland[b]	Glacial Flour, Norway[c]	Lithosphere[d]	Lithosphere[e]	Crust[f]
SiO_2	59.12	67.45	59.19	55.2	59.07	58.7
Al_2O_3	15.34	14.63	15.82	15.3	15.22	15.0
Fe_2O_3	3.08	1.27	3.41	2.9	3.10	2.3
FeO	3.80	3.13	3.58	5.8	3.71	5.2
MgO	3.49	1.69	3.30	5.1	3.45	4.9
CaO	5.08	3.39	3.07	8.4	5.10	6.7
Na_2O	3.84	3.06	2.05	3.0	3.71	3.1
K_2O	3.13	3.55	3.93	1.9	3.11	2.3
TiO_2	1.05	0.41	0.79	1.7	1.03	1.2
MnO	0.12	0.04	0.11	0.3	0.11	0.12
P_2O_5	0.30	0.11	0.22	0.2	0.30	
H_2O	1.15	0.79	3.02	1.7	1.30	
CO_2			0.07	0.4	0.35	

[a] From Clarke and Washington, 1924.
[b] From Sederholm, 1925.
[c] From Goldschmidt, 1954.
[d] From Poldervaart, 1955[b].
[e] From Clarke and Washington, 1924.
[f] From Ahrens, 1965.

Table 3-3

	"Igneous" Rocks					Sedimentary Rocks			Deep-Sea Sediments	
			Granitic Rocks							
	Ultrabasic	Basaltic Rocks	High Calcium	Low Calcium	Syenites	Shales	Sandstones	Carbonates	Carbonate	Clay
1 Hydrogen H	A	A	A	A	A	A	A	A	A	A
2 Helium He	B	B	B	B	B	B	B	B	B	B
3 Lithium Li	0.X	17.	24.	40.	28.	66.	15.	5.	5.	57.
4 Beryllium Be	3.	1.	2.	3.	1.	3.	0.X	0.X	0.X	2.6
5 Boron B	A	5.	9.	10.	9.	100.	35.	20.	55.	230.
6 Carbon C	6.	A	A	A	A	A	A	A	A	A
7 Nitrogen N	A	20.	20.	20.	30.	A	A	A	A	A
8 Oxygen O	A	A	A	A	A	B	B	B	B	A
9 Fluorine F	100.	400.	520.	850.	1200.	740.	270.	330.	540.	1300.
10 Neon Ne	B	B	B	B	B	B	B	B	B	B
11 Sodium Na	4200.	18,000.	28,400.	25,800.	40,400.	9600.	3300.	400.	20,000.	40,000.
12 Magnesium Mg	204,000.	46,000.	9400.	1600.	5800.	15,000.	7000.	47,000.	4000.	21,000.
13 Aluminum Al	20,000.	78,000.	82,000.	72,000.	88,000.	80,000.	25,000.	4200.	20,000.	84,000.
14 Silicon Si	205,000.	230,000.	314,000.	347,000.	291,000.	73,000.	368,000.	24,000.	32,000.	250,000.
15 Phosphorus P	220.	1100.	920.	600.	800.	700.	170.	400.	350.	1500.
16 Sulfur S	300.	300.	300.	300.	300.	2400.	240.	1200.	1300.	1300.
17 Chlorine Cl	85.	60.	130.	200.	520.	180.	10.	150.	21,000.	21,000.
18 Argon Ar	B	B	B	B	B	B	B	B	B	B
19 Potassium K	40	8300	25,200.	42,000.	48,000.	26,600.	10,700.	2700.	2900.	25,000.
20 Calcium Ca	25,000.	76,000.	25,300.	5100.	18,000.	22,100.	39,100.	302,300.	312,400.	29,000.
21 Scandium Sc	15.	30.	14.	7.	3.	13.	1.	1.	2.	19.
22 Titanium Ti	300.	13,800.	3400.	1200.	3500.	4600.	1500.	400.	770.	4600.
23 Vanadium V	40.	250.	88.	44.	30.	130.	20.	20.	20.	120.
24 Chromium Cr	1600.	170.	22.	4.1	2.	90.	35.	11.	11.	90.
25 Manganese Mn	1620.	1500.	540.	390.	850.	850.	X0.	1100.	1000.	6700.
26 Iron Fe	94,300.	86,500.	29,600.	14,200.	36,700.	47,200.	9800.	3800.	9000.	65,000.
27 Cobalt Co	150.	48.	7.	1.0	1.	19.	0.3	0.1	7.	74.
28 Nickel Ni	2000.	130.	15.	4.5	4.	68.	2.	20.	30.	225.
29 Copper Cu	10.	87.	30.	10.	5.	45.	X.	4.	30.	250.
30 Zinc Zn	50.	105.	60.	39.	130.	95.	16.	20.	35.	165.
31 Gallium Ga	1.5	17.	17.	17.	30.	19.	12.	4.	13.	20.
32 Germanium Ge	1.5	1.5	1.3	1.3	1.4	1.6	0.8	0.2	0.2	2.
33 Arsenic As	0.5	2.	1.9	1.5	0.05	13.	1.	1.	1.	13.
34 Selenium Se	0.05	0.05	0.05	0.05		0.6	0.05	0.08	0.17	0.17
35 Bromine Br	1	3.6	4.5	1.3	2.7	4.	1.	6.2	70.	70.
36 Krypton Kr	B	B	B	B	B	B	B	B	B	B
37 Rubidium Rb	0.2	30.	110.	170.	110.	140.	60.	3.	10.	110.
38 Strontium Sr	1.	465.	440.	100.	200.	300.	20.	610.	2000.	180.
39 Yttrium Y	0.X	21.	35.	40.	20.	26.	40.	30.	42.	90.
40 Zirconium Zr	45.	140.	140.	175.	500.	160.	220.	19.	20.	150.
41 Niobium Nb	16.	19.	20.	21.	35.	11.	0.0X	0.3	4.6	14.
42 Molybdenum Mo	0.3	1.5	1.0	1.3	0.6	2.6	0.2	0.4	3.	27.
43 Technetium Tc								C	C	C
44 Ruthenium Ru								C		C
45 Rhodium Rh								D	D	D
46 Palladium Pd	0.12	0.02	0.00X	0.00X	0.0X	0.07	0.0X	D	D	D
47 Silver Ag	0.06	0.11	0.051	0.037	0.13	0.3	0.0X	0.0X	0.0X	0.11
48 Cadmium Cd	0.X	0.22	0.13	0.13	0.13	0.1	0.0X	0.035	0.0X	0.42
49 Indium In	0.01	0.22	0.0X	0.26	0.0X	0.1	0.0X	0.0X	0.0X	0.08

At. no.	Element	Symbol	1	2	3	4	5	6	7	8	9	10
50	Tin	Sn	0.5	1.5	1.5	3.	X.	6.0	0.X	0.X	0.X	1.5
51	Antimony	Sb	0.1	0.2	0.2	0.2	0.X	1.5	0.0X	0.2	0.15	1.0
52	Tellurium	Te	D	D	D	D	D	D	D	D	D	D
53	Iodine	I	0.5	0.5	0.5	0.5	0.5	2.2	1.7	1.2	0.05	0.05
54	Xenon	Xe	B	B	B	B	B	B	B	B	B	B
55	Cesium	Cs	0.X	1.1	2.	4.	0.6	5.	0.X	0.X	0.4	6.
56	Barium	Ba	0.4	330.	420.	840.	1600.	580.	X0.	10.	190.	2300.
57	Lanthanum	La	0.X	15.	45.	55.	70.	92.	30.	X.	19.	115.
58	Cerium	Ce	0.X	48.	81.	92.	161.	59.6	92.	11.5	35.	345.
59	Praseodymium	Pr	0.X	4.6	7.7	8.8	15.	24.	8.8	1.1	3.3	33.
60	Neodymium	Nd	0.X	20.	33.	37.	65.	24.	37.	4.7	14.	140.
61	Promethium	Pm	C	C	C	C	C	C	C	C	C	C
62	Samarium	Sm	0.X	5.3	8.8	10.	18.	6.4	10.	1.3	3.8	38.
63	Europium	Eu	0.X	.8	1.4	1.6	2.8	1.0	1.6	0.2	0.6	6.
64	Gadolinium	Gd	0.X	5.3	8.8	10.	18.	6.4	10.	1.3	3.8	38.
65	Terbium	Tb	0.X	.8	1.4	1.6	2.8	1.0	1.6	0.2	0.6	6.
66	Dysprosium	Dy	0.X	3.8	6.3	7.2	13.	4.6	7.2	0.9	2.7	27.
67	Holmium	Ho	0.X	1.1	1.8	2.0	3.5	1.2	2.0	0.3	0.8	7.5
68	Erbium	Er	0.X	2.1	3.5	4.0	7.0	2.5	4.0	0.5	1.5	15.
69	Thulium	Tm	0.X	0.2	0.3	0.3	0.6	0.2	0.3	0.04	0.1	1.2
70	Ytterbium	Yb	0.X	2.1	3.5	4.0	7.0	2.6	4.0	0.5	1.5	15.
71	Lutetium	Lu	0.X	0.6	1.1	1.2	1.1	0.7	1.2	0.2	0.5	4.5
72	Hafnium	Hf	0.6	2.0	2.3	3.9	11.	2.8	3.9	0.3	0.41	4.1
73	Tantalum	Ta	1.0	1.1	3.6	4.2	2.1	0.8	0.0X	0.0X	0.0X	0.X
74	Tungsten	W	0.77	0.7	1.3	2.2	1.3	1.8	1.6	0.6	0.X	X.
75	Rhenium	Re	D	D	D	D	D	D	D	D	D	D
76	Osmium	Os	D	D	D	D	D	D	D	D	D	D
77	Iridium	Ir	D	D	D	D	D	D	D	D	D	D
78	Platinum	Pt	D	D	D	D	D	D	D	D	D	D
79	Gold	Au	0.006	0.004	0.004	0.004	0.00X	0.00X	0.00X	0.00X	0.00X	0.00X
80	Mercury	Hg	0.0X	0.09	0.08	0.08	0.0X	0.4	0.03	0.04	0.0X	0.X
81	Thallium	Tl	0.06	0.21	0.72	2.3	1.4	1.4	0.82	0.0X	0.16	0.8
82	Lead	Pb	1.	6.	15.	19.	12.	20.	7.	9.	9.	80.
83	Bismuth	Bi	D	0.007	D	D	0.01	D	D	D	D	D
84	Polonium	Po	E	E	E	E	E	E	E	E	E	E
85	Astatine	At	E	E	E	E	E	E	E	E	E	E
86	Radon	Rn	E	E	E	E	E	E	E	E	E	E
87	Francium	Fr	E	E	E	E	E	E	E	E	E	E
88	Radium	Ra	E	E	E	E	E	E	E	E	E	E
89	Actinium	Ac	E	E	E	E	E	E	E	E	E	E
90	Thorium	Th	0.004	4.	8.5	17.	13.	12.	1.7	1.7	X.	7.
91	Protactinium	Pa	E	E	E	E	E	E	E	E	X.	E
92	Uranium	U	0.001	1.	3.0	3.0	3.0	3.7	0.45	2.2	0.X	1.3
93	Neptunium	Np	F	F	F	F	F	F	F	F	F	F
94	Plutonium	Pu	F	F	F	F	F	F	F	F	F	F

a * In some cases, only order of magnitude estimates could be made. These are indicated by the symbol X.

A: These elements are the most important constituents of the biosphere, hydrosphere, and atmosphere. Oxygen is also the most important element of the lithosphere, whereas carbon is important in sedimentary rock.

B: The rare gases occur in the atmosphere in the following amounts (volume per cent): He, 0.00052; Ne, 0.0018; A, 0.93; Kr, 0.0001; Xe, 0.000008. He is produced by radioactive decay of U and Th but is also lost to outer space. A^{40} is produced by the radioactive potassium 40 and is the major isotope of argon in the atmosphere.

The argon and helium contents of rocks will vary with their age owing to the effect of radioactive decay.

The estimated rare-gas contents of igneous rocks are (in cc per gm of rock): He, 6×10^{-6}; Ne, 7.7×10^{-8}; A, 2.2×10^{-4}; Kr, 4.2×10^{-9}; Xe, 3.4×10^{-10}.

C: These elements do not occur naturally in the Earth's crust.

D: The data for these elements are missing or unreliable.

E: All these elements are present as radioactive nuclides in the decay schemes of U and Th.

F: These elements occur naturally only as a consequence of neutron capture by uranium.

TUREKIAN AND WEDEPOHL, TABLE 2
Geological Society of America Bulletin, volume 72

tively, to the corners of a triangle, the corners of a tetrahedron, the corners of an octahedron, the corners of a cube, and the densest packing. The coordinations that commonly exist between a cation and anions can be altered depending on special conditions in an environment, such as elevated pressures and temperatures. For example, for silicon in quartz, tridymite, and cristobalite, the coordination between Si and O is 4, but under sufficient pressure the coordination can be 6, and instead of the silica minerals mentioned, the high-pressure, dense polymorph stishovite would exist; or, as is thought to occur in some of the structural zones of the Earth's interior, $MgSiO_3$ instead of having Si and O with a coordination of 4 would have a denser crystallographic structure in which three fourths of the Si ions would have a coordination of 4 and one fourth would have a coordination of 6 (structure of garnet); or, in zones of greater pressure all the ions of Si would have (according to some hypotheses) a coordination of 6 (structure of ilmenite) (*Chem. Eng. News,* 1967; Schubert, Turcotte and Oxburgh, 1970). One must also recognize the effect of intermediary orbitals between the nucleus of an atom and its exterior electrons, orbitals that tend to mask the force with which the nucleus holds its interior electrons, an effect that results in the concept of effective charge. As will be seen below, these concepts concerning the size and charge of an ion have been used to try to derive general rules concerning the factors that influence the distribution of elements during the crystallization of a magma and during the formation of some chemical and biochemical sediments.

With respect to a starting point for study of the distribution of the chemical elements in the solid phases during the crystallization of a magma, extensive compilations of good analyses of naturally occurring igneous rocks and those produced in the laboratory under preselected conditions of temperature, pressure, and element concentration have existed for many years. From these compiled data one could calculate and derive, with a good degree of confidence, the average concentrations of major elements ($> 1\%$), minor elements ($< 1\%$ but $> 0.1\%$), trace elements ($< 0.1\%$), and their variations (limits) in igneous rocks of the many classes that can crystallize during the cooling of a magma. Detailed analyses of elements in minerals from which we derive the geochemical distribution in rocks as a function of mineralogic phase also exist. By putting these data together with those on the element distribution in sedimentary and metamorphic rocks (and minerals) and giving proper weight to the relative chemical contributions of each basic group of rocks of the Earth's crust, geoscientists have arrived at some chemical compositions that are supposed to be representative of the composition of the lithosphere (see Table 3-2). On the basis of his studies of chemical equilibrium, crystallography, petrology, personal observations and analyses, and data similar to those presented in Tables 3-2, 3-3 and 3-4, Goldschmidt (1937) proposed his classical general rules to explain the distributions of the elements; these rules are especially

Table 3-4   Abundance of the Elements in the Continental Crust, Together with Basalt and Granite Averages (from Taylor, 1964)

Atomic Number	Element	Crustal Average (ppm)	Basalt Average	Granite Average
1	H			
2	He			
3	Li	20	10	30
4	Be	2.8	0.5	5
5	B	10	5	15
6	C	200	100	300
7	N	20	20	20
8	O	46.4%		
9	F	625	400	850
10	Ne			
11	Na	2.36%	1.94	2.77
12	Mg	2.33%	4.5	0.16
13	Al	8.23%	8.76	7.70
14	Si	28.15%	24.0	32.3
15	P	1050	1400	700
16	S	260	250	270
17	Cl	130	60	200
18	A			
19	K	2.09%	0.83	3.34
20	Ca	4.15%	6.72	1.58
21	Sc	22	38	5
22	Ti	0.57%	0.90	0.23
23	V	135	250	20
24	Cr	100	200	4
25	Mn	950	1500	400
26	Fe	5.63%	8.56	2.70
27	Co	25	48	1
28	Ni	75	150	0.5
29	Cu	55	100	10
30	Zn	70	100	40
31	Ga	15	12	18
32	Ge	1.5	1.5	1.5
33	As	1.8	2	1.5
34	Se	0.05	0.05	0.05
35	Br	2.5	3.6	1.3
36	Kr			
37	Rb	90	30	150
38	Sr	375	465	285

Table 3-4 (Continued)

Atomic Number	Element	Crustal Average (ppm)	Basalt Average	Granite Average
39	Y	33	25	40
40	Zr	165	150	180
41	Nb	20	20	20
42	Mo	1.5	1	2
43	Tc			
44	Ru			
45	Rh			
46	Pd			
47	Ag	0.07	0.1	0.04
48	Cd	0.2	0.2	0.2
49	In	0.1	0.1	0.1
50	Sn	2	1	3
51	Sb	0.2	0.2	0.2
52	Te			
53	I	0.5	0.5	0.5
54	Xe			
55	Cs	3	1	5
56	Ba	425	250	600
57	La	30	10	40
58	Ce	60		
59	Pr	8.2		
60	Nd	28		
61	Pm	—		
62	Sm	6.0		
63	Eu	1.2		
64	Gd	5.4		
65	Tb	0.9		
66	Dy	3.0		
67	Ho	1.2		
68	Er	2.8		
69	Tm	0.48		
70	Yb	3.0		
71	Lu	0.50		
72	Hf	3	2	4
73	Ta	2	0.5	2.5
74	W	1.5	1	2
75	Re			
76	Os			
77	Ir			

Table 3-4 (Continued)

Atomic Number	Element	Crustal Average (ppm)	Basalt Average	Granite Average
78	Pt			
79	Au	0.004	0.004	0.004
80	Hg	0.08	0.08	0.08
81	Tl	0.45	0.10	0.75
82	Pb	12.5	5	20
83	Bi	0.17	0.15	0.18
84	Po			
85	At			
86	Rn			
87	Fr			
88	Ra			
89	Ac			
90	Th	9.6	2.2	17
91	Pa			
92	U	2.7	0.6	4.8

Source. Reprinted from *Geochimica et Cosmochimica Acta,* **28**, 1280-1281 (1964).

applicable to the igneous rock minerals with their dominantly ionic bonding, implying which are the factors that control such distributions. The Goldschmidtian rules may be given as follows:

1. When there are two ions available to enter the crystal lattice of a forming mineral, and they have the same valence (electric charge) and essentially the same ionic radius, they can enter into the crystal lattice with equal ease, other factors, such as their concentrations, being similar.

2. When there are two ions with about the same ionic radius but with different electric charges, the ion with the greater charge will preferentially enter the crystal lattice. One can consider as an example the crystallization of the plagioclase feldspars and the fact that the first solid phases that crystallize from a basic magma will contain more calcium and less sodium than the magma. Although the ionic radii of sodium (0.95 Å) and calcium (0.99 Å) are almost equal, the electric charge of calcium (2+) is greater than that of sodium (1+), and because of this, the calcium enters the crystal lattice of the forming plagioclase preferentially and is enriched in the first plagioclase phases that crystallize from a magma.

3. When there are two ions with the same charge, but with different ionic radii, the smaller ion will preferentially enter the crystal lattice of a forming mineral. If one considers the crystallization of olivine from a basic magma

as an example, it is found that the first solid phase of olivine to form is enriched in magnesium (forsterite) because although magnesium and iron have
the same charge (2+), the magnesium (0.65 Å) has an ionic radius smaller than
that of iron (0.76 Å).

Goldschmidt recognized that there were other norms that complemented
his general rules. For example, an element can be substituted for another of
different charge, but only when there is a coupled substitution so that electrical neutrality in a crystal lattice is maintained. In the case of the plagioclase
feldspars, for each ion of Ca (2+) that interchanges for one of Na (1+), there
is a substitution of an ion of Al (3+) for one of Si (4+). With respect to size,
an element may substitute extensively for another if the charges of the two
elements are equal and their ionic radii do not differ by more than about 15%;
if ionic radii differ by more than 15%, such substitution will be very limited.
It is obvious that the degree of substitution will be limited if the bonds that
are formed between the cations and the anions have very distinct ionic–covalent characteristics, but in all cases, the strongest bond is the bond that will
form preferentially.

Although Goldschmidt's rules based on the importance of ionic size and
charge and their relations with coordination number and electrical balance in a
crystal lattice serve to explain and predict many cases of element substitution,
as much for major elements as for minor and trace elements, they have limitations when there is a significant covalent bond contribution in the crystal lattice of a mineral.

Because of this, Ringwood (1955a, 1955b) proposed the complementary
use of the concept of electronegativity in order to understand the distributions
of the chemical elements that could not be explained completely with the
Goldschmidtian rules, especially when the minerals being investigated had high
percentages of covalent bonding. As was noted before, electronegativity describes the tendency of an ion to form covalent bonds (and/or ionic bonds)
and is related to the ionization potential and the affinity of an element for
electrons. A partial list of the common mineral- and rock-forming elements
and their corresponding electronegativities is presented in Table 3-5, together
with the approximate percentages of ionic bonding component that they have
when they are bonded with oxygen; also included are the values for the first
ionization potentials of these elements. Krauskopf (1967) gives a rather complete listing of the electronegativities of the elements. In Table 3-6, it is seen
how the ionic contribution of a bond changes as the electronegativity changes.
Therefore, a less electronegative (or more electropositive) element will substitute for or replace a more electronegative (or less electropositive) element
when the difference in electronegativity between them is more than 0.1 because the strongest bond, one more ionic in character, will be formed. Thus,

Table 3-5 *Electronegativities of some Elements that Comprise Minerals in Common Rocks (compiled from Mason, 1966 and Krauskopf, 1967). Related parameters are also included*

Ion	Electron-negativity[a]	Approximate percent of Ionic Bond	Ionization Potential	kcal/gram-atom
Cs^+	0.7	89	3.893	89
Rb^+	0.8	87	4.176	97
K^+	0.8	87	4.339	100
Na^+	0.9	83	5.138	118
Li^+	1.0	82	5.390	125
Ba^{2+}	0.9	84	9.95	115
Sr^{2+}	1.0	82	10.98	125
Ca^{2+}	1.0	79	11.82	137
Mg^{2+}	1.2	71	14.97	174
Be^{2+}	1.5	63	18.14	210
Al^{3+}	1.5	60	28.31	220
B^{3+}	2.0	43	37.75	290
Sc^{3+}	1.3	65	24.64	200
La^{3+}	1.14	77	19.2	147
$Ce^{3+}-Ho^{3+}$	1.07–0.91	73–75	19.7	160
$Er^{3+}-Lu^{3+}$	0.89–0.85	76		
V^{3+}	1.6 (1.35)	57	29.6	235
Mn^{2+}	1.5 (1.4)	72	15.7	170
Cr^{3+}	1.6	53	32.1	250
Zn^{2+}	1.7	63	17.89	208
Sn^{2+}	1.8	73	14.52	
Pb^{2+}	1.8 (1.6)	72	14.97	170
Fe^{2+}	1.8 (1.65)	69	16.24	185
Ni^{2+}	1.8 (1.7)	60	18.13	220
Co^{2+}	1.8 (1.7)	65	17.3	200
Fe^{3+}	1.9 (1.8)	54		245
Ag^+	1.9	71	7.574	175
Cu^+	1.9 (1.8)	71	7.723	177
Cu^{2+}	2.0	57	20.28	235
Au^+	2.4	62	9.223	212
Si^{4+}	1.8	48	44.95	270
C^{4+}	2.5	23	64.19	370
P^{5+}	2.1	35	64.74	320
N^{5+}	3.0	9	97.43	450
Se	2.4			360
S	2.5			385
O	3.5			530
I	2.5			375
Cl	3.0			460
F	4.0			605

[a]The electronegativity figures represent the elements in octahedral coordination with oxygen and are on a scale of 0.7 to 4.0 by Pauling (1960).

Table 3-6 The Variation of the Percentage of Ionic Character of a Single Chemical Bond with Electronegativity (from Dyna-Slide, 1962)

Difference in Electronegativity	Percentage Ionic Character
0.1	0.5
0.2	1.0
0.3	2.0
0.4	4.0
0.5	6.0
0.6	9.0
0.7	12.0
0.8	15.0
0.9	19.0
1.0	22.0
1.1	26.0
1.2	30.0
1.3	34.0
1.4	39.0
1.5	43.0
1.6	47.0
1.7	51.0
1.8	55.0
1.9	59.0
2.0	63.0
2.1	67.0
2.2	70.0
2.3	74.0
2.4	76.0
2.5	79.0
2.6	82.0
2.7	84.0
2.8	86.0
2.9	88.0
3.0	89.0
3.1	91.0
3.2	92.0

combining this concept with the knowledge of ionic radii and electric charges, Ringwood (1955a) proposed his own rules to explain the distribution and substitution of the elements, taking into account that the stronger bond will generally dominate in the competition for a site in the crystal lattice of a mineral:

1. For a cation and two different anions, the bond with the larger anion is more covalent (or less ionic).

2. For an anion and two different cations, the bond with the smaller cation is more covalent (or less ionic).

3. For two ions with approximately equal sizes, but with different electric charges, the ion with the greater charge forms a covalent bond that is more covalent (or less ionic).

4. The ions of metals from the middle sections of the long periods in the periodic table form more covalent bonds (with anions) than the cations with similar charges and sizes from the first two or three groups of the periodic table.

In spite of the formulation of the concepts of Goldschmidt and Ringwood based on ionic charge, ionic size, and electronegativity, all the problems related to controls of distribution and substitution of the chemical elements in rock originating from the crystallization of a magma have not been resolved. The difficulties arise when the concept developed on the basis of ionic charge and ionic size and that one based on electronegativity work against each other. This happens with the first series transition elements (Sc to Zn) for which ionic sizes and electronegativities result in opposite predictions for element distribution and substitution. This noncorrespondence of predictions also occurs when an ion of greater charge replaces an ion of lesser charge, but with the maintenance of electrical neutrality in the atomic lattice being achieved by a coupled substitution when in the same crystal, an element with a lesser charge replaces an element of greater charge; this does not follow the general rules. In order to overcome these complications, Nockolds (1966) proposed that the three principal factors (ionic size, ionic charge, and electronegativity) be expressed in a single function that would not result in the dicotomous predictions. This function was denominated "bonding energy" by Nockolds and was defined in general terms by the expression:

Bonding energy$_{A-B}$ = Covalent energy$_{A-B}$ + Ionic resonance energy$_{A-B}$

Nockolds (1966) quantified the bonding energy by the equation:

$$\text{Bonding energy}_{M-O} = \frac{11.8\,(S_m + 5.5)}{R} - 5.6 + \frac{26.5\Delta^2}{R}$$

in which S_m = stability ratio of the molecule M−O (M = metal, O = oxygen)

Δ = difference in electronegativity between M and O

R = length of the bond

Table 3-7 shows various characteristics of 42 cations when bonded with oxygen in a 6 coordination and terminates with a figure for the total relative bonding energy. The bonding energy varies with the coordination between ions and it is supposed that it also varies in distinct physical–chemical environments,

Table 3-7 Bonding Energies of X–O Bonds (from Nockolds, 1966)[a]

Bond, X–O	Effective r_X (Å)	Effective r_X (Å)[b]	Average Effective r_X (Å)	R_{X-O} (Å)	Covalent Bonding Energy (kcals/mole)	Ionic Resonance Bonding Energy (kcals/mole)	Total Single Valence Bonding Energy	Ionic Resonance Bonding Energy (%)	Ionic Character from $E_O - E_X$ (%)	Relative Total Bonding Energy
Si–O	0.71	0.70	0.70	1.71	58	37	95	(39)	(45)	380
Ge–O	0.94	0.84	0.89	1.92	55	29	84	(35)	(41)	336
Ti^{IV}–O	0.96	0.98	0.97	1.98	48	39	87	(45)	(51)	348
Zr–O	1.00	0.98	0.99	2.12	44	48	92	(52)	(61)	368
Sn^{IV}–O	1.09	0.96	1.02	2.11	47	28	75	(37)	(43)	300
B–O	0.54	0.57	0.55	1.51	67	37	104	(36)	(41)	312
Al–O	0.76	0.79	0.77	1.88	49	51	100	(51)	(60)	300
Ga–O	0.91	0.88	0.89	1.94	53	35	88	(40)	(47)	264
Sc–O	0.90	0.93	0.91	2.07	44	56	100	(56)	(66)	300
Cr^{III}–O	0.92	0.92	0.92	1.93	50	37	87	(43)	(49)	321 ($d^3 + 60$)
V^{III}–O	0.95	0.93	0.94	1.99	48	38	86	(44)	(51)	299 ($d^2 + 41$)
Fe^{III}–O	0.97	0.94	0.95	1.94	51	31	82	(38)	(43)	246
Ti^{III}–O	0.95	0.97	0.96	2.02	47	42	89	(47)	(55)	290 ($d^1 + 23$)
As^{III}–O	1.04	0.99	1.01	1.94	55	23	78	(29)	(35)	234
Y–O	1.07	1.03	1.05	2.29	38	56	94	(60)	(70)	282
In–O	1.12	1.04	1.08	2.15	46	33	79	(42)	(50)	237
Sb^{III}–O	1.28	1.06	1.17	2.22	46	23	69	(33)	(39)	207
La–O	1.21	1.15	1.18	2.41	37	55	92	(60)	(71)	276
Tl^{III}–O	1.32	1.16	1.24	2.27	44	24	68	(35)	(43)	204

BiIII–O	1.43	1.16	1.29	2.38	42	23	65	(35)	(41)	195
Be–O	0.58	0.61	0.59	1.73	53	59	112	(53)	(61)	224
Mg–O	0.88	0.88	0.88	2.08	42	59	101	(58)	(68)	202
Ni–O	0.98	0.95	0.96	1.98	50	34	84	(40)	(47)	1.97 $(d^8 + 29)$
Co–O	0.99	0.93	0.96	2.00	49	34	83	(41)	(47)	183 $(d^7 + 17)$
FeII–O	1.01	0.95	0.98	2.03	48	35.5	83.5	(43)	(50)	178 $(d^6 + 11)$
Zn–O	1.01	0.96	0.98	2.05	48	37	85	(44)	(51)	170
MnII–O	1.06	1.00	1.03	2.18	43	44	87	(51)	(60)	174
Ca–O	1.15	1.11	1.13	2.40	34	66	100	(66)	(77)	200
Cd–O	1.19	1.16	1.17	2.25	42	36	78	(46)	(53)	156
SnII–O	1.25	1.12	1.18	2.27	44	32	76	(42)	(50)	152
HgII–O	1.28	1.23	1.25	2.25	44	26	70	(37)	(43)	140
Sr–O	1.30	1.25	1.27	2.56	31	64.5	95.5	(68)	(79)	191
PbII–O	1.41	1.30	1.35	2.42	41	28	69	(41)	(47)	138
Ba–O	1.48	1.44	1.46	2.76	28	62	90	(69)	(80)	180
Li–O	0.83	0.81	0.82	2.11	36	75	111	(68)	(77)	111
CuI–O	1.05	0.98	1.01	2.08	46	33	79	(42)	(47)	79
Na–O	1.10	1.08	1.09	2.40	31	69	100	(69)	(79)	100
AgI–O	1.28	1.25	1.26	2.31	40	31	71	(44)	(50)	71
K–O	1.46	1.40	1.43	2.77	25	65	90	(72)	(82)	90
TlI–O	1.55	–	1.55	2.62	37	37	74	(50)	(60)	74
Rb–O	1.58	1.54	1.56	2.90	23	62	85	(73)	(82)	85
Cs–O	1.77	1.72	1.74	3.10	21	60	81	(74)	(83)	81

Source. Reprinted from *Geochimica et Cosmochimica Acta*, 30, 270–271 (1966).

[a]All values for standard sixfold coordination.

[b]r_X in oxide as calculated by Povarennykh (1955, Table 3, p. 152).

although the ions may have the same coordination. It should also be understood that other types of bonding can exist such as the Van der Waal bond, and these add their effects to the total importance of the electrical balance in a crystal. The figures calculated from the above equation correspond to metals with a valence of 1 and because of this, in order to determine the total relative bonding energies, the quantity resulting from such a calculation must be multiplied by the valence of the elements being studied in the event that their valences are greater than 1; when the resulting number is divided by the coordination number of the element in the crystal-atomic environment being considered, the bonding energy per bond is determined. With the total relative bonding energies thus calculated, two rules that Nockolds formulated can be applied to the understanding of element distribution and substitution:

1. When two cations with the same charge are capable of being substituted into a crystal lattice, the cation with the greater total relative bonding energy will preferentially enter the lattice.
2. When two cations of different ionic charge and in a situation of coupled substitution are capable of being substituted into a crystal lattice, the substitution of the element that has the greatest sum of total relative bonding energies will take place preferentially.

The use of the calculations of bonding energy represents a step forward in resolving various problems related to the distribution of the chemical elements and the elemental substitutions that could take place in the igneous rocks and their mineralogic components, but it does not resolve some distributional irregularities related to the transition elements. As we remember, the transition elements are those in which a quantum sublevel *d* below an exterior sublevel is being filled with electrons. In general, all the electrons of a given quantum sublevel have the same energy and the same tendency, availability, and ability to form a bond. In the case of the *d*-electrons of the transition elements, this is not completely correct. As the rules related to the development of the extranuclear structure of an element indicate, the *d*-electrons of the transition elements tend to occupy available orbitals singularly with parallel spin in order to minimize the effect of electrical repulsion, with the orbital filling being completed later when the electrons enter with opposite spin in order to form pairs. But when an ion of a transition metal is placed in an electrostatic field of ions around it, the so-called degeneracy or equality of energy of the electrons of the *d*-orbitals of an atom or isolated ion is lost in a manner that depends on the symmetry of the anions around the atom or ion and the force of the field produced by them. The result of the loss of degeneracy or equality of energy is a division of the quantum sublevel *d* into suborbitals of different energies according to the crystal electric fields developed by the different sym-

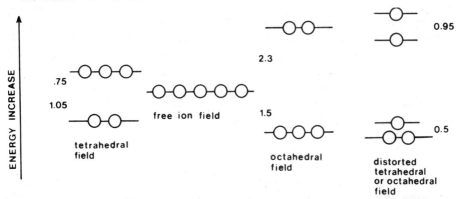

Figure 3-3. Diagrammatic representation of energy-level splitting of the *d*-orbitals resulting from differences in crystal geometries. (Modified from Burns, Clark, and Fyfe, 1964.)

metries of the anions (see Figure 3-3). If the division between the orbitals of the quantum sublevel (and hence the resulting electron energies) is small, the *d*-electrons can fill both suborbitals as in the case of degeneracy, that is, in singular form with parallel spins and later pairing with electrons of opposite spins. But when the division is large, and with the resulting distribution of energy, the complete filling of the suborbitals of the quantum sublevel of lesser energy is favored before electrons enter the suborbitals of the quantum sublevel of greater energy. The magnitude of this division depends on the force of the anionic electric field and the valence of the central cation. When the *d*-suborbitals of lesser energy are preferentially filled, the *d*-electrons are stabilized with an ion of a transition metal, something that does not happen in the case of random filling of the *d*-orbitals. This increase of bonding energy is called the crystal field stabilization energy (CFSE). Bethe (1929) developed this crystal field theory to explain the loss of the degeneracy of atomic orbitals in an electrostatic field around a central ion in a crystal, but there was a delay of about 30 years before his theory was applied to geochemical problems.

Burns, Clark, and Fyfe (1964) used the crystal field theory to explain the chemical behavior of the transition elements of the first series of the Mendeljeff periodic table (Sc, Ti, V, Cr, Mn, Fe, Co, Ni, Cu and Zn) that could not be explained by the general principles of Goldschmidt or of Ringwood. In their work these scientists considered that compounds of transition metals are essentially ionic and that the anions around a central metal ion act as point charges in a crystal lattice. The specific chemical bond between the central metal and the anions was ignored, and only the effects of various electrostatic fields that resulted from different geometries and distances of the point charges of the atomic orbitals of the transition metal ion were taken into account. Burns, Clark, and Fyfe (1964) explained that when there are one, two, or three d electrons (as is the case for Ti^{3+}, V^{3+}, and Cr^{3+}, respectively), two forces act: one which by the interactions between electrons causes the electrons to reach the greatest number of orbitals possible so that each orbital may be occupied singly and with parallel spin; and another which causes the division of the d-orbitals and thus favors the occupation of one group of suborbitals over another, thus creating a system in which there exists the crystal field stabilization energy. However, with 4, 5, 6, or 7 electrons (Cr^{2+} and Mn^{3+}, Mn^{2+} and Fe^{3+}, Fe^{2+} and Co^{3+}, and Co^{2+}, respectively) there is an election of the electron configuration. The transition metal ions that have 8, 9, or 10 d-electrons (Ni^{2+}, Cu^{2+}, and Zn^{2+}, respectively) can establish a single configuration because the d electrons fill the orbitals completely. Burns and his colleagues followed their 1964 report with a series of papers that culminated with a book by Burns (1970) on the topic.

Although various aspects of the factors that influence the distribution and dispersion of the chemical elements in igneous rocks and their mineral components have already been investigated, studies of the subject continue in an effort to refine our concepts and perhaps more thoroughly interpret our observations. For example, Urusov (1970) presented some mathematical expressions to quantify the effect of the difference in the character of a chemical bond on isomorphism for two situations: (1) when the isomorphic components have the same atomic structure but different types of bonding (different degree of ionic bond) and (2) when the structure and the character of the bonding of the components are both different. Urusov made some calculations and arrived at an equation that shows the strong quadratic dependency of the heat of mixing, ΔH_m, on the difference in the character of the bonding of the isomorphous components ($\Delta \epsilon$) of a system:

$$\Delta H_m = 65 \ X_1 X_2 Z^2 \ \frac{5A(\Delta \epsilon)^2}{R} + \frac{n(\Delta R)^2}{(R)}$$

in which ΔH_m = heat of mixing (kcal/mole)
$\ X_1$ = mole fraction of one component
$\ X_2$ = mole fraction of a second component
$\ Z$ = formal charge of the ion
$\ A$ = Madelung constant of the crystal
$\ \Delta \epsilon$ = difference in the degree of the ionic character of bonding between the components
$\ R$ = the shortest interatomic distance
$\ \Delta R$ = the difference in the interatomic distances R_2 and R_1 when $R_2 > R_1$ in the crystals of the components
$\ n$ = the coordination number

The numerical coefficients in the equation are used if R is measured in ang-

Table 3-8 The Relation between Field Forces and the Gel Point of Fusions of Silica with Some Metals (from Hess, 1971), and the Heats of Fusion of the Metals

Element	Field Force Z/a^2	Gel Point (mole % SiO_2)	Heat of Fusion (kcal/gram-atom)
K^{1+}	0.13	44	0.55
Na^{1+}	0.18	44	0.62
Li^{1+}	0.23	44	0.72
Pb^{2+}	0.30	41	1.22
Ca^{2+}	0.35	41	2.10
Mn^{2+}	0.41	35	3.50
Fe^{2+}	0.44	33	3.67
Co^{2+}	0.44	28	3.64
Ni^{2+}	0.46	13	4.21

Source. Reprinted from *Geochimica et Cosmochimica Acta,* 35, 302 (1971).

stroms and H_m in kilocalories per mole. From the equation it is understood that if the components are structurally similar (they have the same Madelung constant and coordination number n), there is a polar isomorphism: the smaller ion replaces the larger ion more easily than the larger ion replaces the smaller ion because the mixing energy has a minimum value when $X_1 \rightarrow 0$ and $X_2 \rightarrow 1$, that is, $R \rightarrow R_2$ (solid solution with the larger ion as the principal component). Nonetheless, it is known that polarity of this type is not important if the components have different bonding characteristics. Therefore, when the components are sufficiently distinct in their bonding character and in their structure, the solubility of the more covalent component is greater than in the reverse case. In this way Urusov (1970) arrived at an additional polarity rule: "the solubility of a covalent compound in a less covalent compound is greater than in the opposite situation, or, the atoms which form the more covalent character bonds (generally the chalcophile or siderophile elements) enter more easily into the structures formed by atoms with the more ionic character bonds (usually the lithophile elements) than in the opposite case." It is known that the chalacophile and siderophile elements dispersed in the silicate minerals correspond to the greater portion of these elements in the crust of the Earth, although other factors exist that influence the dispersion of the chemical elements.

In 1971 Hess studied a polymer model of pseudomagmas of silicates and concluded that the cations of elevated field strength formed the more polymerized fusions and these are characterized by liquid immiscibility in magma compositions of elevated silica contents. Therefore the cation with an elevated

Li
1 IV 0.68
VI 0.82

Be
2 III 0.25
IV 0.35

Table 3-9 Ionic Radii Based on the Criteria of Radius Ratio

Na
1 IV 1.07
V 1.08
VI 1.10
VII 1.21
VIII 1.24
IX 1.40

Mg
2 IV 0.66
V 0.75
VI 0.80
VIII 0.97

K	Ca	Sc	Ti	V	Cr	Mn	Fe
1 VI 1.46	2 VI 1.08	3 VI 0.83	2 VI 0.94	2 VI 0.87	2 VI L 0.81	2 VI L 0.75	2 IV H 0.71
VII 1.54	VII 1.15	VIII 0.95	3 VI 0.75	3 VI 0.72	H 0.90	H 0.91	VI L 0.69
VIII 1.59	VIII 1.20		4 V 0.61	4 VI 0.67	3 VI 0.70	VIII 1.01	H 0.86
IX 1.63	IX 1.26		VI 0.69	5 IV 0.44	4 IV 0.52	3 V 0.66	3 IVH 0.57
X 1.67	X 1.36			V 0.54	VI 0.63	VI L 0.66	VIL 0.63
XII 1.68	XII 1.43			VI 0.62	5 IV 0.43	H 0.73	H 0.73
					6 IV 0.38	4 VI 0.62	
						6 IV 0.35	
						7 IV 0.34	

Rb	Sr	Y	Zr	Nb	Mo	Tc	Ru
1 VI 1.57	2 VI 1.21	3 VI 0.98	4 VI 0.80	2 VI 0.79	3 VI 0.75	4 VI 0.72	3 VI 0.76
VII 1.64	VII 1.29	VIII 1.10	VII 0.86	3 VI 0.78	4 VI 0.73		4 VI 0.70
VIII 1.68	VIII 1.33	IX 1.18	VIII 0.92	4 VI 0.77	5 VI 0.71		
X 1.74	X 1.40			5 IV 0.40	6 IV 0.50		
XII 1.81	XII 1.48			VI 0.72	V 0.58		
				VII 0.74	VI 0.68		
					VII 0.79		

Cs	Ba	La–Lu	Hf	Ta	W	Re	Os
1 VI 1.78	2 VI 1.44		4 VI 0.79	3 VI 0.75	4 VI 0.73	4 VI 0.72	4 VI 0.71
VIII 1.82	VII 1.47		VIII 0.91	4 VI 0.74	6 IV 0.50	5 VI 0.60	
IX 1.86	VIII 1.50			5 VI 0.72	VI 0.68	6 VI 0.60	
X 1.89	IX 1.55			VIII 0.77		7 IV 0.48	
XII 1.96	X 1.60					VI 0.65	
	XII 1.68						

Fr	Ra	Ac–Lw
	2 VIII 1.56	
	XII 1.72	

La	Ce	Pr	Nd	Pm	Sm	Eu
3 VI 1.13	3 VI 1.09	3 VI 1.08	3 VI 1.06	3 VI 1.04	3 VI 1.04	2 VI 1.25
VII 1.18	VIII 1.22	VIII 1.22	VIII 1.20		VIII 1.17	VIII 1.33
VIII 1.26	IX 1.23	4 VI 0.86	IX 1.17			3 VI 1.03
IX 1.28	XII 1.37	VIII 1.07				VII 1.11
X 1.36	4 VI 0.88					VIII 1.15
XII 1.40	VIII 1.05					

Ac	Th	Pa	U	Np	Pu	Am
	4 VI 1.08	4 VIII 1.09	3 VI 1.12	2 VI 1.18	3 VI 1.09	3 VI 1.08
	VIII 1.12	5 VIII 0.99	4 VII 1.06	3 VI 1.10	4 VI 0.88	4 VIII 1.03
	IX 1.17	IX 1.03	VIII 1.08	4 VIII 1.06	VIII 1.04	
			IX 1.13			
			5 VI 0.84			
			VII 1.04			
			6 II 0.53			
			IV 0.56			
			VI 0.81			
			VII 0.96			

Source. Reprinted from *Geochimica et Cosmochimica Acta*, **34**, 952–953 (1970).

(from Whittaker and Muntus, 1970)

B	C	N	O	F
3 III 0.10 IV 0.20	–	–	2 II 1.27 III 1.28 IV 1.30 VI 1.32 VIII 1.34	1 II 1.21 III 1.22 IV 1.23 VI 1.25

Al	Si	P	S	Cl
3 IV 0.47 V 0.56 VI 0.61	4 IV 0.34 VI 0.48	5 IV 0.25	2 IV (1.56) VI (1.72) VIII (1.78) 6 IV 0.20	1 IV (1.67) VI (1.72) VIII (1.65) 5 III 0.20 7 IV 0.28

Co	Ni	Cu	Zn	Ga	Ge	As	Se	Br
IVH 0.65 VIL 0.73 H 0.83 VIL 0.61 H 0.69	2 VI 0.77 3 VIL 0.64 H 0.68	1 II 0.54 2 IVsq 0.70 V 0.73 VI 0.81	2 IV 0.68 V 0.76 VI 0.83 VIII 0.98	3 IV 0.55 V 0.63 VI 0.70	4 IV 0.48 VI 0.62	5 IV 0.42 VI 0.58	2 VI (1.88) VIII (1.90) 6 IV 0.37	1 VI (1.88) VIII (1.84) 7 IV 0.34

Rh	Pd	Ag	Cd	In	Sn	Sb	Te	I
VI 0.75 VI 0.71	1 II 0.67 2 IVsq 0.72 VI 0.94 3 VI 0.84 4 VI 0.70	1 II 0.75 IVsq 1.10 V 1.20 VI 1.23 VIII 1.38 3 IVsq 0.73	2 IV 0.88 V 0.95 VI 1.03 VII 1.08 VIII 1.15 XII 1.39	3 VI 0.88 VIII 1.00	2 VIII 1.30 4 VI 0.77	3 IVpy 0.85 V 0.88 5 VI 0.69	4 III 0.60	I VI (2.13) VIII (1.97) 5 VI 1.03

Ir	Pt	Au	Hg	Tl	Pb	Bi	Po	At
VI 0.81 VI 0.71	2 IVsq 0.68 4 VI 0.71	3 IVsq 0.78	1 III 1.05 2 II 0.77 IV 1.04 VI 1.10 VIII 1.22	1 VI 1.58 VIII 1.68 XII 1.84 3 VI 0.97 VIII 1.08	2 IVpy 1.02 VI 1.26 VIII 1.37 IX 1.41 XI 1.47 XII 1.57 4 VI 0.86 VIII 1.02	3 V 1.07 VI 1.10 VIII 1.19	4 VIII 1.16	

Gd	Tb	Dy	Ho	Er	Tm	Yb	Lu
VI 1.02 VII 1.12 III 1.14	3 VI 1.00 VII 1.10 VIII 1.12 4 VI 0.84 VIII 0.96	3 VI 0.99 VIII 1.11	3 VI 0.98 VIII 1.10	3 VI 0.97 VIII 1.08	3 VI 0.96 VIII 1.07	3 VI 0.95 VIII 1.06	3 VI 0.94 VIII 1.05

Cm	Bk	Cf	Es	Fm	Md	No	Lw
VI 1.06 VIII 1.03	3 VI 1.04 4 VIII 1.01	3 VI 1.03					

field strength has a greater affinity for free oxygen ions than a cation with a lesser field strength. Hess defined the field strength by the relation: Z/a^2 in which Z is the ionic charge and a is the sum of the ionic radii of the cation and oxygen. In the laboratory Hess added the metallic oxide of the cation being investigated to a fusion of silica and determined the gel points for several elements (see Table 3-8).

Finally, it should be noted that with the development of concepts to explain the distribution and dispersion of the elements in natural materials, there are complimentary studies that result in greater accuracy in the measurements of the parameters used in theoretical calculations. As an example one may cite the work of Whittaker and Muntus (1970) on the ionic radii that should be used in geochemistry when the investigations deal with silicates. These authors observed that the new radii (see Table 3-9) represent an important improvement over the values used previously in research on silicates. The values presented in Table 3-9 correspond to a coordination of 6 for the elements, with a radius for oxygen of 1.32 Å, and to a coordination of 6, with an ionic radius for fluorine of 1.25 Å. Whittaker and Muntus recognized that their values are notably greater than those that were used up to the time of their publication (Goldschmidt, 1926; Ahrens, 1952; Pauling, 1960), but were less than the values published by Shannon and Prewitt (1969, 1970).

Studies on the distribution and dispersion of the elements in natural and artificially prepared materials are fundamental to contined advances in geochemistry and must be continued with both practical and theoretical investigations. These alone can further our bases for understanding, explaining, and predicting such distributions and dispersions in the natural environments represented by rock systems.

REFERENCES

Ahrens, L. H., 1952, The use of ionization potentials – 1. Ionic radii of the elements, *Geochim. Cosmochim. Acta,* **2,** 155–169.

Ahrens, L. H., 1964, The significance of the chemical bond for controlling the geochemical distribution of the elements, in *Physics and Chemistry of the Earth,* Part 1, Vol. 5 (Ahrens, Press, and Runcorn, eds.), Pergamon, New York, pp. 1–54.

Ahrens, L. H., 1965, *Distribution of the Elements in Our Planet,* McGraw-Hill, New York, 110 pp.

Bethe, H., 1929, Splitting of terms in crystals, Ann. Phys., **3,** 133–206.

Brett, R., 1971, The earth's core: Speculation of its chemical equilibrium with the mantle, *Geochim. Cosmochim. Acta,* **35,** 203–222.

Burns, R. G., 1970, *Mineralogical Applications of Crystal Field Theory,* Cambridge University Press, England, 224 pp.

Burns, R. G., Clark, R. H., and Fyfe, W. S., 1964, Crystal-field theory and application to problems in geochemistry, in *Chemistry of the Earth's Crust,* Vernadsky Centennial Symposium, Vol. 2 (Vinogradov, ed.), Moscow, pp. 88–106.

Burns, R. G., and Fyfe, W. S., 1966, Distribution of elements in geological processes, *Chem. Geol.,* 1, 49–56.

Burns, R. G., and Fyfe, W. S., 1967, Trace element distribution rules and their significance, *Chem. Geol.,* 2, 89–104.

Burns, R. G., and Fyfe, W. S., 1967, Crystal-field theory and the geochemistry of transition elements, in *Researches in Geochemistry,* Vol. 2, (Abelson, ed.), Wiley, New York, pp. 259–285.

Chem. Eng. News, 1967, Chemistry and the solid earth, 45, 1A–50A.

Clark, S. P. Jr., and Ringwood, A. E., 1964, Density distribution and constitution of the mantle, *Rev. Geophys.,* 2, 35–88.

Clarke, F. W., and Washington, H. S., 1924, The composition of the earth's crust, *U.S. Geol. Surv. Profess. Paper,* 127, 117 pp.

Curtis, C. D., 1964, Applications of the crystal field theory to the inclusion of trace transition elements in minerals during magmatic differentiation, *Geochim. Cosmochim. Acta,* 28, 389–403.

Elson, J., 1969, An empirical bond energy equation, *Chem. Commun.,* 448–450.

Fyfe, W. S., 1964, *Geochemistry of Solids – an Introduction,* McGraw-Hill, New York, 199 pp.

Goldschmidt, V. M., 1926, Geochemische Verteilungsyesetze der Elemente VII. Die Gesetze der Kristallochemie, *Skrfler Norske Videnskaps-Akad. Oslo, I, Mat. Naturv. Kl.,* 2, 5–116.

Goldschmidt, V. M., 1937, The principles of distribution of chemical elements in minerals and rocks, *J. Chem. Soc.,* 655–673.

Goldschmidt, V. M., 1954, *Geochemistry,* Clarendon, Oxford, 730 pp.

Green, J., 1959, Geochemical table of the elements for 1959, *Geol. Soc. Am. Bull.,* 70, 1127–1184.

Hess, P. C., 1971, Polymer model of silicate melts, *Geochim. Cosmochim. Acta,* 35, 289–306.

Krauskopf, K. B., 1967, *Introduction to Geochemistry,* McGraw-Hill, New York, 721 pp.

Mason, B., 1966, *Principles of Geochemistry,* Wiley, New York, 329 pp.

Mulliken, R. S., 1934, A new electronaffinity scale; together with data on valence states and on valence ionization potentials and electron affinities, *J. Chem. Phys.,* 2, No. 11, 782–793.

Nockolds, S. R., 1966, The behavior of some elements during fractional crystallization of magma, *Geochim. Cosmochim. Acta,* 30, 267–278.

Pauling, L., 1960, *The Nature of the Chemical Bond,* 3rd edition, Cornell University Press, Ithaca, New York, 644 pp.

Poldervaart, A. ed., 1955a, *Crust of the Earth, Geol. Soc. America Special Paper,* 62, 762 pp.

Poldervaart, A., 1955b, Chemistry of the earth's crust, *Geol. Soc. Am. Special Papers,* 62, 119–144.

Ringwood, A. E., 1955a, The principles governing trace element distribution during magmatic crystallization. Part I. The influence of electronegativity, *Geochim. Cosmochim. Acta,* 7, 189–202.

Ringwood, A. E., 1955b, The principles governing trace element behavior during magmatic crystallization. Part II. The role of complex formation, *Geochim. Cosmochim. Acta,* 7, 242–254.

Ringwood, A. E., 1966a, Chemical evolution of the terrestrial planets, *Geochim. Cosmochim. Acta,* **30,** 41–104.

Ringwood, A. E., 1966b, The chemical composition and origin of the earth, in *Advances in Earth Science* (Hurley, ed.), M.I.T. Press, Cambridge, Mass.

Ringwood, A. E., 1971, Core-mantle equilibrium: Comments on a paper by R. Brett, *Geochim. Cosmochim. Acta,* **35,** 223–229.

Schubert, G., Turcotte, D. L., and Oxburgh, E. R., 1970, Phase change instability in the mantle, *Science,* **169,** 1075–1077.

Schwarcz, H. P., 1967, The effect of crystal field stabilization on the distribution of transition metals between metamorphic minerals, *Geochim. Cosmochim. Acta,* **31,** 503–517.

Sederholm, J. J., 1924, The average composition of the earth's crust, *Bull. Comm. Geol.,* Finlande, **70,** 3–20.

Shannon, R. D., and Prewitt, C. T., 1969, Effective crystall radii in oxides and fluorides, *Acta Cryst.,* **B25,** 925–946.

Shannon, R. D., and Prewitt, C. T., 1970, Revised values of effective ionic radii, *Acta Crys.,* **B26,** 1046–1048.

Tauson, L. V., 1965, Factors in the distribution of trace elements during the crystallization of magmas, in *Physics and Chemistry of the Earth* (Ahrens, Press, Runcorn, and Urey, ed.), Vol. 6, Pergamon, New York, pp. 215–250.

Taylor, S. R., 1964, Abundance of chemical elements in the continental crust: a new table, *Geochim. Cosmochim. Acta,* **28,** 1273–1285.

Turekian, K. K., and Wedepohl, K. H., 1961, Distribution of the elements in some major units of the Earth's crust, *Geol. Soc. Am. Bull.,* **72,** 175–192.

Urusov, V. S., 1970, Dependence on the limits of isomorphous miscibility on the difference in the character of chemical bonding and structure of the components, *Geokhimiya,* **21,** 59–65.

Vinogradov, A. P., 1962, Average contents of chemical elements in the principal types of igneous rocks of the Earth's crust, *Geochemistry,* 641–664.

Whittaker, E. J. W. and Muntus, R., 1970, Ionic radii for use in geochemistry, *Geochim. Cosmochim. Acta,* **34,** 945–956.

4

WEATHERING AND THE FORMATION OF
THE GEOCHEMICAL SAMPLE OF SOIL

The mineralogical phases with their respective distributions of chemical elements are developed according to the physical–chemical conditions of the
environment of formation during the crystallization of the primary rocks, be
they plutonic or effusive. These rocks may be fairly stable while they are at
conditions generally close to those of the original formation, but some chemical adjustments (towards greater stability) can take place by means of solid-
state diffusion if differences in the chemical potentials of the rock components
exist. Such changes tend to establish the optimum chemical equilibrium and
are rather slight and slow if they are compared with those that occur when the
rocks are transported to significantly different environments in which new
chemical equilibria have to be established with the existing physical–chemical
conditions. The most severe physical–chemical changes take place when the
rocks are exposed to the Earth's surface environment where, in general, the
maximum chemical disequilibrium in a system of primary rocks – hydrosphere
– atmosphere – biosphere can be expected. Here the weathering process acts
to establish a chemical equilibrium between the components of the system
and thus is of utmost importance to the geochemist who is working in prospecting because such a process provides him with the samples that may yield
data on local hidden mineralization and/or on the potential existence of major
and minor mineralization in a wide region.

Weathering may act, depending upon environment, with a tripartite effect
that results in three basic products. The physical or mechanical subeffect may
be due to the action of ice (wedging, crushing, plucking), to the growth of
crystals of minerals or ice (wedging, uplift caused by freezing, colloidal plucking), to load discharge by removal of overlying material (expansion with less
pressure), expansion and contraction of varying degrees in different minerals
due to repeated great variations of temperatures that act during a long period
of time (in combination with humidity), and to the activity of organisms (root
action, borings, tracks, and "destructive" man); this subeffect serves to prepare

rocks for chemical and biological attack by opening pathways and exposing more surface area for such attack. As for the chemical subeffect, leaching takes place by means of solution by water, various inorganic acids (H_2CO_3, HNO_3, H_2SO_4), of which carbonic acid is the most important in the majority of weathering environments, and organic (humic) acids; with solution, other chemical processes such as hydration, ionization and hydrolysis, carbonation, oxidation and reduction, and chelation by organic compounds may contribute to the weathering process. The biological–biochemical subeffect works, as already inferred, to prepare rocks for decomposition by its physical destructive action and to liberate and/or bind ions via sequestering by organic compounds in the weathering environment. All the factors mentioned act with greater or lesser intensity and at varying kinetics depending on the five classical factors that create the weathering environment: the climate (precipitation and temperature), the relief (topography), the vegetation, the mother rock, and geologic time.

The three basic products of the disintegration and decomposition are: (1) detrital material that can be eroded and transported to temporary or final sedimentary deposition zones by gravity, waters, winds, and glaciers; (2) soluble material or particulate matter in suspension that enters the hydrologic system and is maintained in the soluble or colloidal suspension form until it reacts chemically to form solids or joins physically to form larger than colloid masses; and (3) detrital material, soluble material, and material in suspension which remain at or close to the weathering site and develop as parts of a residual soil. The stream sediments, residual soils, and sometimes waters are principal samples in geochemical prospecting (in addition, of course, to *in situ* rocks themselves) and may influence greatly the biogeochemical and gaseous samples that recently have become more important as samples in geochemical exploration programs. Goldschmidt (1937) reported the intense geochemical segregation that takes place during the preparation and transport of sedimentological material and presented a general classification for the final products (Table 4-1); however, his classification did not take into account the secondary economic minerals that are formed from the primary minerals and that also serve in the search for unaltered mineral deposits.

The processes that combine to effect rock decomposition have been studied in detail for some time. More than 30 years ago, Goldich (1938) related the decomposition sequences to the susceptibility of the mineralogical components of igneous rocks as represented by the reaction series described by Bowen (1928). As taught in petrology, the Bowen reaction series or reaction principle describes two characteristic reactions that take place during the crystallization of a magma: one, designated "continuous," is a solid solution series (for example, the plagioclase feldspars) in which the first crystals formed continually change their chemical compositions by interaction with the magma without

Table 4-1 Classification of the Principal Sedimentary Products from the Weathering, Erosion, Transportation, and Deposition of Materials from Preexisting Rocks (from Goldschmidt, 1937)

Resistates	Oxidates	Hydrolyzates	Carbonates	Evaporates	Reduzates	Biodates
Si	Fe, Mn	Al, Si, (K)	Ca, Mg	Na, Ca, Mg, B	C, S, HC, S^{2-}	Ca, Mg, Si, P
Quartz	Goethite	Clays	Calcite	Halite	Coal	Calcite
Zircon	Limonite	Boehmite	Dolomite	Gypsum	Petroleum	Chert
Magnetite	Pyrolusite	Bauxite	Aragonite	Anhydrite	Pyrite	Phosphorite
Ilmenite		Black Shales		Epsomite	Sulfur	Aragonite
Rutile				Borates		
Monazite				Calcite		
Cassiterite				Dolomite		
Au, Pt						

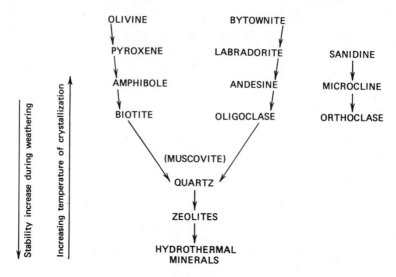

Figure 4-1. The relation between the sequence of crystallization of minerals from a basaltic magma and their stabilities with respect to weathering. (Compiled from Bowen, 1928 and Goldich, 1938.)

changing their basic crystalline structures; and, another designated "discontinuous," (for example, the ferromagnesium minerals) in which the crystals formed initially react with the magma to give a solid with a different crystallographic structure and a distinct chemical composition (Figure 4-1). Goldich (1938) proposed that because minerals such as olivine and calcic plagioclase in a basalt, for example, crystallize at higher temperatures than other common mineralogic phases, they are more out of their stability field (than many of the common igneous rock forming minerals) when they are exposed to the condi-

Table 4-2   Relative Stability of Heavy Minerals in Recent Sedimentary Rocks (from Pettijohn, 1941)

22. Olivine	13. Andalusite	4. Garnet
21. Actinolite	12. Hornblende	3. Monazite
20. Diopside	11. Epidote	2. Tourmaline
19. Hyperstene	10. Kyanite	1. Zircon
18. Sillimanite	9. Staurolite	−1. Rutile
17. Augite	8. Magnetite	−2. Muscovite
16. Zoisite	7. Ilmenite	−3. Anatase
15. Sphene	6. Apatite	
14. Topaz	5. Biotite	

Table 4-3 Stability Indices of Minerals in the Clay-Size Fraction (−2 μm)
(from Jackson, 1956)

Stage of Weathering	Dominant Minerals in the Different Stages
1	Gypsum (also halite, $NaNO_3$, NH_4Cl, etc.)
2	Calcite (also dolomite, aragonite, apatite, etc.)
3	Olivine-hornblende (also pyroxenes, diopside, etc.)
4	Biotite (also glauconite, Fe–Mg chlorite, antigorite, nontronite, etc.)
5	Albite (also anorthite, stilbite, microcline, orthoclase, etc.)
6	Quartz (also cristobalite, etc.)
7	Muscovite (also illite, sericite, Al-Mg chlorite from the weathering of mica)
8	Vermiculite (also hydrated mica, layered 2:1 interstratified silicates)
9	Montmorillonite (also beidellite, saponite, etc.)
10	Kaolinite (also halloysite, etc.)
11	Gibbsite (also boehmite, allophane, etc.)
12	Hematite (also goethite, limonite, pyrolusite, etc.)
13	Anatase (also zircon, rutile, ilmenite, leucoxene, corundum, etc.)

tions at the Earth's surface; thus, they are more susceptible to the rigors of weathering. Also, with respect to the plagioclase feldspars found in igneous rocks, the more calcic phases are more susceptible to weathering breakdown because they are more outside their stability field with respect to temperature than the sodic plagioclase phases. When laboratory investigations were made on the relative rates of mineral decomposition, the resulting data corroborated the concept put forth by Goldich and since then, with these data and empirical observations on rocks in various stages of weathering, reports were presented on the relative susceptibility of many minerals during the weathering of a mother rock (Table 4-2), on the use of minerals in soils as a general measure of intensity of weathering (Table 4-3), and on "quantitative measurements" of this intensity (Table 4-4). Goldich (1938) suggested a method that could be used to establish relative losses or gains of chemical elements during various steps of weathering, basing it on a minimum loss of Al_2O_3 during leaching (because of its greater relative increase in the products of physical–chemical–biological breakdown processes and on the small quantity present in waters deriving from zones of much chemical activity): using the ratio between the quantity of Al_2O_3 in the mother rock and the quantity of Al_2O_3 in the altered

Table 4-4   Weathering Potential Index (WPI) of Selected Minerals (from Loughnan, 1969)

$$WPI = \frac{100 \times \text{moles } (Na_2O + K_2O + CaO + MgO - H_2O)}{\text{moles } (Na_2O + K_2O + CaO + MgO + SiO_2 + Al_2O_3 + Fe_2O_3)}$$

Mineral	WPI	Mineral	WPI	Mineral	WPI
Forsterite	66	Talc	29	Quartz	0
Olivine	54	Nepheline	25	Sillimanite	0
Wollastonite	50	Anorthite	25	Muscovite	−10.7
Enstatite	50	Epidote	23	Analcite	−17
Diposide	50	Biotite	22	Pyrophyllite	−20
Tremolite	40	Leucite	17	Kaolinite	−67
Augite	39	Albite	13	Boehmite	−100
Hornblende	36	Orthoclase	12	Gibbsite	−300

Source. *Chemical Weathering of the Silicate Minerals,* American Elsevier Publishing Company, Inc.

Table 4-5a   Analysis of a Quartz-Feldspar Biotite Gneiss and Its Weathering Products (%). Column I Represents Fresh Rock, and II, III and IV Represent Gradually Increasing Degrees of Weathering of the Mother Rock (from Goldich, 1938)

Oxide	I	II	III	IV
SiO_2	71.54	68.09	70.30	55.07
Al_2O_3	14.62	17.31	18.34	26.14
Fe_2O_3	0.69	3.86	1.55	3.72
FeO	1.64	0.36	0.22	2.53
MgO	0.77	0.46	0.21	0.33
CaO	2.08	0.06	0.10	0.16
Na_2O	3.84	0.12	0.09	0.05
K_2O	3.92	3.48	2.47	0.14
H_2O	0.32	5.61	5.88	10.39
Others	0.65	0.56	0.54	0.58
Totals	100.07	99.91	99.70	100.11
Minerals				
Quartz	30	40	43	25
K-Feldspar	19	18	13	1
Plagioclase	40	1	1	?
Biotite + chlorite	7	t	t	0.2
Hornblende	1	0	0	t
Magnetite, ilmenite, oxides	1.5	5	2	6
Kaolinite	0	36	40	66

Table 4-5b General Calculations of Gains and Losses of Chemical Elements During Weathering (%) from Data Given in Table 4-5a (from Krauskopf, 1967)

Oxide	I	III	A	B	C
SiO_2	71.48	70.51	55.99	−15.49	−22
Al_2O_3	14.61	18.40	14.61	0	0
Fe_2O_3	0.69	1.55	1.23	+0.54	+78
FeO	1.64	0.22	0.17	−1.47	−90
MgO	0.77	0.21	0.17	−0.60	−78
CaO	2.08	0.10	0.08	−2.00	−96
Na_2O	3.84	0.09	0.07	−3.77	−98
K_2O	3.92	2.48	1.97	−1.95	−50
H_2O	0.32	5.90	4.68	+4.36	+1360
Others	0.70	0.54	0.43	−0.27	−39

Source. *Introduction to Geochemistry,* with permission of McGraw-Hill Book Company. Copyright © 1967 by McGraw-Hill, Inc.

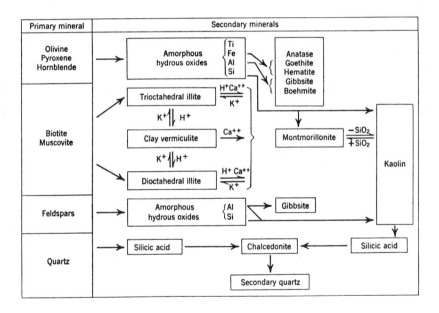

Figure 4-2. Weathering sequences of primary igneous rock-forming minerals. (From Fieldes and Swinedale, 1954.)

rock, the chemical analyses of the altered rock can be normalized and the relative gains and losses of the elements after weathering can be calculated and the results presented in percent (Table 4-5). In Table 4-5 some of the principle

Table 4-6 Behavior of Selected Ore Minerals during Oxidation (from NASA, 1968)

Element	Primary Minerals	Oxidized/Secondary Minerals	Mobility (Relative Rate of Dispersion)
Fe	Siderite, silicates Hematite Pyrite Pyrrhotite	Limonite Limonite Limonite Hydrated ferric silicate	Moderate Slow Fast action of $Fe_2(SO_4)_3$ as oxidizing agent Fast
Cu	Many sulfides	Sulfates	One of the most readily transported, migrates, forms supergene sulfide zone/upper oxidized zone
Zn	Sphalerite (oxides, sulfides)	Smithsonite Calamine	Sulfates and chloride very soluble. Carbonate, silicate difficult; zinc is generally mobile, easily oxidized in presence of pyrite which is common; migrates, forms supergene deposits
Pb	Galena and other mixed sulfides	Cerussite Anglesite	Slight mobility in oxidized zones (somewhat mobile in presence of ferric sulfate and chlorides)
Au	Gold Amalgam Tellurides		Slight mobility in form of chloride, precipitates fast in reducing conditions
Ag	Argentite Native silver Telluride Minor in other sulfides/sulfosalts	Cerargyrite	Mobile in presence of $Fe_2(SO_4)_3$ Less than Cu, Zn, more than gold
Pt	Platinum		Highly resistant to oxidation
Pd	Sperrylite (Pt, As)		Highly resistant to oxidation
Hg	Cinnabar	Calomel Hg_2Cl_2 Oxychlorides	Not easily attacked, not very mobile in H_2SO_4 acid, chlorides, or alkaline sulfides

Element	Primary minerals	Secondary minerals	Mobility
Cd	Sphalerite Greenockite	Greenockite	As mobile as zinc
Ni/Co	Sulfides Arsenides	Arsenates	Very mobile in sulfate solutions
Cr	Chromite	Chromates	Very insoluble
Mn	Manganosiderite $(Mn, Fe)CO_3$ Oxides, sulfides	Pyrolusite (MnO_2) Hydrated oxides and silicates	More soluble than Fe; less easily precipitated; easily transported as sulfate or bicarbonate
Sn	Cassiterite		Very resistant, immobile
W	Scheelite $(CaWO_4)$ Wolframite $(Fe, Mn)WO_4$		Very resistant, highly immobile
V		Vanadinite and other vanadates	Complex oxysalts mobile in groundwater
U	Uraninite	Hydrated oxides and many salts	Very mobile, easily oxidized
Mo	Molybdenite	Molybdite Molybdic ocher Wulfenite	Not very mobile in solution Limited transport and enrichment by colloidal processes
Bi	Bismuthinite (Bi_2S_3) Native Bi Pb–Br sulfites	Bismite $Bi_2O_3 \cdot 3H_2O$ Bismuthite $(Bi_2O_3 \cdot CO_2 \cdot H_2O)$	Like lead, is difficult to dissolve, no great mobility
As	Arsenopyrite $(FeAsS)$/and other arsenides	Arsenolite As_2O_3 Hydrated arsenates	Mobile in presence of Na, K salts. Arsenic is fairly mobile (more than antimony)
Sb	Stibnite Antimonides	Cervantite (Sb_2O_4) Other hydrated oxides	Considerably less mobile than arsenic

mineralogical changes that take place during weathering are also given; such changes are presented graphically in Figure 4-2. Theoretical studies, laboratory investigations, and information such as that in Table 4-5 and Figure 4-2, which in general show clearly that depending on the minerals that comprise a rock, the rock textures (shape and arrangement of minerals), the climate, the drainage, and the time during which weathering proceeds, there is an order of elemental release and, hence, mineral and rock decomposition. The elements Ca, Mg, and Na, for example, respond most rapidly to decomposition, followed in rate of loss by K and Si; all these major elements can be solubilized and carried from the site of leaching out of the immediate hydrologic system. Also, as has been pointed out by other authors, during weathering there is a general reorganization of the chemical components towards more open crystallographic structures than were present in the primary minerals of the system. As is seen from Tables 4-2, 4-3, and 4-4, researchers worked principally with silicates, oxides and hydroxides, and carbonates in their studies on weathering, but often did not consider to any degree the sulfide and other economically very important minerals. However, NASA (1968) published information on the comportment of several economic primary minerals during weathering attack as well as on the common secondary products formed from them (Table 4-6); this latter fund of information serves to complete the consideration of the weathering panorama from the point of view of geochemical prospecting.

Perhaps the two chemical properties of waters that most influence the different chemical reactions produced during the weathering of a rock are the pH and the Eh. The pH is equal to the negative logarithm of the hydrogen ion concentration (more precisely hydrogen ion activity) of a solution and is a measure of its acidity or alkalinity on a scale ranging from 0 (acidic) to 14 (alkaline), with 7 being the pH value for neutral solutions. The Eh is a measure of the oxidation potential [loss of electron(s) from a cation or capture of electron(s) by an anion] or reduction potential [capture of electron(s) by a cation or loss of electron(s) from an anion]. The potential of an element for reduction or oxidation is measured in volts (or millivolts), comparing its potential with that of hydrogen which has the potential for reduction according to the equation $H_2 = 2H^+ + 2\,\bar{e}$, arbitrarily set at 0.00 volts when all the reactants are at unit activity in the measuring system and the reaction takes place at standard conditions of 25°C and 1 atm pressure.

The pH and Eh are linked by the Nernst relation:

$$Eh \text{ (in volts)} = \frac{0.059}{2} \log \frac{[H^+]}{P_{H_2}} = -0.059 \text{ pH} - \frac{0.059}{2} \log P_{H_2}$$

In order to establish the limiting relations of pH and Eh in natural environments (presence of water), the limits are taken as the reactions by which H_2O is oxidized or reduced. In the case of reduction, that is, the evolution of H_2 from H_2O, the Eh–pH link is expressed as:

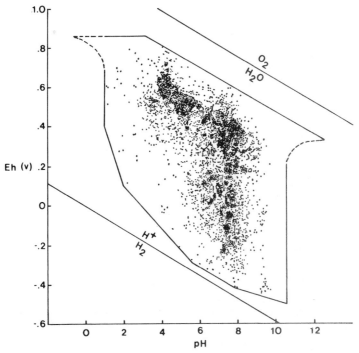

Figure 4-3. Distribution of Eh–pH measurements in natural aqueous environments. The environments investigated include meteoric waters, carbonaceous swamps, soils, subsurface waters, mine waters, fresh water lakes and rivers, fresh water sediments, near-shore marine sediments, seawater, open sea marine sediments, evaporites, geothermal waters, and connate waters. (Reproduced from Bass Becking, Kaplan, and Moore, 1960, *J. of Geol.,* **68**, 243–284.)

$$Eh = -0.059 \, pH - 0.0295 \log P_{H_2}$$

When the partial pressure of hydrogen, P_{H_2}, is equal to 1,

$$Eh = -0.059 \, pH - 0.0295 \log (1)$$

and therefore

$$Eh = - \, 0.059 \, pH$$

For the oxidation of water, $2H_2O = O_2 + 4H^+ + 4e^-$, the Eh–pH relation is given as

$$Eh = E^0 + \frac{0.059}{4} \log \frac{[P_{O_2}] \, [H^+]^4}{[H_2O]^2}$$

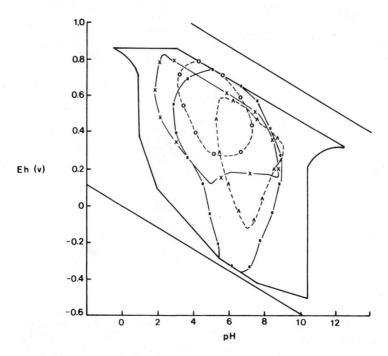

Figure 4-4. Approximate distribution of Eh and pH for meteoric waters, soils, shallow subsurface waters, and oxidized mine waters in environments related to the weathering process. (○) Meteoric waters; (●) soils; (Λ) shallow subsurface waters; and (X) oxidized mine waters. (Reproduced from Bass Becking, Kaplan and Moore, 1960, *J. of Geol.*, **68**, 243–284.)

in which E^0 is the standard potential which is equal to 1.23 volts. Therefore,

$$Eh = 1.23 + \frac{0.059}{4} \log \frac{P_{O_2}}{[H_2O]^2} + \frac{4(0.059)}{4} \log [H^+]$$

$$= 1.23 + 0.015 \log \frac{P_{O_2}}{[H_2O]^2} - 0.059 \text{ pH}$$

For the decomposition of water for a dilute solution in which $[H_2O] = 1$ and $P_{O_2} = 1$, the equation is simplified to

$$Eh = 1.23 - 0.059 \text{ pH}$$

Plots of Eh against pH can be made using the two limiting equations and there results a field of values bounded by two straight and parallel lines (Figure 4-3).

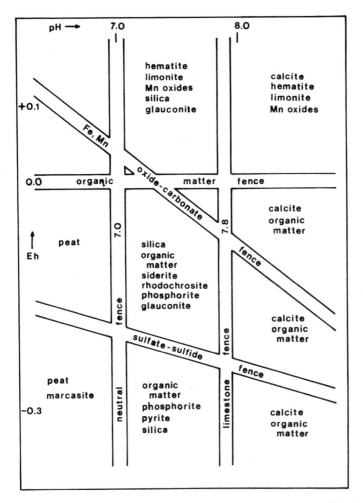

Figure 4-5. Sedimentary associations in relation to environments with specific pH and redox potential. (Reproduced from Krumbein and Garrels, 1952, *J. Geol.,* **60**, 1–33.)

The values for pH and Eh in almost all the natural environments of the Earth's surface fall between the limiting values indicated by the two calculated straight line plots. Bass Becking, Kaplan and Moore (1960) have measured the limits of pH and of oxidation–reduction potential (Eh) in natural environments (Figure 4-3), and the diagrammatic data for natural values measured for environments characteristic of meteoric waters, soils, subsurface waters, and oxidized mine waters are presented in Figure 4-4. The data of Bass Becking, Kaplan, and Moore, based on real measurements, correspond well with those of Krumbein and Garrels (1952) which were derived from theoretical calculations and reviewed in their famous fence diagram for sedimentary materials (Figure 4-5).

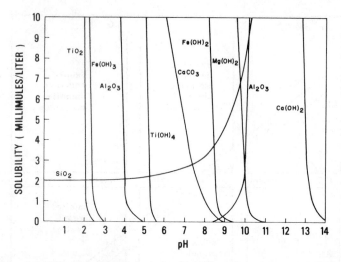

Figure 4-6. Solubility changes with variation in pH for some components liberated by chemical weathering. (Reproduced from Loughnan, 1969, *Chemical Weathering of the Silicate Minerals,* Elsevier, New York, 154 pp.)

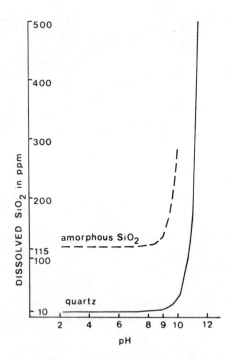

Figure 4-7. Solubility of crystalline and amorphous silica in relation to pH (at 25°C).

Each one of these factors has an influence on the solubility of minerals and the field of stability of minerals or their ions or component compounds. For example, laboratory experiments have been performed from which it can be readily understood how the solubility of some compounds liberated by weathering changes with changes of the pH of the environment (Figure 4-6). Silica has a rather constant solubility between pH values of approximately 2 and 9, but when the pH exceeds 9, the solubility increases greatly (Figure 4-7); Figure 4-7 also shows that the crystallographic form influences the initial solubility of a mineral or species being studied and that amorphous silica as opal (or disordered cristobalite) is about 20 times as soluble initially as quartz, but that at a pH greater than 9, the two minerals act similarly with respect to the rate of solution and the quantity of SiO_2 in solution.

Another example of the pH influence on solubility is given by the calcium carbonates, calcite and aragonite, which are soluble at acidic pH values and show diminishing solubilities with increases of pH until they precipitate and are essentially insoluble at a pH of approximately 8.9; in this case also, there is a crystallographic effect because aragonite is 16% more soluble than calcite. Alumina shows another aspect of the effect of pH with respect to solubility in that it is almost insoluble between pH values of 4 and 10, values not generally found in common natural environments. Another interesting comparison can be made of the solubilities of ferrous and ferric hydroxides; $Fe(OH)_3$ is practically insoluble at a pH greater than 3 but very soluble in environments with an acidity less than pH 3; on the other hand, $Fe(OH)_2$ is essentially insoluble at a pH greater than 9.4 but very soluble at pH values less than 9.4. When the soil scientist or geochemist has to consider the genesis of laterites, latosols, or bauxites, for example, the relations between the solubilities of the minerals in a mother rock and their decomposition products with the pH assume great importance.

Obviously several factors influence the pH of an environment and can cause large seasonal differences in the acidity or alkalinity of waters active in weathering. The climate, for example, influences the pH by the effect of temperature; warmer water can retain less gas than colder water, and because the interaction between CO_2 and H_2O in the system

$$H_2CO_3 \rightleftharpoons H^+ + HCO_3^- \rightleftharpoons H^+ + CO_3{}^{2-}$$

results in the commonly most important control of pH, it is obvious that with a higher temperature, an environment may have a higher pH. However, in a situation with elevated temperature (and other favorable conditions) sufficient to increase the quantity of vegetation in decomposition, the decomposition may result in an addition of CO_2 to the environment with a resulting lowering of pH. It should be clear than that opposing geochemical reactions can be

Table 4-7 Standard Potentials (in volts) for Some Geochemically Interesting Reactions

Each half-reaction is given with the reduced form of the element at the left and its oxidized form at the right. Thus, strong reducing agents are towards the upper part of the table and strong oxidizing agents below. Some reactions react differently with the acidity or basicity of the test solutions so that the table includes some reactions that go only in an alkaline environment. In geochemistry, the voltages more reducing than the hydrogen electrode have the negative (−) sign and those more oxidizing than the hydrogen electrode have the positive (+) sign; this is opposite to the signs used in many chemistry texts.

Potentials in acidic solutions

$Li = Li^+ + e^-$	−3.03
$K = K^+ + e^-$	−2.93
$Ca = Ca^{2+} + 2e^-$	−2.87
$Na = Na^+ + e^-$	−2.71
$Mg = Mg^{2+} + 2e^-$	−2.37
$Th = Th^{4+} + 4e^-$	−1.90
$Al = Al^{3+} + 3e^-$	−1.66
$Mn = Mn^{2+} + 2e^-$	−1.18
$V = V^{2+} + 2e^-$	−1.18
$Si + 2H_2O = SiO_2 + 4H^+ + 4e^-$	−0.86
$Zn = Zn^{2+} + 2e^-$	−0.76
$Cr = Cr^{3+} + 3e^-$	−0.74
$Fe = Fe^{2+} + 2e^-$	−0.44
$Cr^{2+} = Cr^{3+} + e^-$	−0.41
$Cd = Cd^{2+} + 2e^-$	−0.40
$Ti^{2+} = Ti^{3+} + e^-$	−0.37
$Co = Co^{2+} + 2e^-$	−0.28
$V^{2+} = V^{3+} + e^-$	−0.26
$Ni = Ni^{2+} + 2e^-$	−0.25
$Sn = Sn^{2+} + 2e^-$	−0.14
$Pb = Pb^{2+} + 2e^-$	−0.13
$H_2 = 2H^+ + 2e^-$	0.00
$H_2S = S + 2H^+ + 2e^-$	+0.14
$Sn^{2+} = Sn^{4+} + 2e^-$	+0.15
$Cu^+ = Cu^{2+} + e^-$	+0.16
$S^{2-} + 4H_2O = SO_4^{2-} + 8H^+ + 8e^-$	+0.16
$H_2SO_3 + H_2O = SO_4^{2-} + 4H^+ + 2e^-$	+0.17
$Ag + Cl^- = AgCl + e^-$	+0.22
$As + 2H_2O = HAsO_2 + 3H^+ + 3e^-$	+0.25
$2Hg + 2Cl^- = Hg_2Cl_2 + 2e^-$	+0.27
$U^{4+} + 2H_2O = UO_2^{2+} + 4H^+ + 2e^-$	+0.33
$Cu = Cu^{2+} + 2e^-$	+0.34
$V^{3+} + H_2O = V^{2+} + 2H^+ + e^-$	+0.34

Table 4-7 (Continued)

Potentials in acidic solutions (continued)

$S + 3H_2O = H_2SO_3 + 4H^+ + 4e^-$	+0.45
$Cu = Cu^+ + e^-$	+0.52
$2I^- = I_2 + 2e^-$	+0.54
$HAsO_2 + 2H_2O = H_3AsO_4 + 2H^+ + 2e^-$	+0.56
$Pt + 4Cl^- = PtCl_4^{2-} + 2e^-$	+0.73
$Fe^{2+} = Fe^{3+} + e^-$	+0.77
$2Hg = Hg_2^{2+} + 2e^-$	+0.79
$Ag = Ag^+ + e^-$	+0.80
$Hg = Hg^{2+} + 2e^-$	+0.85
$NO + 2H_2O = NO_3^- + 4H^+ + 3e^-$	+0.96
$VO^{2+} + H_2O = VO_2^+ + 2H^+ + e^-$	+0.99
$Au + 4Cl^- = AuCl_4^- + 3e^-$	+1.00
$Fe^{2+} + 3H_2O = Fe(OH)_3 + 3H^+ + e^-$	+1.06
$HgS = S + Hg^{2+} + 2e^-$	+1.11
$2H_2O = O_2 + 4H^+ + 4e^-$	+1.23
$Mn^{2+} + 2H_2O = MnO_2 + 4H^+ + 2e^-$	+1.23
$2Cr^{3+} + 7H_2O = Cr_2O_7^{2-} + 14H^+ + 6e^-$	+1.33
$2Cl^- = Cl_2 + 2e^-$	+1.36
$Pb^{2+} + 2H_2O = PbO_2 + 4H^+ + 2e^-$	+1.46
$Au = Au^{3+} + 3e^-$	+1.50
$Mn^{2+} = Mn^{3+} + e^-$	+1.51
$Mn^{2+} + 4H_2O = MnO_4^- + 8H^+ + 5e^-$	+1.51
$Au = Au^+ + e^-$	+1.68
$Ni^{2+} + 2H_2O = NiO_2 + 4H^+ + 2e^-$	+1.75
$Co^{2+} = Co^{3+} + e^-$	+1.82
$2SO_4^{2-} = S_2O_8^{2-} + 2e^-$	+2.00
$2F^- = F_2 + 2e^-$	+2.87

Potentials in Basic Solutions

$Mg + 2OH^- = Mg(OH)_2 + 2e^-$	-2.69
$U + 4OH^- = UO_2 + 2H_2O + 4e^-$	-2.39
$Al + 4OH^- = Al(OH)_4^- + 3e^-$	-2.35
$Mn + 2OH^- = Mn(OH)_2 + 2e^-$	-1.55
$Zn + 2OH^- = Zn(OH)_2 + 2e^-$	-1.25
$SO_3^{2-} + 2OH^- = SO_4^{2-} + H_2O + 2e^-$	-0.93
$Se^{2-} = Se + 2e^-$	-0.92
$Fe + 2OH^- = Fe(OH)_2 + 2e^-$	-0.88
$H_2 + 2OH^- = 2H_2O + 2e^-$	-0.83
$V(OH)_3 + OH^- = VO(OH)_2 + H_2O + e^-$	-0.64
$Fe(OH)_2 + OH^- = Fe(OH)_3 + e^-$	-0.56
$Pb + 3OH^- = Pb(OH)_3^- + 2e^-$	-0.54
$S^{2-} = S + 2e^-$	-0.48

Table 4-7 (Continued)

Potentials in Basic Solutions (Continued)

$2Cu + 2OH^- = Cu_2O + H_2O + 2e^-$	-0.36
$Cr(OH)_3 + 5OH^- = CrO_4^{2-} + 4H_2O + 3e^-$	-0.13
$NH_3 + 9OH^- = NO_3^- + 6H_2O + 8e^-$	-0.12
$Cu_2O + 2OH^- + H_2O = 2Cu(OH)_2 + 2e^-$	-0.08
$Mn(OH)_2 + 2OH^- = MnO_2 + 2H_2O + 2e^-$	-0.05
$Hg + 2OH^- = HgO \ (red) + H_2O + 2e^-$	$+0.10$
$Mn(OH)_2 + OH^- = Mn(OH)_3 + e^-$	$+0.10$
$Co(OH)_2 + OH^- = Co(OH)_3 + e^-$	$+0.17$
$PbO \ (red) + 2OH^- = PbO_2 + H_2O + 2e^-$	$+0.25$
$4OH^- = O_2 + 2H_2O + 4e^-$	$+0.40$
$Ni(OH)_2 + 2OH^- = NiO_2 + 2H_2O + 2e^-$	$+0.49$
$MnO_2 + 4OH^- = MnO_4^- + 2H_2O + 3e^-$	$+0.57$

Source. Compiled from Mason, 1966; Krauskopf, 1967; and Flaschka et al. 1969; these have the original data from Latimer, 1952; Sillen, 1964).

competing, with the dominance of one or another affecting the final determination of pH. The mother rock that is being weathered is important because the meteoric and subsurface waters that come into contact with it react according to the composition of the rock and have pH values determined to a great degree by the mother rock; for example, water that comes into contact with limestones or dolomites can hydrolyze, resulting in a basic pH. Finally, there are competitions between various chemical reactions, some already mentioned, that determine the pH of the waters of an environment, and due to the complexity that exists in common natural environments, the effect of each competing reaction cannot be isolated and measured precisely in order to perfectly understand the controls of pH in a natural system.

As for the Eh, it is known that many chemical species are more or less stable and more or less mobile, depending on the reducing-oxidizing conditions of an environment. One of the better examples of the above characteristics can be given by the Fe^{2+} ion, which is mobile and stable in certain level reducing environments but which reacts rapidly in oxidizing environments to form Fe^{3+} which precipitates in the presence of water as $Fe(OH)_3$ (at a pH greater than 3). Another example is the U^{6+} ion, which is stable and mobile in environments where it is not able to combine with other species, but under reducing conditions (-0.5 volts) the ion changes to U^{4+} which is rather insoluble and enters into solid phases. Standard conditions numerical potentials have been determined for a great number of reactions (Table 4-7) and from these one may establish qualitatively and quantitatively how competing elements are going to react in an oxidizing-reducing situation. It is well known that the reduced form of a species in reaction has sufficient chemical energy to reduce the oxidized form of another species that has a greater oxidation potential. Thus, for the reactions

$$Mn^{2+} = Mn^{3+} + e^- \qquad E^\circ = 1.51 \text{ volts}$$

and

$$Fe^{2+} = Fe^{3+} + e^+ \qquad E^\circ = 0.77 \text{ volts}$$

the Fe^{2+} reduces the Mn^{3+}. With Table 4-7 and the measurement of the Eh of an environment, it is possible to get a good idea of the possible ions present and their valence states. However, in natural environments, there are competing reactions and a reaction that goes in the laboratory under controlled conditions and away from the influence of competing reactions may not take place or may take place incompletely in natural environments. In spite of this, the calculations and concepts based on measurements of Eh in an environment are most important in geochemistry because they give us an idea of how certain chemical species may be expected to act under different physical–chemical–biological conditions. Krauskopf (1967) gave a good example of the utility of such measurements: if we measure a redox potential of 0.5 volts in an aqueous system, we can determine quantitatively the ionic form of Fe dissolved in the water; if the solution is acidic, the selection will be between Fe^{2+} and Fe^{3+}. The standard potential for the reaction $Fe^{2+} = Fe^{3+} + e^-$ is 0.77 volts, and from the measurement made (0.5 volts) we know that the environment is more reducing; thus, one might expect that the Fe^{2+} will be the dominant ion and this can be checked by using a calculation based on the Nernst relation:

$$0.5 = 0.77 + \frac{0.059}{1} \log \frac{[Fe^{3+}]}{[Fe^{2+}]}$$

which transposes to

$$\log \frac{[Fe^{3+}]}{[Fe^{2+}]} = - \frac{0.27}{0.059} = -4.58$$

and

$$\frac{[Fe^{3+}]}{[Fe^{2+}]} = 10^{-4.58} = 2.6 \times 10^{-5}$$

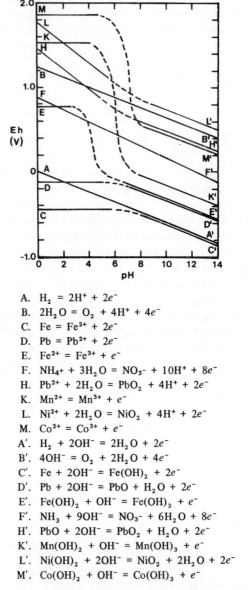

A. $H_2 = 2H^+ + 2e^-$
B. $2H_2O = O_2 + 4H^+ + 4e^-$
C. $Fe = Fe^{2+} + 2e^-$
D. $Pb = Pb^{2+} + 2e^-$
E. $Fe^{2+} = Fe^{3+} + e^-$
F. $NH_4^+ + 3H_2O = NO_3^- + 10H^+ + 8e^-$
H. $Pb^{2+} + 2H_2O = PbO_2 + 4H^+ + 2e^-$
K. $Mn^{2+} = Mn^{3+} + e^-$
L. $Ni^{2+} + 2H_2O = NiO_2 + 4H^+ + 2e^-$
M. $Co^{2+} = Co^{3+} + e^-$
A'. $H_2 + 2OH^- = 2H_2O + 2e^-$
B'. $4OH^- = O_2 + 2H_2O + 4e^-$
C'. $Fe + 2OH^- = Fe(OH)_2 + 2e^-$
D'. $Pb + 2OH^- = PbO + H_2O + 2e^-$
E'. $Fe(OH)_2 + OH^- = Fe(OH)_3 + e^-$
F'. $NH_3 + 9OH^- = NO_3^- + 6H_2O + 8e^-$
H'. $PbO + 2OH^- = PbO_2 + H_2O + 2e^-$
K'. $Mn(OH)_2 + OH^- = Mn(OH)_3 + e^-$
L'. $Ni(OH)_2 + 2OH^- = NiO_2 + 2H_2O + 2e^-$
M'. $Co(OH)_2 + OH^- = Co(OH)_3 + e^-$

Figure 4-8. Variation of redox potential with changes of pH for some selected reactions. (Reproduced from Mason, 1966, *Principles of Geochemistry,* by permission of John Wiley & Sons, Inc.) Copyright © 1966 by John Wiley & Sons, Inc.

From this we know that under standard conditions the concentration of Fe^{2+} is almost 40,000 times as great as that of Fe^{3+}.

Table 4-8 Ionic Potentials of Some Common Elements, Z/r (from Mason, 1966 and Barth, 1962)[a]

Element (Ion)	Ionic Potential	Element (Ion)	Ionic Potential
Cs^+	0.60 (0.61)	Th^{4+}	3.9 (4.2)
Rb^+	0.68 (0.67)	Ce^{4+}	4.3 (4.2)
K^+	0.75 (0.71)	Fe^{3+}	4.7
Na^+	1.0	Zr^{4+}	5.1 (4.6)
Li^+	1.5 (1.3)	Be^{2+}	5.7 (5.9)
Ba^{2+}	1.5 (1.4)	Al^{3+}	5.9 (5.3)
Sr^{2+}	1.8 (1.6)	Ti^{4+}	5.9 (6.3)
Ca^{2+}	2.0 (1.9)	Mn^{4+}	6.7
Mn^{2+}	2.5	Nb^{5+}	7.5 (7.3)
La^{3+}	2.6	Si^{4+}	9.5 (10)
Fe^{2+}	2.7	Mo^{6+}	9.7
Co^{2+}	2.8	B^{3+}	13 (15)
Mg^{2+}	3.0 (2.6)	P^{5+}	14 (15)
Y^{3+}	3.3 (3.4)	S^{6+}	20
Lu^{3+}	3.5	C^{4+}	25 (27)
Sc^{3+}	3.7	N^{5+}	38 (45)

[a]The values in parentheses indicate Barth (1962) values which differ from those given by Mason (1966).

We have already noted the relation between the pH and Eh of an environment and this becomes especially obvious when we consider Figure 4-8 in which it is shown that, in general, the oxidizing potential for a reaction diminishes with higher pH values. Thus for a theoretical approach to geochemical prospecting, some weathering products in an environment can be examined and ideas can be generated on other products that should be present in the system and those that have been carried out of the system. These concepts can be related to the mobilities of the elements and general estimations can be made on how the dispersion haloes of the elements in a study zone may be disposed areally.

Once the ions liberated from solids by the various weathering processes enter the hydrological system, they can react inorganically according to their ionic potentials (the ionic potential of an element is equal to its ionic charge divided by its ionic radius, Z/r). The ionic potential was recognized by Goldschmidt (1937) an an important factor related to the manner in which elements react after being liberated from the mother rock and injected into the sedimentary cycle. The designation "ionic potential" is considered by some geochemists as a misnomer, especially since it can be confused with ionization potential. Perhaps it would be better to denominate the factor Z/r as "hydropotential" or "potential for reaction in aqueous media." A listing of selected Z/r potentials is given in Table 4-8

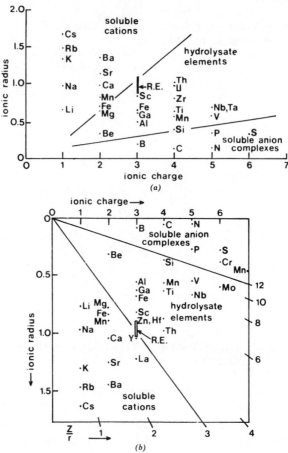

Figure 4-9. Diagrammatic representations that show the *Z/r* relation by plots of ionic charge against ionic radius, and the responses of elements in hydrogeological sedimentary environments. (*a*) Reproduced from Mason, 1966, *Principles of Geochemistry* by permission of John Wiley & Sons, Inc.; Copyright © 1966 by John Wiley and Sons, Inc. (*b*) Reproduced from Barth, 1962, *Theoretical Petrology* by permission of John Wiley & Sons, Inc. Copyright © 1962 by John Wiley & Sons, Inc. The graphs are based on the same data.)

and is represented diagrammatically in Figure 4-9, in which we see that the manner in which ions act with respect to water can be categorized in several reaction fields based on their ionic charges and ionic radii. The less reactive ions in aqueous media retain their ionic characters in solution and have *Z/r* values up to 3; other ions with intermediate *Z/r* potentials (between 3 and 12) tend to bond with (OH)⁻ and are those elements generally associated with the hydrolyzates. Finally, the elements with *Z/r* potentials greater than 12 tend to

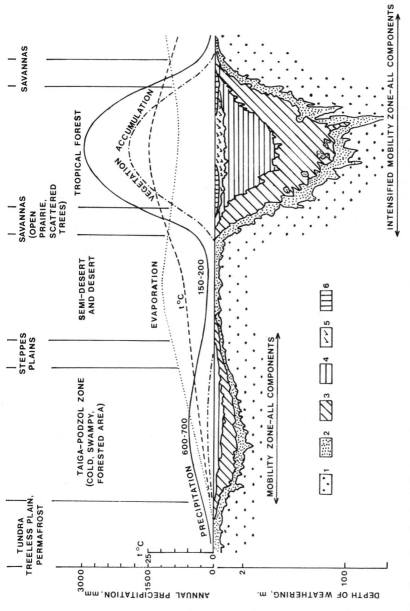

Figure 4-10. Formation of a leached, decomposed weathered crust in tectonically inactive areas: (1) new crust; (2) grus zone, little modified chemically; (3) hydromica-montmorillonite-beidellite zone; (4) kaolinite zone; (5) ochers, Al_2O_3; (6) hard layer $Fe_2O_3-Al_2O_3$. (Reproduced from Strakhov, 1967, *Principles of Lithogenesis*, Oliver & Boyd Ltd. and Consultants Bureau Enterprises, Inc.)

form soluble anionic complexes (with oxygen). As can be seen in Figure 4-9, more than three response fields can be represented by the ionic valences, ionic radii, and corresponding Z/r potentials in order to attempt to better detail the reactivities of the elements with respect to water.

All these factors and others not mentioned enter into play during the weathering of a mother rock and its resulting important products: sediment and soil. If the geochemist recognizes the dominance of one or more of these in a prospecting zone, he can judge which element(s) could be a pathfinder or otherwise important in the search for ore and generalize where the so-called geochemical anomalies may be found with respect to a given type of deposit due to the different mobilities of the elements as a function of the physical–chemical environment.

The residual soil is the geochemical sample that is most often used to detail the location of hidden mineralization once a zone of economic interest is localized by a stream sediment survey or other geological aspect. As is the case in a general consideration of weathering, the climate, by virtue of its influence on precipitation, evaporation, temperature, and the accumulation of vegetation, is the factor that predominates in determining the vertical extent of weathering (Figure 4-10) and the physical, chemical, and biological characteristics of a soil. With a knowledge of the climate (or of the paleoclimate), the geochemist can begin his planning for a prospecting program with a good concept of where in a soil profile a sample should be taken for maximum utility and efficiency in the exploration project. There are several definitions of what a soil is, depending on how it is studied (from the point of view of edaphology, agronomy, biology, civil engineering, etc.) and there are as many classification of soils as there are definitions. In 1967, FitzPatrick proposed the use of 74 descriptive modifiers (with subscript numeration to represent the thicknesses of the described horizons) in order to pigeon-hole 41 subclasses of soils that represent 23 great classes; on the other extreme, attempts were made to eliminate some of the classificatory confusion that existed in a field of research with a geotechnical definition of soil and rock to be used in engineering works, basing it on the degree of weathering, consistency and degree of fracture, and using the adjectives fresh, weathered and decomposed, hard and soft, and strong and fractured rock. For geochemical prospecting we can use a classification based on the relative amount of precipitation, the temperature, and our observations of vegetation (Figure 4-11), but always with the complementary use of our geological, geomorphological, and hydrogeological knowledge of the study zone; this classificatory scheme will be given later.

For the geochemist who uses the soil in prospecting, the best definition of a soil may be that given by Hawkes and Webb (1962): "a soil is a natural body of mineral and organic constituents, differentiated into horizons, of variable depth, which differs from the material below in morphology, physical

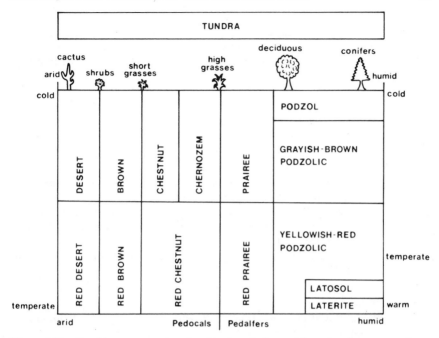

Figure 4-11. Graphic representation showing the relative position of various great soil groups with respect to climate and vegetation. (Modified from Thompson, 1965, *El Suelo y Su Fertilidad,* Editorial Reverte, S.A.)

make-up, chemical properties and composition, and biological characteristics" (Joffe, 1949). Ideally the soil with a classical profile shows three principal horizons (A, B, and C) that can be divided into subhorizons on the basis of characteristics given in the above definition or other more detailed definitions (Table 4-9 and Figure 4-12). Horizon A begins forming from the mother rock, which is designated as horizon C, during the first phase of development of a soil (by the physical alterations and chemical decompositions with a resulting vertical and lateral dispersion of the chemical elements and detrital particles). As the process of soil formation continues, horizon A keeps developing as a consequence of physical, chemical, and biological changes that evolve during the process. In general, the A horizon can be said to suffer eluviation via the loss of material by solution and physical translocation (under the influence of gravity and the movement of precipitation downward vertically or laterally depending on the topography). The material lost by the A horizon, for example, sesquioxides and the clays, can accumulate in the B horizon (zone of illuviation) or can enter into the hydrologic system in solution and/or suspension and be carried out of the immediate zone of soil formation. Loss of material is not the only change in the A horizon. If there is vegetation growing in a

Table 4-9 Description of an Ideal Soil Profile

O: Organic horizons of mineral soils

O_1: Formed or being formed in the upper part of mineral soils on top of the mineral part

O_2: Dominated by fresh or partially decomposed organic matter

O_3: Contains more than 30% organic matter if the mineral fraction contains more than 50% clay, or contains more than 20% organic matter if the mineral fraction does not contain clays; intermediate contents of clay require a proportional content of organic matter

A: Mineral horizons which consist of

A_1: Horizons of accumulation of organic matter formed or being formed on or adjacent to the surface

A_2: Horizons that have lost clay, Fe, or Al, with the resulting concentration (relative) of quartz and other resistant minerals in the silt- or sand-size fractions

A_3: Horizons dominated by A_1 or A_2 but transitional to an underlying B or C horizon

B: Dominant characteristic or characteristics as described in one or more of the following

B_1: An illuvial concentration of clay-size silicates, Fe, Al, or humus, alone or in combination

B_2: Residual concentration of sesquioxides or clay-size silicates, singly or mixed, that have formed by means other than solution and movement of carbonates or other more soluble salts

B_3: Thin layers of sesquioxides sufficient to impart conspicuous dark reddish colors, darker and more intense than those of the overlying and underlying horizons

B_4: An alteration of material from its original condition that obliterates the structure of the original rock, results in the formation of clay-size silicates, free oxides, or both, and has a prismatic structure, blocky or granular if the textures are such that changes in volume accompany changes in humidity

C: Mineral layer or horizon excluding the mother rock that is similar or dissimilar to the material from which the solum has (presumably) formed, is only slightly affected by the pedogenic processes, and has no properties characteristic of A or B

R: Underlying consolidated mother rock − granite, sandstone, limestone.

Source. *U. S. Department of Agriculture Handbook 18,* 1951 and September, 1962 Supplement.

soil, the transfer of nutrients to the different physiological parts of the vegetation takes place and with the death and fall of these parts there can result a return of nutrients, including metals, to the surface; these metals can enter in-

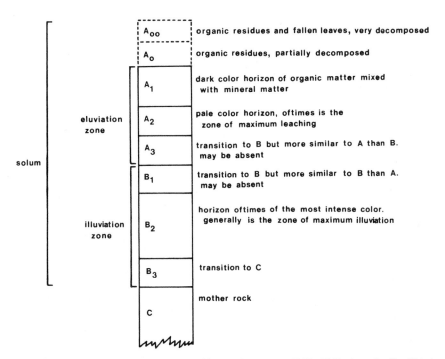

Figure 4-12. Hypothetical soil profile. (From Thompson, 1965, *El Suelo y Su Fertilidad,* Editorial Reverte, S.A.)

to the weathering cycle again or can be strongly retained in an environment by organic complexes that are formed by the decomposition of the organic matter, depending on the conditions that are dominant in the environment. Also, soluble material can be added to the profile from below by means of capillary activity or if the water table rises and falls; thus in subhumid and arid regions the soils have concretions or well-defined layers of salts, especially of calcium carbonate. These last two processes are very important in geochemical prospecting if the geochemist recognizes that a redispersion has taken place due to the actions subsequent to those involved in the initial formation of a soil. Depending on the importance given to detail, a principal horizon can be divided into various subhorizons; from Figure 4-12, it can be seen that the A zone is the one with maximum biological activity and can be arbitrarily presented in several subhorizons according to the necessities of the scientific study that is being made. The same factors that influence the dispersion and distribution of the chemical elements during general weathering influence the process of soil formation, but in a certain sense are intensified by the development of each horizon and subhorizon with its own physical, chemical, and biological characteristics: pH; Eh; state of organic material with respect to quantity, decomposition, complexes, and its own content of metalliferous and nonmetalliferous elements; mineralogy (especially of the resistates, hydrolyzates, and colloids, and clays with their distinct capacities for ion exchange); position with respect to the permanent or seasonal water table level; and degree of maturity. Generally one can work well in soil geochemical prospecting using as a base a simple and common profile such as that presented by Hawkes and Webb (1962). Very important is the fact that the number of studies that clearly detail soil profiles in an area under investigation are increasing depending on the experience of the geochemical prospecting team and its realization that comparative data on soils are very desirable for future exploration programs.

As for a classification of soils that can be applied effectively and comfortably in geochemical prospecting, a general one can be used for zonal soils and then extended to include azonal or intrazonal soils. Zonal soils are defined as "those soils that extend over great areas (or zones) that are limited by geographic and geologic characteristics; they have well-developed properties; they are situated on mother rock that has been *in situ* for a sufficient period of time so as to allow the active climatic forces and biological factors to be developed to their maximum influence" (USDA, 1951). Azonal soils do not have their profiles well developed even though they are found in regional areas where zonal soils have developed well; this is very often due to local relief, but also may be caused by the presence of a different class of mother rock or to the youth and state of immaturity of a soil, as for example, lithosols over solid rocks, regosols on nonconsolidated material, and alluvial soils on active flood plains. The soils denominated "intrazonal" are those that reflect the effect of mother rock or of local relief that overcomes the effect of climate and of vegetation during its development, as for example, clay- or peat-type soil, which is formed as the result of imperfect or impeded drainage (Bunting, 1967). Table 4-10 presents a classification of soils representative of the various existing classifications (with more or less detail), and in Table 4-11 some general data are given on soil profiles, vegetation, climate, drainage, and other factors that influence the development of some of the great groups of soils cited in Table 4-10.

During geochemical exploration, one works with many classes of soils, and because of this, it is convenient to now consider the characteristics of some of these classes from the point of view of prospecting to emphasize the importance of knowing and understanding the different dispersion processes and evaluating them before and during the reconnaissance and detailed programs. Initial consideration is given to the zonal soils that are formed in humid areas where there is an average of more than approximately 65 cm of annual precipitation, a flora of high grasses and forests in temperate and tropical zones, and low-order plants and shrubs in cooler zones. These soils are called "pedalfers" for their notable contents of Al and Fe oxides.

The latosols or lateritic soils are formed in warm areas ($> 25°C$), tropical and temperate, where there is much rain, high humidity, a dense forest or savannah vegetation, and good drainage. However, in the same areas, there exist subenvironments (deserts, swamps, sites of strong relief, and rugged topography with permanent snow, etc.) that lack the characteristics mentioned and thus originate different classes of soils in the very latosolic or lateritic region. In general the alteration of mother material is deep and intense. When the mother rock is basic or less siliceous, the latosols or lateritic soils are formed better by the loss of the bases, the alkalis, and the combined silica (due to the decomposition of the feldspars) to the waters of the hydrologic system and by the accumulation of clays in the B horizon (from the reconstitution of the remains of the weathered feldspars) and hydrated oxides of Fe and Al. The B horizon, rich in clays and colored reddish-brownish-yellow because of the sesquioxides of Fe, may attain thicknesses up to 100 meters. In these soils the A_1 horizon is generally poorly developed, but the A_2 horizon shows good development. It is in the upper part of the latosols and lateritic soils that the resistates and the stable oxides (for example, of Cr, Ti, and Ni) are accumulated. The stability factor can be understood if one studies the solubilities of the components of a rock with respect to pH (Figure 4-6); it is clearly seen that there is a decreasing order of stability of $TiO_2 > Fe_2O_3 > Al_2O_3 > SiO_2$, a relation that explains the presence or lack of some components in a soil. When a soil of this class develops (especially on ferromagnesian alkaline rocks and on neutral rocks) with the complete hydration of the silicates and the "induralization" (Bunting, 1967) of silica and the hydrated oxides of Fe and Al with the resulting formation of limonite and gibbsite, the process is called "ferralitization" and the soil is called "ferralitic." In a more acid environment, there is a dispersion of silica, possibly in a colloidal form of the monomer, but the positive sesquioxides (hydroxides of Fe^{3+}, Al, and Cr^{3+} and hydrated dioxides of Ti, Zn, and Th, for example) remain, and instead of forming ferralitic soils, they form kaolinitic soils if there is an imperfect drainage so that there is a wet zone and silica is not completely leached from the system. If, at a subsequent time, the process of desilicification occurs, bauxite can form (on acidic mother rock). Under the conditions of continuous humidity, the latosolic or lateritic soils contain more Al than Fe because the Fe is more mobile; Fe tends to accumulate at the water table level by the formation of concretions, thin lenses, or soft compact masses that harden with dehydration, thus forming laterites, sometimes pisolitic in character. Often the concretions or lenses may enclose very white quartz sand that was trapped during their development. Figure 4-13 shows schematically a sequence of weathering in a tropical zone where the effect of climate and vegetation on the weathering products is clearly seen. One of the final weathering products in latosolic or lateritic areas is laterite. Laterite itself is not considered a soil but rather a specific

Table 4-10 The Great Groups of World Soils (from Lopez-Ritas, 1967)

Order	Suborder	Great Soil Groups
Zonal soils		
Pedocals	1. Soils of the cold zone	1. Tundra
	1. Pale soils of arid regions	2. Desert soils
		3. Red desert soils
		4. Sierozens
		5. Brown soils
		6. Reddish-brown soils
	2. Dark soils of semiarid subhumid regions (perhaps with humid grasses)	7. Chestnut soils
		8. Reddish chestnut soils
		9. Chernozem soils
		10. Prairee soils
		11. Reddish prairee soils
	3. Transition soils of prairee-forest zones	12. Degraded chernozem soils
		13. Noncalcareous brown soils
Pedalfers	4. Pale podzolized soils from treed regions	14. Podzol soils
		15. Brown podzol soils
		16. Grayish-brown podzol soils
	5. Lateritic soils from temperate and tropical forested regions	17. Yellow podzol soils
		18. Red podzol soils

Intrazonal soils

1. Halomorphic soils (saline and alkaline soils from arid regions with imperfect drainage and littoral deposits)

2. Hydromorphic soils from bogs, swamps, and permeable and plains zones

3. Calomorphic zones

Azonal soils

19. Yellowish-brown lateritic soils
20. Reddish-brown lateritic soils
21. Laterites

1. Saline soils; Solonchak
2. Solonetz soils
3. Soloth soils
4. Wiesenboden soils
5. Alpine meadow soils
6. Swamp soils
7. Semiswamp soils
8. Planosols
9. Muddy podzolic soils
10. Lateritic soils
11. Brown forest soils
12. Rendzina soils

1. Lithosols
2. Alluvial soils
3. Sandy (dry) soils

Table 4-11 General Data on Soil Profiles, Vegetation, Climate, Drainage, and other Factors that Influence the Development of Some of the Great Soil Groups (from Thompson, 1965)

Great Soil Group	Typical Profile	Native Vegetation	Climate	Natural Drainage	Other Factors
Tundra	Layers of dark brown peat over mottled gray horizons with oxide; substrata of material always frozen	Lichen, muskeg, flowering plants and shrubs	Humid and cold	Reduced	
Podzol	A layer of leaves and humus, acidic, an A_1 horizon well defined and of a dark-gray color, a well-defined A_2 whitish-gray horizon, a B_1 horizon colored brown or dark coffee, and a B_2 horizon of a yellowish brown color; very acidic	Conifer forest or mixture of broad-leafed conifer forest	Cold–temperature, except in some zones where it is temperate; humid	Good	
Latosol (Laterite)	Surface of red brown soil, B horizon is intense red, primary material red or with reticulate mottle, weathering very deep	Tropical forest and savannah vegetation	Tropical; humid to dry; large to moderate rains	Good	
Reddish and yellowish-brown lateritic soil	Brown or reddish-brown clay soil, friable, over a yellowish-brown or red fractured granular clay; acid to neutral; deep substratum with reticulate mottle in certain places	Forest with tropical rain, broadleafed and always green	Humid tropical to dry; high to moderate rainfall	Good	
Chernozem	Black or dark gray soil to a depth of 90 to 120 cm passing to a more pale or whitish color because of the accumulation of calcareous material	High, mixed prairee grasses	Temperate to cold; subhumid	Good	
Desert	Pale gray or pale brown gray, scarce organic matter, very much mixed with limestone material	Scarce bushes	Temperate to cold; arid	Good to imperfect	

Soil	Profile	Vegetation	Climate	Drainage	Topography
Lithosol	Incomplete solum or without clear morphologic expression, consists of recently weathered masses of hard rock	Depends on climate	All climates; more in desert climates and less in the humid tropical climate	Good to excessive	Rugged topography, primary material generally consolidated
Regosol (includes dry sands)	Essentially undeveloped, in sands, material dragged by glaciers; loess, etc.	Scanty grasses or dwarf forests the major part lacking vegetation	Humid to arid, cold to hot	Excessive	Rugged topography
Alluvial	Profile slightly developed, part of organic matter is accumulated; stratified	Scanty grasses or dwarf forests the major part lacking vegetation	All climates except the coldest	A wide range from deficient to good	Deposited by current activity
Humic gley	Organic–mineral horizons, dark mineral color, of moderate thickness situated on mineral-gley horizons	Swampy forests or grassy swamps	Temperate or cold; humid to subhumid	Deficient	Drainage deficient
Bog soil	Brown, dark brown, or black peat, or land covered by residue on a peat material	Swampy forest, or reeds and grasses	Cold to tropical and generally humid	Very low	Drainage deficient, area covered by water most of time
Solonchak	Gray, thin, saline crust in the fine granular surface, grayish soil, cracked and saline; the salts may be concentrated above or below	Scarce growth of halophyte grasses, shrubs, and some trees	Generally subhumid to arid and can be warm or cold	Deficient or imperfect	Deficient drainage with evaporation of water, and capillary salt accumulations

Source. *El Suelo y Su Fertilidad*, Editorial Reverte. S. A.

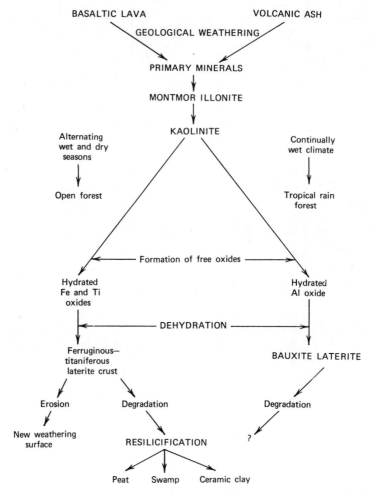

Figure 4-13. Weathering sequence of basaltic lavas in the Hawaiian Islands. (From Sherman, 1952, as presented by Loughnan, 1969, *Chemical Weathering of the Silicate Minerals,* Elsevier, New York.

state of the weathering process during the formation of a soil; it may contain up to 80% of hydrated Fe oxides and is incapable of sustaining vegetation, because of which a hard top layer may be present instead of a soil layer. In the search for mineralization by means of geochemical prospecting, one may use the A horizon for the sampling because the resistates and stable oxides accumulate there, but it is also possible that there can be metals dispersed in and trapped by the clays of the B horizon. However, the dense flora acts to return nutrient and other metals (and nonmetals) to the surface as dead parts fall to the ground, with the result that the chemical elements enter once more into the weathering cycle with the possibility of being trapped by organic compounds in the A horizon or leached again to the lower part of the soil. In a geochemical exploration program then, it is always fundamental to make a pilot study to determine where in the accessible profile and in what size fraction would the greatest contrast in concentration of the elements being sought, be encountered.

Passing from the warm areas to cooler and temperate areas where there are high humidities due to low evaporation and where a forest vegetation (dominated by conifers in podzol environments and by wide-leafed trees in environments of podzolic soils) grows, podzol soils are formed if there is good drainage with the corresponding strong leaching of the A horizon and illuviation in the B horizon. In the temperate climate with high humidity, an accumulation of humic acids is formed resulting in strong leaching and transfer of Fe and Al oxides towards the B horizon. The upper part of the profile is rich in organic material due to the mixing of residues by organisms and may present an upper layer of 10- to 12-cm thickness at the top of a forest soil; this organic material decomposes to organic acids that combine with Fe and Al, eluviating these elements from the A horizon. The A_1 horizon is generally absent in podzols, but the A_2 horizon is well developed and very marked by the white or gray appearance that is characteristic of a podzol. A part of the organic residues accumulates in the B_1 horizon and gives it a dark color that changes to a more reddish or yellowish color in the B_2 horizon. Obviously the podzol is an acidic soil in which the clays and humus are saturated with respect to hydrogen. The areas where podzols form best are characterized by long winters and cool, short summers. Bunting (1967) describes the podzolization process in five steps:

1. Decalcification of the mother rock.
2. Formation and migration of clay mineral complexes by argilluviation.
3. Formation of free oxides by intense weathering of silicates and destruction of clays (silicification).
4. Transfer or cheluviation of Fe and Al by organic acids.
5. Precipitation in the B horizon.

The thicknesses of the horizons that are developed by podzolization vary more with the mother rock and the local site than with the climate and vegetation. Of course, podzol regions may have nonpodzolic soils depending on the mother rock, strong relief and topography, and specific subsoil conditions (permafrost). Millar, Turk, and Foth (1965) describe the ideal podzol soil profile as follows: "the humus horizon at the surface (O_2) is thin or moderately thick, the A_1 is thin or lacking; the eluviated leached zone (A_2) is strongly developed, brown or reddish brown in color and very often cemented by organic compounds and Fe oxides; there is a secondary maximum concentration of organic material in the B zone; the texture is often sandy; the A_2 horizon has a white or gray aspect due to the strong leaching and the accumulation of quartz." Because there is a strong dispersion of metals during the formation of the podzol soils, it is convenient to evaluate the B horizon for the geochemical sample in a prospecting program as well as the A_2 horizon and thus determine from what level the soil sample should be taken and which will be the size fraction of that sample which will give the best data for the proposed study plan.

The podzolic soils are transitional between podzol soils (and areas) and those of more temperate and drier areas. They are developed in a subhumid or semiarid and mesothermal environment where there is a native vegetation of broadleafed forests or forests with a mixture of broad-leafed and conifer trees. The podzolic soils are neutral or acidic (pH <5) but are not as acid as the podzol soils, with the result that leaching is not as strong. In general, the A_1 horizon is thin and of a dark color caused by a mixture of organic and mineral material. However, the A_2 horizon is well developed and is a grayish-brown in notable contrast to the ash gray color of the A_2 horizon of the podzol soil. Because of the moderate acidity of the podzolic soil, there is a translocation of Fe and Al from the A horizon to the B horizon, leaving the A horizon relatively enriched in silicon. There are more clays in the B_2 horizon and this horizon often has a blocky structure that grades into a more fragmented structure in the B_3 horizon, which itself is of a lighter color. With respect to geochemical prospecting, the greatest dispersion of metals in the podzolic soils should be encountered in the B horizon, perhaps 30 to 60 cm below the surface, but the resistant (to weathering) minerals can accumulate at a higher level in the soil profile.

More towards the higher latitudes, there are tundra-type (or arctic, subarctic, polar) soils. These have their profiles lesser developed because of the low temperatures and consequent limitation of biological activity. The subsoil may be frozen throughout the entire year (permafrost) with the result that the drainage is impeded. The native vegetation consists of a covering comprised of about 20% grasses, mosses, and weeds, alive and dead (Millar, Turk, and Foth, 1965). Such chemical weathering as may take place is weak and of short duration during the summer seasons when there is a partial thawing. Because of the thawing and refreezing, the cryogenic forces move and mix the existing soil; the detrital particles may be moved laterally by the pressure of active refreezing between the permafrost and the layer that is refreezing, an action that serves to expose new parts of the permafrost for later thawing and movement, especially if there is a slight topographic dip (greater than about 2 degrees). If there is good drainage and a significant vegetation cover in a general tundra zone, acids are formed that serve to hydrolyze the minerals present in the same way as they would in a podzolic-soil environment. The majority of tundra soils

are acidic, but some are humic and a few are saline and often occur with rego-
lith- and gley-type soils.

Zonal soils in subhumid and arid regions receive less than 65 cm of precipitation annually and support a native vegetation of high grasses or shrubs (gram-
ineas and desert plants); these soils are called pedocals because of the presence
of a calcareous level in one of the principal horizons. This calcareous accumulation indicates the depth of penetration of calcium-bearing waters.

The group of chernozem soils is developed in temperate and subtropical regions where the precipitation is approximately 35 to 75 cm annually. They
are characterized by a notable concentration of calcium carbonate in the lower
part of the B horizon that is located from 60 to 90 cm from the surface; as is
the case with other soils of subhumid or arid zones, the calcareous layer is
found higher in the soil profile where there is less precipitation. In chernozem
regions, there is a steppe vegetation of tall grass of or mixed tall and short
prairee grasses that grow in a subhumid climate with hot, early summers,
drought in the late summer and autumn, and generally dry and snow-free winters. The soil profile is dominated by a dark and relatively thick A horizon of
a humus–mineral mixture, but due to the limited precipitation, there is less
weathering, leaching, and floral growth and less transfer of clays and sesquioxides from the A horizon to the B horizon than in the opposite situation in humid zones, thus leaving the B horizon generally light or dark brown in color.
The chernozems most often do not present the A_2 subhorizon. They are
formed better on mother rock rich in lime and are neutral or slightly alkaline;
this factor, combined with the presence of clay minerals with high ion-exchange capacities, does not facilitate the dispersion of metals. Thus, for geochemical prospecting in chernozem soils, one may work with samples from the
A horizon in order to obtain the best contrasts and reliable results, but always
with the reservation that a pilot study be made to determine if the above gen-
eralization is correct for the particular study zone and to determine the size
fraction of the soil sample that should be used in the prospecting zone. It
should be noted that although the calcareous layer originates mostly from surface reactions, it may also originate from below via capillary activity. Caliche
or calcrete can be deposited in important amounts from a subsurface water
table reaction as the water table reaches a given level and then falls or rises seasonally, or if there is a slow but continuous fall of the water table level, this
could result in deposits similar to the famous "tosca" of the Argentine Pam-
pean area. Goudie (1972) has recently presented an excellent paper in which
he gives more than 300 chemical analyses of calcrete deposits and discusses
their origins. Such information may be useful to geochemists who expect to
work in pedocal soil areas.

The desert soils may have only slightly developed profiles due to the minimum humidity prevalent in deserts and the consequent limitation on the proc-

esses of weathering, leaching, and growth of vegetation. The native vegetation consists of grasses (very dry areas), shrubs, and cactus (in areas with more humidity), plus the short-lived perennial plants that grow in the spring or after rains. Because of the rapid decomposition of dead vegetation, there is a low content of organic material in desert soils. Often there are accumulations of salts at the surface or at a few centimeters depth, especially where the soil is formed from a calcareous mother rock. The detrital material from the mother rock is present throughout the entire soil profile, generally in sand size in the upper section and in coarser size downwards; it is sometimes present with silt and clay but with minor sesquioxides in the B horizon. Depending on the climatic alterations in the general environment, the surface may have a hard crust and/or layers of gravel. The colors of desert soils vary from reddish in warm zones to light gray in cold zones. As is the case in all major soil regions, there are variations of a soil type that are due to other factors such as topography (presence of pediments, mesas, buttes, etc.). With respect to geochemical prospecting in desert soils, it must be remembered that there is generally little chemical dispersion of metals except when there is vegetation with roots that reach a B horizon, for example, where there is more humidity because of the lack of penetration of the heat of the day or where the vegetation roots penetrate deeply towards the water table level in the search for humidity. In these cases there can be a transfer of metals from below to above by means of the vegetation parts that are continually added to the surface with the death of the flora, thus entering again into the weathering cycle. If there is a B zone developed in the desert soil and it contains clays and silts, these must be considered as potentially good sites for obtaining informative geochemical samples.

Much information on the classes of soils in a given region and their physical, chemical, and biological characteristics can be obtained in provincial and federal offices and in private institutions or those related to universities. In addition, by its widely extended program of technical assistance, the United Nations Organization for Food and Agriculture (FAO) has published many regional and individual country reports on the recognition and classification of soils. By combining the general data with such detailed data as may be available, the geochemist can develop a firm basis on which the first phases of a geochemical exploration program can be planned both for the reconnaissance phase and the follow-up detailed phase. An excellent example of the utility of available information that provided a basis for a geochemical exploration program can be found in the report by O'Brien and Romer (1971) in which these authors describe the discovery of the Navan deposit of base metals in the Irish Republic. In 1968 the Irish Institute of Agriculture published a series of maps of trace metals present in stream sediments to assist agricultural development; for O'Brien and Romer these maps plus the geological information they had al-

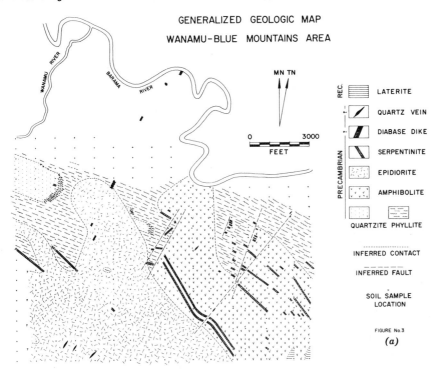

GENERALIZED GEOLOGIC MAP

WANAMU-BLUE MOUNTAINS AREA

Figure 4-14. Example of the utility of geochemistry of soils in geologic mapping: (a) generalized geologic map; (b) geochemical map of Ni in the soils. (From Kilpatrick, 1969; reprinted from *Quart. Colo. School Mines,* **64** No. 1, by permission of the Colorado School of Mines. Copyright © 1969 by the Colorado School of Mines.)

ready assembled were sufficient to allow them to obtain permission for prospecting in a zone located centrally over a possible fault contact between Lower Carboniferous limestones and older rocks. The Navan deposit was localized by geochemical prospecting with soils analyzed for Cu, Pb, and Zn, using a hot acid attack on the samples prior to analysis. In the reconnaissance phase of the program, an anomaly of Pb–Zn was easily picked out and gave values of 200 to 5000 ppm versus a background value of 70 ppm Zn and 50 ppm Pb, with prospect dimensions of more than 450 meters in length SE-NW and 240 meters in width, estimated from the elevated Zn values. It was proven that there was no contamination involved in the sampling by taking a second suite of samples at a depth of 60 cm; these confirmed the results of the initial reconnaissance and gave values of up to 25,000 ppm Zn. The detailed prospecting with soil samples taken from a 60-cm depth on a grid of 30 by 30 meters localized the mineralized outcrop and extended the dimensions of the deposit to more than 915 meters along its strike by 305 meters in width. The discovery was confirmed by an induced polarization study

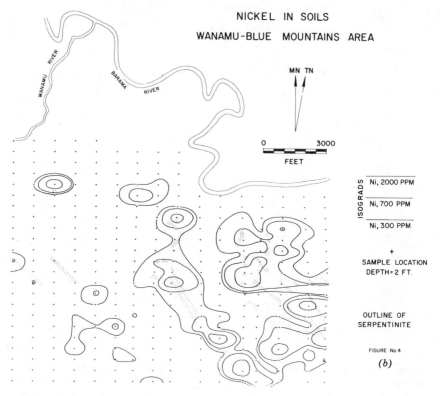

Figure 4-14. (b)

and the test drillings proved the richness of the deposit. The drill-hole samples also showed that the sulfide components of the ore completely lacked any sign of oxidation either at depth or at a few feet from the surface, and there were no oxidation stains in any of the fractures or joints of the ore host rock. O'Brien and Romer believe that the Lower Carboniferous strata in the study zone are in their present condition effectively impermeable to the filtration of water. Because the major part of the chronostratigraphically equivalent limestones of the Lower Carboniferous limestones of Ireland are covered by glacial material (till), with few outcrops of mother rock, or on low plains, one would think that geochemical prospecting by the method described in the publication by O'Brien and Romer can be used effectively in the search for base-metal-bearing ores in limestones of Ireland or in similar physical, chemical, and biological environments.

Research into the geochemistry of soils is not limited to prospecting but also has applications in other fields, for example, in epidemiological studies

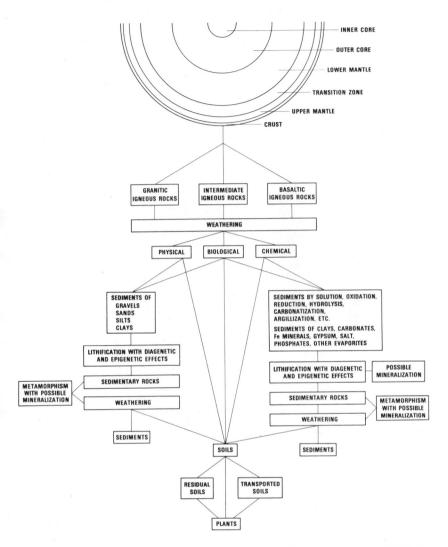

Figure 4-15. The relation between weathering and the various classes of materials that serve as samples in geochemical prospecting programs.

(which will be treated in the chapter on geochemistry in health and pollution studies) and in the geological mapping of zones covered by residual soils and vegetation. Kilpatrick (1969) presented an excellent example of this latter application of soil geochemistry; in the Wanamu-Blue Mountains of the Northwest district, Guyana, the rock underlying the soil is composed of a complex assemblage

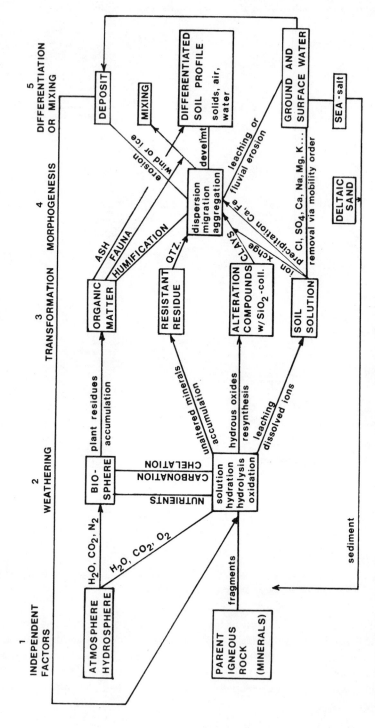

Figure 4-16. A scheme of soil genesis and profile development (From Yaalon, 1960.)

of mafic and pelitic metamorphic rocks (phyllites, quartzites, amphibolites, and epidiorites) intruded by thin, tabular bodies of serpentinites in the form of dikes. The residual soils of the area, sampled at a depth of about 60 cm at intervals of 152 meters along traverse lines 304 meters apart, contain anomalous elevated concentrations of Ni, Cr, Co, and Mn; the anomalies of these metals in the soils closely follow the locations of the serpentinite bodies across the study zone and thus can be used in the geologic mapping of these bodies hidden by the overlying material (Figure 4-14).

We conclude this section with two graphical representations, one which illustrates the generalizations between weathering and the formation of various sample types for geochemical prospecting (Figure 4-15), and another which presents a scheme of the genesis of soils and the development of their profiles (Figure 4-16). What should be very clear is that the geochemical soil sample represents an environment with well-defined physical, chemical, and biological parameters, but parameters that can vary locally in a wide study region. The value of the data that comes from the soil analyses depends not only on the numbers that result from the analyses, but also on the geological observations in the field and good planning of the exploration program before going to the field.

REFERENCES

Barth, T. W., 1962, *Theoretical Petrology*, Wiley, New York, 416 pp.

Bass Becking, L. G. M., Kaplan, I. R., and Moore, D., 1960, Limits of the natural environment in terms of pH and oxidation–reduction potentials, *J. Geol.*, **68**, 243–284.

Bear, F. E. (ed.), 1964, *Chemistry of the Soil*, 2nd edition, Reinhold, New York, 515 pp.

Bowen, N. L., 1928, *The Evolution of the Igneous Rocks*, Princeton University Press, Princeton, 334 pp.

Bunting, B. T., 1967, *The Geography of Soil*, 2nd edition, Hutchinson, London, 213 pp.

Carroll, D., 1970, *Rock Weathering*, Plenum Press, New York, 203 pp.

Cloke, P. L., 1966, The geochemical application of Eh–pH diagrams, *J. Geol. Educ.*, **XIV**, 140–148.

Fieldes, M., and Swinedale, L. D., 1954, Chemical weathering of silicates in soil formation, *New Zealand J. Sci. Technol.*, **36B**, 140–154.

FitzPatrick, E. A., 1967, Soil nomenclature and classification, *Geoderma*, **1**, 91–106.

Flaschka, H. A., Barnard, A. J., and Sturrock, P. E., 1969, *Quantitative Analytical Chemistry*, Vol. 1 Barnes & Noble, New York, 594 pp.

Garrells, R. M., and Christ, C. L., 1965, *Minerals and Equilibria*, Harper & Row, New York, 450 pp.

Goldich, S. S., 1938, A study in rock weathering, *J. Geol.* **46**, 17–58.

Goldschmidt, V. M., 1937, The principles of distribution of chemical elements in minerals and rocks, *J. Chem. Soc.*, 655–673.

Goldschmidt, V. M., 1954, *Geochemistry*, Clarendon Press, Oxford, 730 pp.

Goudie, A., 1972, The chemistry of world calcrete deposits, *J. Geol.,* 80, 449–463

Hansuld, J. A., 1966, Eh and pH in geochemical prospecting, Geological Survey of Canada, Dept. of Energy, Mines and Resources, Paper 66-54, *Proceedings, Symposium on Geochemical Prospecting,* pp. 172–187.

Hawkes, H. E., and Webb, J. S., 1962, *Geochemistry in Mineral Exploration,* Harper & Row, New York, 415 pp.

Jackson, M. L., 1956, Soil chemical analysis – advanced course, Author's publication, Professor of Soils, University of Wisconsin, College of Agriculture, Dept. of Soils, Madison, Wisconsin.

Joffe, J. S., 1949, Pedology, 2nd edition, Rutgers Univ. Press, New Brunswick, 662 pp.

Kilpatrick, B. E., 1969, Nickel, chromium, and cobalt in tropical soils over serpentinites, *Quart. Colo. School Mines,* 64, 323–332.

Krauskopf, K. B., 1967, *Introduction to Geochemistry,* McGraw-Hill, New York, 721 pp.

Krumbein, W. C., and Garrels, R. M., 1952, Origin and classification of sediments in terms of pH and oxidation–reduction potential, *J. Geol.,* 60, 1–33.

Larsen, E. J., 1964, The new classification, *Soil Conservation,* 30, No. 5, 99–102.

Latimer, W. M., 1952, *The Oxidation States of the Elements and Their Potentials in Aqueous Solution,* 2nd edition, Prentice-Hall, New York, 392 pp.

Lopez-Ritas, J., 1967, *El Diagnostico de Suelos y Plantas,* Mundi-Prensa, Madrid, 267 pp.

Loughnan, F. C., 1969, *Chemical Weathering of the Silicate Minerals,* Elsevier, New York, 154 pp.

Malyuga, D. M., 1964, *Biogeochemical Methods of Prospecting,* Consultants Bureau, New York, 205 pp.

Mason, B., 1966, *Principles of Geochemistry,* Wiley, New York, 329 pp.

Millar, C. E., Turk, L. M., and Foth, H. D., 1965, *Fundamentals of Soil Science,* 4th edition, Wiley, New York, 491 pp.

NASA, 1968, *Application of Biogeochemistry to Mineral Prospecting – A Survey,* Special Publication 5056, National Aeronautics and Space Administration, Washington, D. C., 134 pp.

O'Brien, M. V., and Romer, D. M., 1971, Tara geologists describe Navan discovery, World Mining, International edition, June, 38–39.

Pettijohn, F. J., 1941, Persistence of heavy minerals and geologic age, *J. Geol.,* 49, 610–625.

Schasscha, E. B., Appelt, H., and Schatz, A., 1967, Chelation as a weathering mechanism – I. Effect of complexing agents on the solubilization of iron from minerals and granodiorite, *Geochim. Cosmochim. Acta,* 31, 587–596.

Sillen, L. G., 1964, Stability constants of metal-ion complexes. Sec. I.: Inorganic ligands, *Chem. Soc. London, Spec. Publ.* 17.

Strakov, N. M., 1967, Principles of lithogenesis, vol. 1, Consultants Bureau, New York, and Oliver & Boyd, Edinburgh & London, 245 pp.

Thompson, L. M., 1965, *El Suelo y Su Fertilidad,* 3rd edition, Editorial Reverte S. A., Barcelona, 407 pp.

Yaalon, D. H., 1960, Some implications of Fundamental concepts of pedology in soil classification, Trans. Int'l. Congress Soil Sci., 7th, 16, 119–123.

5

SAMPLES FOR
EXPLORATION GEOCHEMISTRY

The sample used in geochemical prospecting is that one which can be analyzed for the elements or ions useful in the search for mineral deposits (of metals, nonmetals, or hydrocarbons). This does not lessen the role of other indicators of potential areas of economic ore deposits such as magnetics, gravity, electric conductivity (and resistivity), seismology, aerial photos, or in a sense, all classes of remote sensing, and of course the most important indicators in the final consideration, the geological and geochemical observations and fundamental knowledge.

Each general sample has its own advantage with respect to the information it can provide in addition to the geochemical data, the area it represents, and the ease with which it can be collected, transported, and prepared for laboratory analyses. The papers published from the last four symposia on geochemical exploration and prospecting (Ottawa, Canada, 1966; Denver, Colorado, 1968; Toronto, Canada, 1970; London, England, 1972) show a significant range of sample classes (Table 5-1) from the most common to very ingenious ones, some of which we will consider below.

The sample that provides the most data for locating, evaluating, and estimating the volume of a deposit, and for interpreting its genesis, is the mother rock, with the designation "mother" being used here to include all rock classes although their genesis may be distinct. The sample of mother rock provides us with geochemical data representative of the collection site; if it is a spot sample, it represents only the point of collection, but if it is a channel sample or a general sampling of an entire outcrop, it represents the area sampled. Obviously the geological observations in the outcrop zone of the mother rock allow the projection of a wide area that might be represented geochemically by the sample, but always with the reservation that what appears to be a relatively homogeneous rock or series of rocks perhaps is not so geochemically.

The quantity of one or various indicator elements related to the mineralization increases in the fresh host rock (and/or component minerals) approaching

Table 5-1   Samples Used in Geochemical Prospecting

Fresh host rock
Weathered host rock
Minerals separated from the host rock
Ore
Minerals separated from the ore

Stream sediments
Alluvium
Minerals separated from stream sediments and alluvium (magnetic and nonmagnetic heavy minerals)
Ultrasonic concentrates of material films (fines) adhered to detrital fragments
Lake sediments

Soils
Humus related to soils
Bacteria related to soils

Glacial materials
Peat

Snow
Stream waters
Lake waters
Subsurface waters
Spring waters

Air
Soil gases
Rock gases
Vapors
Aerosols

Vegetation (flora)
Animals (fauna)
Man

Source.   Compiled from the reports presented at the first four international symposia on geochemical prospecting (Ottawa, Canada, 1966; Denver, Colorado, 1968; Toronto, Canada, 1970; London, England, 1972).

a mineralized zone, an observation reported in many investigations and applied in lithogeochemical prospecting programs.

In rock altered by hydrothermal activity, the aureoles of mineralogical or chemical element dispersion can be studied and used in an attempt to delineate the extension of the alteration zone and the intensity of alteration in subzones, thus indicating where the geochemical prospecting should be intensified.

In a well-executed study, Hansen and Kerr (1971) used x-ray diffraction analysis to semiquantitatively evaluate silicification, feldspathification, and sericitization in areas of porphyry mineralization by measuring the amounts of quartz, potassium feldspars, and sericite in hydrothermally altered rocks. In the areas where these workers did their research (Kingman, Arizona — sericitization associated with Cu-Mo mineralization; Okanogan County, Washington — potassium feldspathization and silicification associated with Mo mineralization; Boulder Batholith, Jefferson County, Montana — concentric pneumatolitic zonation associated with Cu-Mo mineralization; and, Copper Creek, Arizona — alteration associated with Cu mineralization in test drilling cores), the facies alteration of potassium silicates correlated well with the Cu and Mo mineralization, so much so that a plot of alteration values could be used to delimit the extension of mineralization and indicate the most probable geologic-geographic trend for the occurrence of ore. However, the authors stated that a study of each mineralized area is necessary to establish exploration norms because the intensity of alteration usually varies in its degree due to differences in rock types and their geographic–geologic location. Without doubt, the utility of alteration contours is a technique that merits further investigation and subsequent application in the exploration and exploitation phases of the location and development of the porphyry type of mineralized deposits (mesothermal to hypothermal).

Specific minerals separated from a mother rock can be geochemically analyzed and provide indications as to the proximity of a vein or other type of ore deposit. For example, Hamil and Nackowski (1971) studied the distribution of various elements in magnetite from quartz monzonite intrusives in the search for sulfide minerals of the Basin and Range Province, Utah and Nevada; they found that elevated contents of Ti and Zn in the magnetite correlated well with important Pb-Zn mineralization and that low contents of Ti and Zn in the magnetite correlated well with important Cu mineralization associated with the parent intrusive. In the cupriferous area of Michigan, Ruotsala, Nordeng and Weege (1969) found that Mn in accessory calcite in lava flows bearing native Cu mineralization correlated more with an exploitable tenor of Cu than other elements evaluated (Co, Cu, Cr, Ni, Mg, Fe and Hg). After compiling available data from the United States and the Soviet Union, Beus (1969) reported that the concentration of niobium pentoxide in muscovite from columbite-bearing pegmatites is significantly greater than its concentration in muscovite from pegmatites without columbite. Biotite is another mica that has been investigated with the purpose of using it in geochemical prospecting. In a preliminary study of the distribution of some elements in biotites from a quartz monzonite with contact metasomatic W-Mo-Cu mineralization (Pine Creek, California), Darling (1971) found that the biotite from a pluton directly related to ore deposits of W, contained more F, Zn, Cu, and

Li and less Na and Y than the biotite from a non-ore-bearing pluton. He also observed that the biotite directly from (or near) mineralized contacts contained more Na, Y, and Sr than the biotite from nonmineralized contacts; nearing the ore, the contents of B, Bi, Cu, Sc, Ag, Sr, Y, Cr, and V in biotite may increase, whereas that of F may decrease. Studies on this class of samples are somewhat limited in number but are being carried forth; although the samples indicate a real potential for use in the search for ore, they cannot be considered as principal type samples for geochemical prospecting (at this time).

Unfortunately the mother rock often does not outcrop sufficiently to permit a systematic sampling or it does not outcrop in areas where weathering acts to fragment and/or dissolve it physically, chemically, and biologically with the result that the rock is hidden beneath a soil and/or vegetation cover of varying thickness. In such a situation, the stream sediment (or alluvium) can be used as a reconnaissance sample representative of the weathering products of materials in upstream areas. An evaluation is made of the meaning of the data from the stream sediments according to the knowledge of drainage basin pattern and of the regional geology, thus defining or delimiting zones of major interest for detailed or intensified geochemical prospecting that would follow the general reconnaissance. In these general explorations by means of stream sediments, the sampling density used varies with the geologic, topographic, and geomorphologic conditions of the study areas, the type of deposit being sought, and the time and economic limitations of a program; in general, it would be between one sample every 400 meters to one sample every kilometer or more. In their report on geochemical mapping in Southern Madagascar, Borucki and Ratsimbazafy (1969) divided the topographic maps of the area into quadrants (of 4 km^2) and placed a sampling site in each quadrant based on the following norms, which were appropriate for the prospecting zone:

1. In the center of the quadrant.
2. Immediately below the confluence of two (or more) streams.
3. In streams with their sources in the same quadrant.
4. At sites of relatively easy access.

In other studies different norms have been established for sampling and these depend on the geologic environment, the experience of the director of a program, and the information he wants from the samples. For example, Bolviken and Sinding-Larsen (1972) carried out a lead exploration project in southern Norway and collected duplicate stream sediment samples from some 3000 sites near roads; the collection stations were at least 30 meters from each road site and duplicate samples were approximately 10 meters away. Such duplicate sampling allowed an estimation of statistical parameters for the metal distribution within each sample site. Of extreme interest in this report is the fact that for Pb, the duplicate sampling proved to be adequate and necessary; a Pb anomaly could have been missed if just one sample had been taken from each location since only one of the duplicates showed an anomalous value. The followup of this anomaly led to the discovery of a Pb deposit.

The subsample of the stream sediment or alluvium to be used must be carefully selected in order to obtain data that result in the maximum concentration contrast with respect to background value for the elements to be studied. This selection of the most effective subsample in the prospecting program cannot be made on the basis of results obtained in other study zones, but must be established for the particular region of interest by means of a pilot or orientation study. An investment of time and funds in a pilot study could result in a savings of both in a major project and also provide an evaluation of the subsample or subsamples that could lead to discovery of a greater number of anomalous zones for subsequent detailed studies. In many programs researchers have worked satisfactorily with data from the geochemical analysis of the −80 mesh fraction of stream sediments or alluvium, but in other studies, geochemical information from sediments fractions of greater or lesser sizes has been used to achieve the best results; such fraction selections were determined by pilot study data. Also, in the size fractions analyzed with orientation aims, it is possible that a further separation of heavy minerals, and/or magnetic and nonmagnetic heavy minerals, would indicate the most efficient sample for a given exploration program. In a paper published in 1971, Huff compared exploration methods in alluvium for porphyry copper deposits in the Lone Star district (desert) of the Gila Mountains, Arizona; he collected seven classes of subsamples from the heads of arroyos that drain known copper deposits and found anomalous values for copper at distances of about 8 to 11.4 km from the cupriferous deposits. The subsamples that Huff evaluated included:

1. The −80 mesh fraction of fine alluvium from the sides of arroyos.

2. The −80 mesh fraction of coarser alluvium from the beds of arroyos.

3. The −30 to +80 mesh fraction of the coarse alluvium.

4. Assemblages of hydrothermally altered pebbles.

5. Ultrasonic concentrates of fine films of material adhered to coarse detrital grains.

6. Magnetic heavy minerals from the −30 mesh fraction.

7. Nonmagnetic heavy minerals from the −30 mesh fraction.

Of the seven classes of subsamples compared, the ultrasonic concentrates and the nonmagnetic heavy minerals showed an elevated range of copper contents and the higher values clearly identified the anomalous arroyos. Huff believes that these two subsamples may be especially good for tracing anomalies over a long distance or for localizing anomalies with a wide sample spacing; however, the obvious disadvantage in using these subsamples is the work necessary in their preparation for analysis. The subsamples of silt, fine sand, and hydrothermally altered pebbles also clearly indicated the anomalous arroyos and have the advantage of being easily collected.

The distribution of Cu and Zn has been studied in alluvial magnetites from the coarse fraction of sediment in various zones of Central Ecuador and has been compared with data on the Cu and Zn distribution in the sediment −80 mesh fraction (de Grys, 1970). De Grys reported background values for Cu between 0 and 100 ppm for both classes of samples; but for Zn, the background values were 0 to 200 ppm for the −80 mesh fraction of stream sediments and 100 to 300 ppm for the magnetites of the coarse fraction of the sediments. In the Chancha area (disseminated mineralization of pyrite, chalcopyrite, covellite, and molybdenite) the dispersion trains for the Cu and Zn in the two sample types (−80 mesh fraction and magnetite) can be traced over equal distances (1 to 10 km) and either of the two samples could have been used in the geochemical reconnaissance study. In the San Fernando (silicification and gossan) and San Bartholome (veins with sphalerite, galena, and sulfides of Sb, As, and Bi) areas, the mean concentration of Zn in the alluvial magnetites was 1000 ppm, but there were no Cu anomalies. For two other areas (Molleturo chalcopyrite, sphalerite, galena and other sulfides, and Gualleturo galena and sphalerite in quartz veins) the mineralized zones were detected with the −80 mesh fraction since the magnetites showed anomalies only near the mineralization. From the examples studied, de Grys (1970) concluded that there was little advantage in analyzing the magnetites instead of the −80 mesh fraction, although she believes that the investigation of the content of trace elements in magnetites could be useful to resolve the genesis of this mineral in different areas of mineralization.

In Brazil, gravels have been used as a new type of sample in geochemical prospecting. According to Goncalves (1971) the field method followed consisted of selecting stream gravels that showed signs of hydrothermal alteration, Fe_2O_3 staining, veining of calcite, quartz or any other mineral that indicates the existence of geologic activities related to mineralization, signs indicative of fault or fracture origins, and any other characteristic that can serve as an aid in the interpretation of the genesis of a sample. The Brazilian geochemists collected between 5 and 10 samples of stream gravels in each river (or stream) and examined and fragmented them; the most representative pieces of about 5 cm^3 were selected for further investigation. Once in the laboratory, the samples were separated into groups according to characteristics that indicated that they were derived from the same rock type. This was followed by a macroscopic description of the fragments for the composition of the rock, type of alteration, or whatever other feature for which it was collected. After this, chemical analyses were made. Once the presence of an element in anomalous quantities was detected, an attempt was made to determine the provenance of the anomalous gravels for future detailed geochemical programs.

As was described in Chapter 4, the residual soil is generally used in the detail stage of exploration geochemistry. The soil represents the immediate area of its collection, and its content of metals could give an indication of the relative content of these same metals in the mother rock from which the soil was derived. Although some researchers do not advise the use of soils in prospecting areas where marked creep has been active or where there have been slides or mass movements, an understanding of the topography and of geological processes permits the use of these "moved" soils when the approximate direction from which the soils can be projected and one can determine the possible geometric forms of potential deposits in a zone. As is the case with stream sediments, the size fraction to be used in a study must be carefully selected as well as the soil horizon that will be sampled in the exploration program.

In 1967 Boyle and Dass wrote that the majority of the geochemical reconnaissances (based on soils) carried out in Canada and other regions favored the use of the B soil horizon. This was because the B horizon contains the maximum accumulation of clay minerals and iron and manganese hydroxides, solids that tend to absorb or coprecipitate metals frequently used in prospecting for veins or other types of deposits. However, it is also true that in some areas the concentrations of various trace elements are much lower in the B horizon than in the A horizon. Boyle and Dass cited their work in Cobalt, Ontario, Canada, where the best results in prospecting for known mineralization came from the A soil horizon; the concentrations of the metals Pb, Zn, Cu, As, Sb, Mo Ag, Co, Mn, and Hg are much greater in the A horizon and resulted in better contrasts than did the analytical concentrations in samples from the B horizon. These authors attributed the concentrations of metals in the A horizon to a biochemical factor; the deciduous leaves and conifer needles fall to the ground carrying metals obtained from the B and C horizons and from slightly weathered mother rocks. This fallen material undergoes humification with the resulting formation of products that sequester or chelate the metals. After several years of such activity, an elevated concentration of metals may develop in the A horizon. However, Roy and Paul (1967) gave two examples from India where the A horizon was, in general, less effective in prospecting than the B and C horizons, and they observed that the concept of using the A horizon is good, but that the *modus operandi* of the biochemical mechanism proposed by Boyle and Dass (1967) is very sensitive to various factors, such as very dense vegetation with very deep roots, the availability of metals in soluble forms, the type of vegetation that can accumulate specific metals, and the existence of a reducing environment because in an oxidizing environment the vegetable material decays with the resulting solubilization of minerals. Lapointe (1968) criticized this last factor as being somewhat simplistic and noted that the chemistry of the A horizon is that of the formation of carboxylic salt but mainly of chelation. In addition, the stability of compounds of chelates depends on many factors, and the influence of the reducing factor with the chelation and the solubility of chelates is such that generalizations on the subject cannot be made. Finally, Lapointe commented that laboratory experiments with cellulose and other organic matter, under different conditions, showed that the Eh and pH are not primary factors in the sequestering and selectivity of chelates for specific ions. As has been written and will be repeated in various sections of this book, each geological-geochemical environment has to be treated in a particular way, and criteria derived only from other studies cannot be used to plan a geochemical exploration project.

If the soil in a study area changes from one class to another, it is very probable that the type of sample being used must be changed. The grid spacing used in geochemical soil sampling is very variable and depends on the size of the target being sought and the mobility of the chemical elements being used in the project; the element selection is dependent on the genesis of the potential deposit and the conditions of temperature, pressure, and elemental concentration existent during its formation, and in the actual physical, chemical, and biological environment of the soil. The sample that will give the best data and the optimum grid spacing in an area can be determined scientifically and economically by means of a pilot study, an essential step in exploration geochemistry programs. Directly related to the soils are the flora and fauna. As already noted in the section on the usefulness of flora and fauna in exploration programs, these biogeochemical samples are assuming the importance in geochemical prospecting that they deserve.

Glacial materials (which have been studied extensively in Scandanavia, Canada, Russia, and other regions that were completely or partially covered by glaciers in the past) can provide us with geochemical samples that, to a certain degree, are similar to stream sediments and soils. Generally, the basal till is an acceptable geochemical sample when it is accessible without having to dig or drill through a thick overburden. However, although the cost of drilling to obtain a basal till sample is high, it is possible that the value of a potential deposit may justify the investment. Garret (1971) described a project of the Geological Survey of Canada in which there was drilling through an average of more than 11 meters of glacial overburden plus 1 meter into the mother rock. There was little variation in the contents of the elements (Cu and Zn) analyzed in the glacio-lacustrine sediments overlying the basal till, but in the −80 mesh fraction of the basal till, anomalous levels of Cu and Zn could be traced over distances of 120 to 150 meters from a known deposit in the direction of ice movement. When heavy minerals of the −80 to +230 mesh fraction were analyzed, the elemental dispersion train increased more than 50%; this means that the cost of an exploration program based on heavy-minerals geochemistry could be less because the increased dispersion of the elements would permit the use of a wider prospecting grid, thus requiring fewer samples.

Van Tassel (1969) reported on mechanical and geochemical prospecting by drilling of permanently frozen overburden in the Galena Hill – Keno Hill Yukon, Canada. Movement by solifluction has transported detrital material from areas of Ag mineralization downslope in zones where the overburden consists of more than 30 meters of glacial till and mother rock detritus. The operation proved successful after an inspection of mineralized detrital material complemented by geochemical analyses of sedimentary samples representative of 1.5-meter sections of the drilled matter. The author observed that this type of exploration has been very satisfactory in the sampling of zones of veins to a depth of about 30 meters and that the method cannot be used in areas of excessive water and broken ground.

In addition, the mineralogical analysis of basal till can serve in the geologic mapping done prior to the geochemical prospecting. Larson and Nichol (1971) recognized the potential for mineralization in the contact zones between Devonian sandstone and Carboniferous limestone in the west of Ireland where the rock is covered by glacial deposits. They tried to identify the nature of the rock below the glacial material by the mineralogical analysis of the silt-size fraction of the till. Their results were positive since the mineralogical composition of the till silts seemed to be diagnostic for various rock classes (basic intrusives, micaceous schist, shale, sandstone, quartzite, and granite). With this preliminary exploration step, prospects for geochemical studies could be established.

By definition, "eskers" are subglacial stream deposits generally expressed topographically as rather long winding ridges of crudely bedded gravel and sand. Cachau-Herreillat and La Salle (1971) thought sediments from eskers could be used in a manner similar to the use of samples of ancient drainage patterns. They carried out mineralogical studies on the sand and gravel of the Mattagami esker and geochemical studies on samples from the B horizon of the podzol soil developed on the esker and from the C horizon. The geochemical data from the C horizon samples had some high elemental concentrations, but an interpretation of these data cannot be made as yet because of the lack of information from the mineralogical investigation; without this information it is difficult to ascertain in which minerals the anomalous elements are concentrated. Although it is very possible that the results of this research may not be useful in geochemical prospecting, the concept of using the eskers is good.

Water is one of the most important agents in geologic–geochemical processes active in the preparation and transport of geochemical samples. Also, by itself, water provides another geochemical sample to use in the search for mineralized areas. For example, in the Keno Hill region, Canada, Ag–Pb–Zn ore deposits, one of the samples evaluated in a pilot study was stream water; the cold-extractable base-metal water anomalies generally coincided with the known deposits, as did stream-sediment-sample anomalies (Gleeson, 1966). In another pilot study of a permafrost area in Canada, Allan and Hornbrook (1971) examined various classes of samples from an area of Cu mineralization; the data from lake waters, especially from lakes with one axis longer than 305 meters, were very satisfactory for regional geochemical exploration. Similarly, Nigrini (1971) studied three streams related to sources of Cu, Pb, and Zn (Bathurst, New Brunswick, Canada) and suggested that after being in contact with ore, the subsurface waters are strongly charged with CO_2, H^+, and metal cations; with exposure of these waters to the atmosphere, there is a rapid increase of pH, rapid decreases of P_{CO_2}, Cu, Pb, and Zn, and simultaneous deposition of hydrous oxides of Fe, rich in Cu, Pb, and Zn.

Subsurface waters that pass through sulfide-bearing deposits can oxidize the sulfides and become anomalously charged with sulfate. In such a case, a knowledge of the hydrogeology is necessary to interpret from where the analyzed waters could have accumulated the sulfate and thus indicate zones for more intensified geochemical exploration. In a similar way, the Mo content in subsurface waters, streams, or rivers may be determined and used as a prospecting guide in areas of possible copper porphyry mineralization; depending on the chemical properties of the waters that make contact with porphyry copper deposits, the Mo could be present in anomalous concentrations and thus serve as a pathfinder element. Barakso and Bradshaw (1971) studied the leaching and loss of Mo at the surface environment and concluded that the mobility of Mo is controlled by the pH and Eh of the active waters, although they noted that the amount of Fe oxide and other ions in a system can influence the concentration of Mo by their ability to capture or interact with those distinct ionic forms of Mo that may be present. Horsnail and Elliot (1971) investigated some environmental influences on the secondary dispersion of Mo and Cu in regions of Canada and found that acidic groundwater, especially near oxidizing pyrite, facilitates the mobility of Cu, but that the mobility of Mo is better under slightly alkaline (or basic) conditions; also, these authors reported that the mobilities of Cu and Mo are apparently restricted by chelation with organic compounds, capture by precipitates of Fe and Mn oxides, and precipitation in the proximity of caliche. Two of the better treatments of the effect of pH and Eh on the mobility of Mo (with respect to exploration geochemistry) have been presented by Hansuld (1966a, 1966b).

Subsurface waters that issue forth at the Earth's surface as permanent or seasonal springs have been used by De Geoffroy, Wu, and Heins (1967) as the basis for a regional geochemical study of the Southwest Wisconsin zinc area, via a sampling of 3766 springs in an area of more than 1000 km^2. Of the 56 zinc anomalies found, 26 corresponded to known zinc deposits; test drillings of some of the remaining anomalies confirmed the presence of exploitable zinc ore close to the anomalies. The factors that influence the utility of this method are climatologic, topographic, geologic, and hydrologic in character. According to de Geoffroy, Wu, and Heins, the annual precipitation should be equal to or greater than the annual evapotranspiration; the topography should be such that the less permeable horizons and circulation channels are cut repeatedly by the Earth's surface; the formations of stratified (sedimentary or volcanic) host rock should include numerous horizons with contrasting permeabilities because this facilitates the lateral movement of subsurface waters to their exits as springs; and, the hydrologic and topographic conditions should ensure a sufficient superposition between the oxidation zone and the primary haloes of ore bodies not readily detectable by other than the spring sampling exploration approach. These conditions are satisfied in Wisconsin and the prospecting method by means of spring sampling resulted in a rapid, complete, and economical technique for making an inventory of the mineralogical and hydrological resources, especially in carbonate areas (except when karst conditions exist) where other geochemical methods cannot be used with confidence. De Geoffroy, Wu, and Heins suggested the implementation of geochemical prospecting with springs when the necessary conditions are existent as, for example, in different regions of the eastern third and the northwest of the United States and in Central and Western Europe.

Surface waters have been used in prospecting for uranium. Dyck and Smith (1969) tested the possibility of using radon-222 in these studies and found a general areal correlation between the content of radon or of uranium in surface waters and uraniferous zones. The low cost and the simplicity of the measurement of radon by portable equipment in the field lend themselves well to preliminary geochemical reconnaissance programs. Already described is the work of Meyer (1969) in which the measurements of the contents of uranium in lake waters indicated regional distributional tendencies of uranium that corresponded favorably with geophysical reconnaissance data obtained from aircraft. In an orientation study of epigenetic pitchblende mineralization, McDonald (1971) observed that the content of U_3O_8 in lake waters defined zones of mineralization; he also noted that there was an increase of the specific conductance of lake waters in the proximity of mineralized areas.

Most recently Jonasson and Allan (1972) proposed the use of another hydro-related sample in geochemical prospecting – snow. A comparison of chemical analyses of snow samples from over nonmineralized background sites and from areas of close and narrow veinlet mineralization showed significantly

Table 5-2 Comparative Element Concentration Ranges in Snows from Within and Outside the Ore Zone (Ni and Cu Sulfides Mineralization) of Donaldson Mine Area, Ungava, Quebec, Canada, and in Spring Drainage both Downstream and Upstream from the Ore Zone (from Jonasson and Allan, 1972)

Element	Concentration Range in Snows Inside Ore Zone (ppb)	Concentration Range in Snows Outside Ore Zone (ppb)
Ni	5 — 53	4 — 8
Cu	3 — 34	3 — 6
Zn	4 — 220	16 — 45
Hg	0.5 — 1.1	0.5 — 1.1

	Concentration Range in Spring Drainage Down from Ore Zone (ppb)	Concentration Range in Spring Drainage Up from Ore Zone (ppb)
Ni	2 — 29	2 — 4
Cu	1 — 7	0 — 1
Zn	1 — 4	0 — 2
Hg	0.55 — 1.1	0.55 — 1

higher contents and wider ranges of values of several metals from the snow inside the ore zone (Table 5-2). For both sets of samples (inside to outside of ore zones), there was a definite concentration gradient from ground-level snow (highest) up to surface snow (lowest). Because of this latter observation, the authors believe that the source of the metals (Hg, Cu, and Zn) is from below the snow and not from atmospheric particulates and that the metals are derived from the weathered mineralized rock beneath the soils rather than from the soils proper or plant mull. It is possible that the metals find their respective ways into the fallen snows by two processes: gaseous diffusion of organometallics and solution migration by capillary movement of water across interlocked ice crystal faces. Jonasson and Allan are of the opinion that the metal levels detected in the snows built up in a 1- to 3-month period.

In our consideration of samples for geochemical exploration we pass from solids and liquids to gases. The use of soil air or vapors that emanate from the terrestrial surface is not a new method in geochemical prospecting. For many decades measurements of organic gases in soil atmospheres have been used in the search for hydrocarbons, a technique described in the section on geochemical prospecting for petroleum and natural gas. More recently, the gas radon-222 in soil gases and in waters has been used as a guide in the search for uranium. Stevens, Rouse, and De Voto (1971) described three instances of exploration for uranium in the Western United States with radon-222. Radon-222 moves in solution from the points of its formation (from the decay of

[238]U) by diffusion and by barometric pumping for distances of a few inches to hundreds of feet. The prospecting in the three cases to be cited was done by the determination of radon-222 in soil gas and in the atmosphere itself. In one area (the Denver Basin, Colorado), the magnitude of the anomalous content of radon over the background value correlated inversely with the depth of uranium mineralization; in another zone (Lisbon Valley, Utah), the presence or location of mineralization at depth of from 19 to 2000 feet could not be adequately predicted with the radon anomalies in soil gases; in the third case (Laramie Basin, Wyoming), the content of soil gas radon showed a strong correlation with the uranium content of the area soils. Thus, although it is possible to work with radon concentrations in soil gas or in the atmosphere, grave problems exist in using it without reservation as a pathfinder element in the search for uranium.

Mercury, because of its high volatility and geochemical responses, lends itself as a volatile pathfinder element in the search for various elements (Au, base metals, and other sulfide-forming elements). In his summary on the research in geochemical prospecting being conducted by the United States Geological Survey, Erickson (1969) described the experiments of Vaughan and McCarthy on the measurement of Hg in soil gases and in the atmosphere. Vaughan and McCarthy collected Hg vapor in plastic pyramids placed directly on the Earth's surface, from surface vehicles, and from aircraft that flew at low altitudes. Their experiments with plastic tents suggested that Hg in soil gases delineated a part of the Cortez, Nevada, gold deposit better than Hg in soils directly above the ore body, and this through as much as 30 meters of gravel overburden. Also, measurable quantities of Hg in the atmosphere were detected from a Beaver aircraft specially equipped and flying at altitudes up to 300 meters above the surface; the major concentration of the element was found below 90 meters altitude.

In geochemical prospecting with Hg, the problem of the origin of the Hg affects its utility as a pathfinder element. Trost and Bisque (1971) differentiated between the Hg from a vapor source and ionic Hg (aqueous) by the analysis of two distinct soil horizons, one rich in humus and the other rich in clay. The ratios Hg_{org}/Hg_{clay} with values less than 2 corresponded to an ionic source of Hg and ratios with values greater than 2 corresponded to a vapor source. Barringer (1971) published an abstract regarding new methods for the measurement of volatile elements in which he wrote about the manufacture of optical equipment for continuously measuring the concentrations of Hg and I in the atmosphere and about the possibilities of installing this equipment in a plane for testing; his experiments from a stratopheric balloon have already established the potential for making certain kinds of gas measurements from high altitude, but the meteorological effects and effects of vegetation on the data obtained must yet be investigated and evaluated.

Scientists from the Soviet Union have applied techniques developed in prospecting for hydrocarbons to the search for metalliferous deposits and to assist in structural mapping. Ovchinnikov, Sokolov, Fridman, and Yanitskii (1972) realized that there is a continuous but localized release of significant concentrations of various gases to the atmosphere from the Earth's crust. According to their observations, zones of abyssal fractures and associated cracks are especially important in this respect and the gas stream increases sharply with the presence of metalliferous deposits in such zones. These authors note that gases of different genesis and composition accompany ore deposits throughout their existence and evolution; for example, gaseous phases play an important role in the formation of endogenic ore deposits and may be preserved in a post-ore phase, and gaseous products from the oxidation of ores occur in hypergene conditions in near-surface areas. Such gases may be occluded in the rock and ore (as inclusions, for example) or may enter the network of ore-controlling fractures. It is obvious, therefore, that given the proper equipment for sampling and analyzing such gases (include CO_2, H_2, CH_4, O_2, N_2, He, Rn, A, H_2S, SO_2, F, Cl, I, and Br), the analyses may be used as indicators, both in structural mapping to reveal fractures of different age, origin, and depth and in the search for buried deposits.

Elynson (1972) applied this technique in the search for sulfide deposits by analysis of subsurface air for H_2S, basing his approach on the thesis that blind ore bodies enrich the subsurface atmosphere with sulfur-bearing gases. Gases found at depths of 1.5 to 2 meters in a semidesert area with sulfide ore deposits showed significant concentrations of H_2S over the exposed ore bodies and over those occurring at a depth of 30 meters through an overburden of 1 to 10 meters. Elynson believes that it is possible that the presence of H_2S is related to chemical and biochemical oxidation–reduction processes acting on sulfides and sulfates. Once formed, the H_2S seems to be quite stable in the environments characteristic of secondary enriched sulfide ore bodies in the oxidation zone of sulfide ore deposits (reduced or weakly acid environments). Similarly, Eremeev, Sokolov, Solovov, and Yanitskii (1972) applied helium geochemical surveying to structural mapping and ore-deposit forecasting by sampling gases at 20- to 30-meter depths (thus eliminating atmospheric effects and assuring the sensitivity of the mass spectrometry or ion pump-type detectors used). According to control studies by Soviet scientists, the helium field is stable in time with variations arising only as a result of seismic activity. In the pilot studies made to date on a survey sampling of 1 to 4 samples per 100 km^2 where the basement is exposed or not far from the surface (less than 500 meters), it appears that only rock permeability affects the helium field and that no geological structure-related factors affect the distribution near the surface. The permeability zones through which the helium travels to the surface reach extensive depths and are thought (by the researchers) to be tectonically controlled. Intensities of helium anomalies apparently depend directly on the depth at which the fracture-permeability zone is located. The initial studies show that He-bearing anomalies occur in geological rock units associated directly or indirectly with mobile diagenetic fractures and have been revealed in association with diamond-bearing kimberlite pipes, carbonatites, and hydrothermal gold, lead–zinc, iron, and mercury deposits. In addition to the technique potential in prospecting for ore deposits and in structural mapping (together with other geophysical and geochemical studies), helium surveying may serve in earthquake forecasting. Gas emanation studies in exploration projects for metalliferous ore certainly hold promise for the future and deserve the research importance assigned to them by the scientific community.

The idea of using aerosols in geochemical prospecting from helicopters or fixed-wing aircraft emerged from the fact that the very fine sized particles from outcrops of mineralized rocks or from soils are carried by air currents to the atmosphere, thus forming particulate aerosols (Weiss, 1971). The chemical composition of these aerosols can be analyzed and the nature of the ore deposits from which they were derived can apparently be identified in a very general way. Weiss described the collecting of these atmospheric particulates using aluminum frames around which were wound about 275 meters of nylon twine having a 0.1mm diameter. This collecting device is lowered from an airplane or helicopter flying at altitudes of 30 to 60 meters and collects aerosols for a given time, after which it is replaced by a sister device thus allowing for continuous sampling; geographic position is controlled by aerial photos or navigation techniques. The collectors are stored in plastic bags until they are transported to the laboratory for separation and analysis of the collected particulates. Size distribution of the aerosols and their element concentrations are determined. Obviously this technique has its limitations and cannot be used over surfaces covered by water, snow, or dense tropical vegetation; on the other hand, rather inaccessible areas (deep canyons, mountain peaks, etc.) can be studied in a preliminary reconnaissance. According to estimates by Weiss (1971), with 2 months of good weather, 36,000 km^2 can be covered with flight lines spaced at about 800 meters and at velocities between 160 and 320 km/hour. The initial tests of this geochemical exploration technique were made in South Africa and Southwest Africa and indicated that the method can detect mineralized outcrops, identify their content of metals, and define their position with an accuracy of 2.4 km^2 in areas of tens of thousands of square kilometers.

Whatever may be the type or types of sample one considers using in geochemical prospecting, its potential usefulness in the exploration environment must be evaluated by means of an orientation study of relatively little areal extension; this would indicate not only the most efficient sample or subsample for the project, but also the best sampling grid and/or sample density. Data from geochemical analyses are only indicators of the presence of ore or mineralization in an area studied and do not give us information on the tenor of an ore nor of volume of ore, although they can give us an idea of the general limits of a deposit; other exploration techniques such as induced polarization or electromagnetism can serve to better localize an ore deposit. However, the tenor of the ore, the real extension of the deposit, and the volume determination of ore present, can come only from detailed studies of core samples from test drilling.

REFERENCES

Allan, R. J., and Hornbrook, E. H. W., 1971, Exploration geochemistry evaluation study in a region of continuous permafrost, Northwest territories, Canada, in *Geochemical Exploration* (Boyle, tech. ed.), Special Volume No. 11, Canadian Institute of Mining and Metallurgy, pp. 53–66.

Barakso, J. J., and Bradshaw, B. A., 1971, Molybdenum surface depletion and leaching, *Geochemical Exploration*, Special Volume No. 11, Canadian Institute of Mining and Metallurgy, pp. 78–84.

Barringer, A. R., 1971, Optical detection of geochemical anomalies in the atmosphere (Abstr.), *Geochemical Exploration* (Boyle, tech. ed.), Special Volume No. 11, Canadian Institute of Mining and Metallurgy, pp. 474.

Beus, A. A., 1969, Geochemical criteria for assessment of the mineral potential of one of the igneous rock series during reconnaissance exploration. *Quarterly Colo. School Mines,* **64,** 67–74.

Bolviken, B., and Sinding-Larsen, R., 1972, Sample error and other criteria of consequence for the interpretation of stream sediment data, Handbook, 4th *Intern. Geochem. Exploration Symp., Abstracts,* London, p. 16.

Borucki, J., and Ratsimbazafy, C., 1969, Geochemical mapping of the Horombe Plateau, *Quart. Colo. School Mines,* **64,** pp. 75–88.

Boyle, R. W., and Dass, A. S., 1967, Geochemical prospecting – use of the A horizon in soil surveys, *Econ. Geol.,* **62,** 274–276.

Boyle, R. W., and Dass, A. S., 1968, Geochemical prospecting – use of A horizon in soil surveys, *Econ. Geol.,* **63,** 423.

Cachau-Herreillat, F., and La Salle, P., 1971, The utilization of eskers as ancient hydrographical networks for geochemical prospecting in glaciated areas (Abstr.), *Geochemical Exploration* (Boyle, tech. ed.), Special Volume No. 11, Canadian Institute of Mining and Metallurgy, p. 121.

Darling, R., 1971, Preliminary study of the distribution of minor and trace elements in biotite from quartz monzonite associated with contact-metasomatic W-Mo-Cu ore, California, U.S.A., *Geochemical Exploration* (Boyle, tech. ed.), Special Volume No. 11, Canadian Institute of Mining and Metallurgy, pp. 315–322.

De Geoffroy, J., Wu, S. M., and Heins, R. W., 1967, Geochemical coverage by spring sampling method in the southwest Wisconsin zinc area, *Econ. Geol.,* **62,** 679–697.

de Grys, A., 1970, Copper and zinc in alluvial magnetites from Central Ecuador, *Econ. Geol.,* **65,** 714–717.

Dyck, W. and Smith, A. Y., 1969, The use of radon-222 in surface waters in geochemical prospecting for uranium, *Quart. Colo. School Mines,* **64**, 223–236.

Elynson, M. M., 1972, Gaseous surveys in prospecting for sulfide ore deposits, Handbook, *4th Intern. Geochem. Exploration Symp., Abstracts,* London, p. 21.

Eremeev, A. N., Sokolov, V. A., Solovov, A. P., and Yanitskii, I. N., 1972, Application of helium surveying to structural mapping and ore deposit forecasting, Handbook, *4th Intern. Geochem. Exploration Symp., Abstracts,* London, p. 22.

Erickson, R. L., 1969, *Quart. Colo. School Mines,* **64**, 237–244.

Garrett, R. G., 1971, The dispersion of copper and zinc in glacial overburden at the Louvem Deposit, Val d'Or, Quebec, *Geochemical Exploration* (Boyle, tech. ed.), Special Volume No. 11, Canadian Institute of Mining and Metallurgy, pp. 157–158.

Gleeson, C. F., 1966, The distribution and behavior of metals in stream sediments and waters of the Keno Hill Area, Yukon Territory, Geochemical Survey of Canada, *Proc., Symp. Geochem. Prospecting,* Dept. of Energy, Mines and Research, Paper 66–54, pp. 134–144.

Goncalves, G. N. D., 1971, Amostragem de rolados: un navo coadjuvante na prospeccao geoquimica, *Bol. Soc. Brasil. Geol.,* Bol. Espec. No. 1, 25th Congresso Brasileiro.

Hamil, B. M. and Nackowski, M. P., 1971, Trace element distribution in accessory magnetite form quartz monzonite intrusives and its relation to sulphide mineralization in the Basin and Range Province of Utah and Nevada – A preliminary report, *Geochemical Exploration* (Boyle, tech. ed.), Special Volume No. 11, Canadian Institute of Mining and Metallurgy, pp. 331–333.

Hansuld, J. A., 1966a, Eh and pH in geochemical prospecting, *Proceedings, Symp. Geochem. Prospecting,* Geological Survey of Canada, Dept. of Energy, Mines, and Resources, Paper 66–54, pp. 172–187.

Hansuld, J. A., 1966b, Behavior of molybdenum in secondary dispersion media, *Mining Engr. J.,* **18**, 73–77.

Hansen, D. M., and Kerr, P. F., 1971, X-ray diffraction methods of evaluating potassium silicate alteration in porphyry mineralization, *Geochemical Exploration* (Boyle, tech. ed.), Special Volume No. 11, Canadian Institute of Mining and Metallurgy, pp. 334–340.

Horsnail, R. F. and Elliot, I. L., 1971, Some environmental influences on the secondary dispersion of molybdenum and copper in Western Canada, *Geochemical Exploration* (Boyle, tech. ed.), Special Volume No. 11, Canadian Institute of Mining and Metallurgy, pp. 166–175.

Huff, L. C., 1971, A comparison of alluvial exploration techniques for porphyry copper deposits, *Geochemical Exploration* (Boyle, tech. ed.), Special Volume No. 11, Canadian Institute of Mining and Metallurgy, pp. 190–194.

Jonasson, I. R. and Allan, R. J., 1972, Snow – a sampling medium in hydrogeochemical prospecting, Handbook, *4th Intern. Geochem. Exploration Symp., Abstracts,* 31.

Lapointe, G., 1968, The A horizon in soil surveys: geochemical kit for trace amounts of molybdenum, *Econ. Geol.,* **63**, 572.

Larsson, J. O. and Nichol, I., 1971, Analysis of glacial material as an aid in geological mapping, *Geochemical Exploration,* Special Volume No. 11, Canadian Institute of Mining and Metallurgy, pp. 197–203.

Macdonald, J. A., 1969, An orientation study of the uranium distribution in lake waters, Beaverlodge District, Saskatchewan, *Quart. Colo. School Mines,* **64**, 357–366.

Meyer, W. T., 1969, Uranium in lake water from the Kaipokok Region, Labrador, *Quart. Colo. School Mines,* **64,** 377–394.

Nigrini, A., 1971, Investigations into the transport and deposition of copper, lead and zinc in the surficial environment (Abstr.), *Geochemical Exploration* (Boyle, tech. ed.), Special Volume No. 11, Canadian Institute of Mining and Metallurgy, pp. 235.

Ovchinnikov, L. N., Sokolov, V. N., Fridman, A. I., and Yanitskii, I. N., 1972, Gaseous geochemical methods of structural mapping and search for ore deposits, Handbook, *4th Intern. Geochem. Exploration Symp., Abstracts,* London, pp. 42–43.

Roy, S. S. and Paul, D. K., 1967, Geochemical prospecting – use of A horizon in soil surveys, *Econ. Geol.,* **62,** 1102–1103.

Ruotsala, A. P., Nordeng, S. C. and Weege, R. J., 1969, Trace elements in accessory calcite – a potential exploration tool in the Michigan Copper District, *Quart. Colo. School Mines,* **64,** 451–456.

Stevens, D. N., Rouse, G. E. and De Voto, R. H., 1971, Radon-222 in soil gas: Three uranium exploration case histories in the Western United States, *Geochemical Exploration* (Boyle, tech. ed.), Special Volume No. 11, Canadian Institute of Mining and Metallurgy, pp. 258–264.

Trost, P. B. and Bisque, R. E., 1971, Differentiation of vaporous and ionic mercury in coils, *Geochemical Exploration* (Boyle, tech. ed.), Special Volume No. 11, Canadian Institute of Mining and Metallurgy, pp. 276–278.

Van Tassel, R. E., 1969, Exploration by overburden drilling at Keno Hill Mines Limited, *Quart. Colo. School Mines,* **64,** 457–478.

Weiss, O., 1971, Airborne geochemical prospecting, *Geochemical Exploration* (Boyle, tech. ed.), Special Volume No. 11, Canadian Institute of Mining and Metallurgy, pp. 502–514.

Whitney, P. R., 1973, Variation of heavy metal content with particle size in stream sediments (Abstracts), Geological Society of America, Northeastern Section Meeting, Allentown, p. 238.

6

BIOGEOCHEMICAL
PROSPECTING

For some time biogeochemical prospecting was thought of in terms of knowledge and chemical analyses of plant species or other forms of vegetation. This is generally correct, but before considering this "botanogeochemistry," it is necessary to become acquainted with the branches of biogeoinvestigation that are being developed and review some recent data on the use of other biological features in the search for geoeconomic riches hidden beneath a botanopedological cover of relatively little thickness.

Boyle and Garrett (1970) published a review on geochemical prospecting and reported that in Finland, dogs have been trained (conditioned) to smell out ore-bearing rocks in dispersion trains originating from sulfide deposits and thus assist in localizing the mineral deposits. It is obvious that the olefate organ of dogs can be conditioned to a very fine sensitivity for sulfur dioxide and other gases that may emanate from oxidizing sulfide minerals. This "biocanino-geochemistry" method was successful in Finland, and Canadian mining companies have recently financed a program on experimentation on the utility of dogs for geochemical exploration in Canada.

Another method, partly botanogeochemical and partly remote sensing, but based on spectral reflectance measurements, is being researched for its capability in indirectly establishing the mineralization potential of an area (Canney, 1971). Canney studied the spectral reflectance curves (from 350 to 1100 nm) of *Picea rubens* Sarg (red spruce) and of *Abies balsamea* (L) Mill (balsam fir) from an area of a large show (but of small tenor) of copper–molybdenum on the side of a mountain where the soil and the cited trees concentrated anomalous amounts of Cu and Mo. The spectral reflectance curves of five specimens of each species from this mineralized zone were compared with those from five samples of each species from a nearby nonmineralized zone with other factors such as soil type, soil humidity, and orientation of trees maintained as constant as possible. The spectral reflectance curves of the two groups of red spruce were almost equal in the visible region of the spectrum, but in the near infrared region, the sample group from the mineralized area showed a markedly reduced reflectance. However, the spectral curves of balsam fir showed higher reflectance in all the wavelengths for the samples from the mineralized zone as compared to those from the nonmineralized zone. The data suggest the possibility that multispectral aerial photography techniques could be used to distinguish tree groupings of *Abies balsamea* (L) Mill (balsam fir) with anomalous metal contents from other tree groupings in a zone of interest. However, much more research is necessary to establish the technique because the preliminary study by Canney was not statistically valid due to his very limited sampling.

Insects may also provide geochemical exploration assistance. Two Rhodesian geologists (West and Sutton) have started to use the six-legged, wood-devouring termite to prospect for gold beneath the thick Kalahari Desert sands and grits in Southwestern Rhodesia. Water is essential to termites and they burrow deep into the desert subsurface to get to the water table. Termites grow fungus gardens inside their conical-shaped, earthen-hill homes and assure the garden growth with an intricate system of humidity control by digging as deep as 40 meters (in the Sahara Desert) down to the water table to allow moisture to evaporate into the nest; according to Sutton, termites may dig down 30 to 60 meters and have been known to burrow as deep as 160 meters to find water. Other termites carry damp particles of rock matter from their mines into their structures and pile up cone-shaped hills that may reach heights of 2 meters or more. The detrital mineral particles comprising the hills yield information about the mineralogy and chemical composition of the rocks below. The geologists take 250-gram samples from hundreds of termite hills, sieve them, and make a heavy-mineral separation for microscope examination (for an immediate estimate of gold content) and emission spectrograph analysis for Cu, Zn, Ni, and other elements. In Rhodesia exploration blocks about 2 km on a side are being mapped geochemically and it is expected that geochemical anomalies will manifest themselves either as clusters or linear forms that might be indicative of vein deposits of a similar shape in the subsurface. If the technique gives positive results, it can be used in similar desert or semidesert environments where termites abound (for example, in the Botswana extension of the Kalahari Desert) or perhaps into which termites could be introduced if allowed by the existing ecosystem and thus provide a relatively inexpensive method for prospecting. One problem exists: it is unknown whether the termites burrow straight down or at an angle; in the latter case, the anomaly would be offset from the mineral source below, but limits to the maximum amount of offset can be readily estimated. As with other indirect geochemical prospecting techniques, the exploration system serves to localize zones of maximum interest and only test drilling in these zones can determine the precise position of a vein or vein system below the surface and provide samples to correctly assess the tenor of any ore encountered and its known, probable, and possible extensions.

The above exploration methods do not require geochemical analysis of the biosample, but in a well-conceived project, Warren et al. (1971) analyzed the livers of 96 trout (rainbow and cut-throat) from 47 geographically separated sites in British Columbia, Canada. The livers from 30 locations contained 60 ppm or less of copper (in wet weight of the livers) and those from 44 locations contained 50 ppm or less of zinc. All the livers with more than 50 ppm of Zn came from fish that inhabited waters draining from areas of economic Zn mineralization. Of the 17 locations where the livers contained more than 60 ppm of Cu, 4 have significant Cu mineralization, and 7, because of their known general geology, should be objects of detailed geochemical prospecting; the remaining 6 sites have potential for exploration, but their geology is little known and they have not yet been investigated. As well as initiating another way of carrying out geochemical prospecting, these authors showed us that field work can be rather agreeable unless one is allergic to fish.

The studies above notwithstanding, the biogeochemistry most employed at present is based on botanical specimens. Prospecting with plants can be carried out by one or a combination of the following three methods:

1. A knowledge of the effects of toxic quantities of some elements on the manner of growth (morphology) or the physiological development of the botanical specimens—geobotanical prospecting.

2. A knowledge of indicator plants, that is, plants that grow only where there is a specific element or assemblage of elements present in relatively large amounts. This is related directly to the rock type (or types) from which a soil is developing. Such plants may be universal or local indicators of element presence.

3. A knowledge of accumulator plants. Certain species will accumulate one or more metals if they are present in the growth environment, but such plants do not need these metals for normal development, nor do they suffer physiological or morphological changes if they accumulate what might be considered excessive quantities of the metals.

This prospecting approach is based on the knowledge that plants absorb necessary elements (nutrients) and unnecessary elements and compounds mainly by their primary and secondary root systems in a soil that contains or perhaps does not contain all the nutrients that a plant needs for its proper growth. The important chemical forms in which macro and micro quantities of essential plant elements are absorbed are given in Table 6-1 and the processes of absorption and excretion of the nutrients are presented graphically in Figure 6-1. It should be noted that the quantity of an element in a plant may not be rep-

Table 6-1   Essential Elements to Plants and the Important Chemical Forms in which They are Absorbed (from Sutcliffe, 1962).

Macronutrients		Micronutrients	
Element	Form Absorbed	Element	Form Absorbed
Carbon	Carbon dioxide, bicarbonate	Boron	Borates
		Chlorine	Chlorides
Hydrogen	Water, several anions	Copper	Cations
Oxygen	Water, gaseous oxygen, several anions	Iron	Cations
		Manganese	Cations manganates
Nitrogen	Nitrates, ammonia	Molybdenum	Molybdates
Phosphorus	Phosphates	Zinc	Cations
Sulfur	Sulfates		
Potassium	Cations		
Calcium	Cations		
Magnesium	Cations		

Source.   Reprinted from *Mineral Salt Absorption in Plants,* with permission of Pergamon Press. Copyright © 1962 by Pergamon Press.

resentative of the quantity present in the subsurface because the concentration may derive from pollution. In a paper on the biogeochemistry of lead in Canada, Warren and Delavault (1960) clearly showed that exhaust emissions from gasoline motors are the cause of higher concentrations of lead in vegetation growing close to highways. "Normally," 10 to 100 ppm of Pb are found as background values in vegetation ash, but Warren and Delavault (1960, 1967) measured more than 1000 ppm of Pb in the ash of plants located consistently close to roads and outside zones of significant Pb mineralization. Other classes of metal pollution can come from fertilizers, fungicides, pesticides, rodenticides, coal or wood ash, alteration of the chemical balance of the elements in soils, and industrial pollution (Warren, Delavault and Cross, 1965). The geologist–geochemist in the field must be alert to these possibilities of contamination when planning an exploration program based on plants, soils, or natural waters.

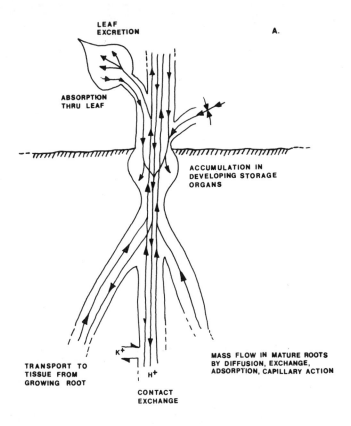

LEAF
EXCRETION

A.

ABSORPTION
THRU LEAF

ACCUMULATION IN
DEVELOPING STORAGE
ORGANS

TRANSPORT TO
TISSUE FROM
GROWING ROOT

K⁺

H⁺

CONTACT
EXCHANGE

MASS FLOW IN MATURE ROOTS
BY DIFFUSION, EXCHANGE,
ADSORPTION, CAPILLARY ACTION

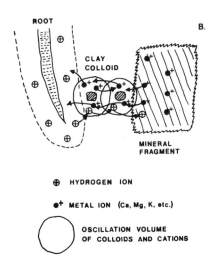

ROOT

B.

CLAY
COLLOID

MINERAL
FRAGMENT

⊕ HYDROGEN ION

●⁺ METAL ION (Ca, Mg, K, etc.)

◯ OSCILLATION VOLUME
OF COLLOIDS AND CATIONS

Figure 6-1. Graphic illustration of absorption and excretion processes in plants. (Top reproduced from Sutcliffe, 1962, *Mineral Salt Absorption in Plants,* with permission of Pergamon Press. Copyright © 1962 by Pergamon Press Ltd. Bottom from Keller and Frederickson, 1952.)

Table 6-2 Physiological and Morphological Changes (Mutability) in Plants that May be Caused by Toxic Quantities (Excesses) of Metals (compiled from Cannon, 1960a and Malyuga, 1964)

Element	Effect
Al	Short and stubby roots; leaf scorch, mottling
B	Dark foliage; marginal scorch of older leaves of high concentrations; shortened internodes; deformed; incomplete development; creeping forms; heavy (slow) pubescence; increased gall production
Cr	Yellow leaves with green veins
Co	White dead patches on leaves
Cu	Dead patches on lower leaves from tips; purple stems; chlorotic leaves with green veins; poorly developed (stunted) roots; creeping sterile forms in some species
Fe	Incompletely developed (stunted) tops; thin or sometimes stubby roots; cell division disturbed in algae, resulting cells greatly enlarged
Mn	Chlorotic leaves; reddish coloration and lesions of stem and petiole; curling and dead areas on leaf margins; distortion of laminae
Mo	Incomplete development (stunting); yellow–orange coloration
Ni	White dead patches on leaves; chlorosis; apetalous sterile forms; abnormal forms
Zn	Chlorotic leaves with green veins; white dwarfed forms; dead areas on leaf tips; roots stunted or poorly developed
U	Abnormal number of chromosomes in nuclei; unusually shaped fruits; sterile apetalous forms; stalked-leaf rosette
Mo, Cu	Unusual development of black bands on the petals of *Eschscholtzia sp.* (black cross)
Pb, Zn	Development of different forms with the "double flower" in the *Eschscholtzia sp.*
Bitumen	Gigantism and deformity; abnormal repeated flowering

With respect to the first method of geochemical exploration by means of vegetation, that is, the knowledge of the effects of toxic quantities of chemical elements on the morphology or the physiological development of botanical species, the reader is referred to Table 6-2 in which the general effects (on macro and micro scales) of excesses of selected metals (plus B and bitumen) on plants are described. The degree of toxicity that can develop is relative and depends on many factors, such as the concentration of ions present in the growth environment with the toxic ion. Scharrer (1955) showed that pure salts of rubidium or strontium are toxic to angiosperms, but that one third of a plant's potassium can be replaced by strontium without there being a toxic effect. Also, this toxicity can be modified by organic compounds that seques-ter certain metals and do not allow the metals to poison the development en-zymes in plants (Bowen, 1966). One of the major problems in this geobotani-cal mineral prospecting is that the geologist is not generally trained in botany to the degree where he can work with confidence and recognize important fea-tures such as chlorotic leaves, incomplete development, dwarf forms, or a fruit in odd forms, all of which may be due to toxicity. During the planning stage of a geochemical exploration program, a botanist can be of great assist-ance to the project director by indicating the flora that will be encountered in the field and the effect of the climate and soil on the growth of the flora. In the Soviet Union a geobotanical institute was founded at the beginning of this century, and since 1946 most Soviet prospecting teams in the field include a botanist who may be a specialist in the flora of the study zone. Because of this, the Soviet programs in biogeochemistry are very advanced; nonetheless, groups in Canada (Fortescue and Hornbrook, 1967, 1969), the United States, Scandanavia, and New Zealand are investing more time and effort in this aspect of biogeochemical prospecting.

Indicator plants are those that have attained a balance between all their re-quirements with respect to nutrients, solar orientation, soil, pH, Eh, humidity, and so on. The nutrients may include one or more sensitive metals, and when an indicator plant is found, this means that the metal needs of the species are available in the growth environment, although this alone is not a sure indica-tion that there is an economic mineral deposit below the soil. However, the absence of an indicator plant does not provide positive proof that the metal that is being sought is not present in the environment in minimum concentra-tions for satisfying the needs of the plant and maintaining its balance with na-ture; the possibility exists that another factor does not allow the plant to grow in an area. Those plants whose presence indicates the availability of a minimum quantity of a specific metal to the roots are called universal or local indicator plants. A listing of selected universal indicator plants, local indicator plants, and indicator plants whose extensions have not yet been evaluated, has been compiled from the works of Cannon (1960a,b, 1971), Malyuga (1964), Bowen (1966), and NASA (1968) and is presented in Table 6-3. These are the plants that may indicate the existence of mineralization in specific geographical zones and environments. Their presence may be the result of the availability of a metal or the effect of an environmental factor affecting the ore, as might be the pH, Eh, some organic or inorganic complex, or another such factor. What-

Table 603   Universal (U) and Local (L) Indicator Plants (compiled from Cannon, 1960a,b; Malyuga, 1964; Bowen, 1966; and NASA, 1968).

	Botanical Name	Common Name	Element	Area Used
(U)	Gypsophila patrini	Karum, Kachim	Cu	U.S.S.R., Rudnyi Altai
(U)	Acrocephalus roberti		Cu	Katanga
(U)	Ocimum homblei	Basil	Cu	Rhodesia
(U)	Merceya latifolia	Copper moss	Cu	Sweden, Montana
(U)	Viscaria alpina	German catchfly	Cu	Norway
(U)	Viola calamineria (lutea)	Zinc violet	Zn	Belgium, Germany
(U)	Crotalarina cobalticola	Rattlebox	Co	Katanga
(U)	Silene cobalticola	Catchfly	Co	Katanga
(U)	Astragalus bisulcatus	Poison vetch, milk vetch	Se	Western U.S.A.
(U)	Astragalus racemosus	Poison vetch	Se	Western U.S.A.
(U)	Astragalus pectinatus	Poison vetch	Se	Western U.S.A.
(U)	Astragalus convertiflorus	Blue poison vetch	Se	Western U.S.A.
(U)	Astragalus preussi	Preuss poison vetch	Se, U	Western U.S.A.
(U)	Astragalus pattersoni	Poison vetch, milk vetch	Se, U	Western U.S.A.
(U)	Oanopsis spp.	Goldenweed	Se	Western U.S.A.
(U)	Aster venustus	Woody aster	Se	Western U.S.A.
(U)	Stanleya spp.	Princeplume	Se	Western U.S.A.
(U)	Xylorrhiza		Se	Western U.S.A.
(U)	Townsendia incana		Se	Western U.S.A.
(L)	Anabasis salsa	Saltwort	Bitumen	Caspian Sea
(L)	Salsola spp.		Bitumen	Caspian Sea
(L)	Allium sp.	Onion	Bitumen	California
(L)	Eurotia ceratoides	Winterfat	B	U.S.S.R.
(L)	Limonium suffruticosum	Statice	B	U.S.S.R.
(L)	Polycarpea spirostylus	Pink	Cu	Australia

(L)	Elsholtzia haichowensis	Elsholtzia	Cu	China
(L)	Elchscholtzia mexicana	California poppy	Cu	Arizona
(L)	Armeria maritima	Thrift	Cu	Scotland
(L)	Erigonum inflatum	Desert trumpet	Gypsum	Western U.S.A.
(L)	Mintzelia spp.	Blazing star	Gypsum	Western U.S.A.
(L)	Betula sp.	Birch	Fe	Germany
(L)	Clusia rosea	Copey clusia	Fe	Venezuela
(L)	Erianthus giganteus	Beardgrass	Pb	Tennessee
(L)	Convolvulus althaeoides	Bindweed	P	Spain
(L)	Astragalus declinatus	Milk vetch	Mo, Cu	Kadzharan, Armenia
(L)	Astragalus sp.	Garbancillo	Se, U	The Andes
(L)	Eriogonum ovalifolium	Eriogonum	Ag	Montana
(L)	Philadelphus sp.	Mock orange	Zn	Washington
(ND)[a]	Alyssum bertolonii		Ni	Italy
(ND)	Alyssum murale		Ni	Georgia (U.S.S.R.)
(ND)	Asplenum adulterium		Ni	Norway
(ND)	Ruta gaveolens		Zn	U.S.A.
(ND)	Astragalus thompsonae	Milk vetch	Se, U	Western U.S.A.
(ND)	Amorpha canescens	False indigo	Pb	U.S.A.
(ND)	Equisetum arvense	Horsetail	Au	Czechoslovakia
(ND)	Lonicera confusa	Honeysuckle	Ag, Au	Australia
(ND)	Drypton sp.		Cu	
(ND)	Gymnocolea acutiloba		Cu	
(ND)	Mielichoferia		Cu	
(ND)	Arenaria setacea		Hg	
(ND)	Thalictrum sp.		Li	
(ND)	Cirsium sp.		Li	
(ND)	Sempervivum soboliferum		Sn	
(ND)	Pluchea quitoc		Sn	

[a]Nondetermined extension.

Table 6-4 *Accumulator Plants for Use in Biogeochemical Prospecting*

Scientific Name	Common Name	Metal Detected																										
		Cu	Zn	Al	Se	Hg	Ge	Fe	Au	B	U	Mn	Ni	Mo	Ag	Sn	W	V	Sr	Co	Mg	Ba	Pt	As	Cr	RE	F	I
Alnus sinuata	Green alder	X	X					X				X																
Salix sp.	Willow	X	X						X			X																
Quercus tubinella	Scrub oak	X																										
Quercus emoryi	Emory oak	X																										
Cercocarpus lediofolius	Mountain mahogany		X																									
Tsuga mertensiana	Mountain hemlock		X																									
Abies amabilis	Balsam fir		X																									
Thuya plicata	Giant cedar		X																									
Piaus contorta	Lodgepole pine	X	X					X	X	X	X	X			X													
Populus trichocarpa	Black cottonwood	X	X					X			X	X																
Larix occidentalis	Larch	X	X							X	X	X																
Populus tremuloides	Aspen	X	X					X	X	X	X	X			X													
Shepherdia canadensis	Sootpolallie		X						X	X	X	X																
Pseudotsuga taxifolia	Douglas fir	X	X					X	X		X	X			X													
Thuja plicata	Western red cedar	X	X					X			X	X																
Tsuga heterophylla	Western hemlock	X	X					X				X																
Juniperus scopularum	Rocky mountain juniper	X	X																									
Juniperus sp.	Juniper							X	X			X																
Pinus monophylla	Pinyon pine										X																	
Pinus ponderosa	Ponderosa pine	X	X					X				X			X													
Pinus albicaulis	White bark pine	X	X																									
Picea engelmanni	Engelmann spruce	X	X					X				X																
Picea sitchensis	Sitka spruce	X	X																									
Picea glacua	White spruce	X	X					X				X																
Juniperus communis	Dwarf juniper	X	X																									
Abies lasiocarpa	Rocky mountain fir	X	X					X					X	X														
Betula glandulosa	Scrub birch	X	X					X					X	X														
Betula occidentalis	Mountain birch	X	X																									

130

Scientific name	Common name				
Betula popyrifera	Silver birch	X	X	X	
Philadelphus lewisii	Syringa orange	X	X	X	
Acer glabrum	Mountain maple	X	X	X	
Populus grandidintata	Apple	X	X		
Betula fontinalis	Poplar	X			
Abies grandis	Western red birch				
Salix scouleriana	Grand fir				
Alnus rubra	Willow				
Acer macrophyllum	Red alder				
Acer circinatum	Broad leaf maple				
	Vine maple				
	Citrus				
Prunus demissa	Choke cherry	X	X		
Cornus nuttallii	Flowering dogwood				
Rhus typhina	Sumac	X			
Sassafras variifolium		X			
Acer saccharinum	Soft maple	X			
Carya cordiformis	Pignut	X			
Carya ovalis obvalis		X			
Carya ovata	Hickory	X			
Cornus florida	Flowering dogwood	X			
Juglans nigra	Black walnut	X			
Juniperus virginiana	Red cedar	X			
Maclura pomifera	Osage orange	X			
Platanus occidentalis	Sycamore	X			
Prunus serotina	Black cherry	X			
Quercus alba	White oak	X			
Quercus borealis maximus	Red Oak	X			
Quercus macrocarpa	Bur oak	X			
Quercus marilandica	Blackjack oak	X			
Quercus muhlenbergii		X			
Quercus palustris	Pin oak	X			
Quercus stellata	Post oak	X			

Table 6-4 Accumulator Plants for Use in Biogeochemical Prospecting

Scientific Name	Common Name	Metal Detected																										
		Cu	Zn	Al	Se	Hg	Ge	Fe	Au	B	U	Mn	Ni	Mo	Ag	Sn	W	V	Sr	Co	Mg	Ba	Pt	As	Cr	RE	F	I
Quercus velutina	Black oak	X																										
Rhus copallina	Sumac	X																										
Robinia pseudoacacia	Black locust	X																										
Sassafras albidum	Nees	X																										
Ulmus alata	Winged elm	X																										
Ulmus americana	American elm	X																										
Ulmus fulva	Slippery elm	X																										
Pine	Pine																	X										
Hickory	Hickory							X				X								X				X				
Acar saccharum	Sugar maple																						X					
Lycopodium fabelliforme	Princess pine			X																								
	Pecan																								X			
	Brazil nut																				X							
	Chestnut																								X			
Quercus wislizeni	Oak													X														
Pseudotsuga douglasii																					X	X						
Afzelia africana				X																								
Rubiacaea sp		X													X								X					
Bahia nitida		X													X								X					
Parinari curatellifolia		X													X								X					
Albizzia zygia		X													X								X					
Lophira alata		X													X								X					
Vitex cuneata		X													X								X					
Parkia oliveri		X													X								X					
Millettia sp.		X													X								X					
Newbouldia lavis		X													X								X					
Trichilia prievriana		X													X								X					
Anacardiaceae sp.		X													X								X					

Sp. indet.	
Quercus douglasii	Blue oak
Artemisia tridentata	Sagebrush
Gaultheria shallon	Salal
Vaccinium parvifolium	Red huckleberry
Arctostaphylos uva-ursa	Kinnikinnick
Pachystima myrsinites	False box
Corylus californica	Hazel
Vaccinium sp.	Blueberry
Echinopanax horridus	Devil's club
Pteridium aquilinum	Bracken
Holodiscus discolor	Ocean spray
Prosopis juliflora	Mesquite
Astragalus sp.	Milkvetch
Oryzopsis hymenoides	Indian ricegrass
Atriplex confertifolia	Shadscale
Vaccinium vitis idaea	
Ledum palustre	
Equisetum sp.	Horsetails
Vaccinium ovalifolium	
Symphoricarpos racemosus	Wax berry
Artemisia trifida	Sagebrush
Viola calaminaria	Violet
Thlaspi calaminaria	
Amorphia canascens	
Viola tricolor	Violet
Thlaspi alpestre	
Asparagus officinalis	
Linaria vulgaris	Butter and eggs
	Red clover
Zea mays	Corn
Euphorbia masculata	
Solidago sp.	Golden rod

Table 6-4 Accumulator Plants for Use in Biogeochemical Prospecting

Scientific Name	Common Name	Metal Detected																											
		Cu	Zn	Al	Se	Hg	Ge	Fe	Au	B	U	Mn	Ni	Mo	Ag	Sn	W	V	Sr	Co	Mg	Ba	Pb	As	Cr	RE	F	I	Li
Equisetum arvense	Horsetail	X																											
Lobelia inflata		X																											
Lobelia syphillitica		X																											
Plantago lanciolata	Plantain	X																											
Pycnanthemum flexuosa		X																											
Ambrosia artemisiifolia	Ragweed	X																											
Amelanchier canadensia	Shad bush	X																											
Smilacina racemosa	Solomon seal	X																											
Tomanthera auriculata		X																											
Viola sagittata	Violet	X																											
Daucus carota	Wild carrot	X																											
Achillea millefolium	Yarrow	X																											
Physalis sp.	Ground cherry	X																											
Celtis occidentalis	Hackberry	X																											
Persicaria hydropiper	Smartweed	X																											
Ambrosia coronopifolia	Ragweed	X																											
Ambrosia elatior	Ragweed	X																											
Ambrosia trifida	Giant ragweed	X																											
Andropogon furcatus	Big bluestem	X																											
Andropogon scoparius	Little bluestem	X																											
Apocynum sibiricum	Dogbane	X																											
Gnaphalium obtusifolium	Cudweed	X																											
Belianthus mollis	Sunflower	X																											
Liatrus aspera		X																											
Desmodium sessilifolium		X																											
Phytolacca decandra	Pokeweed	X																											
Salvia pitcheri		X																											
Sorghastrum nutans	Indian grass	X																											

Symphoricarpos orbiculatus Buckbush
Tridens flavis
Vernonia interior Ironweed
Amelanchier alnifolia Saskatoon
Clover
Beets
Beans

Calluna rulgaris

Sweet leaf
Club moss
Meadow beauty
Shortia
Galax
White clover
Sweet clover

Polycarpia spirostyles
Thymus serpyllum
Viola hirto Violet
Tobacco
Woody aster
Devil's paint brush
Chick pea

Astragalus racemosus Two grooved poisonvetch
Oonopsis condensata
Astragalus bisulcatus
Xylorhiza parri
Astragalus pectinatus
Astragalus grayii
Aplopappus fromantii
Stanleya pinnata
Gutierrezia sarothrae

Arctostaphylos viscida Dandelion

Table 6-4 Accumulator Plants for Use in Biogeochemical Prospecting

Scientific Name	Common Name	Metal Detected																											
		Cu	Zn	Al	Se	Hg	Ge	Fe	Au	B	U	Mn	Ni	Mo	Ag	Sn	W	V	Sr	Co	Mg	Ba	Pb	As	Cr	RE	F	I	Li
Adenostoma fasciculatum																							X						
Vivia americana														X															
Medicagohispida														X															
Lotus corniculatus														X															
Trifolium repens														X															
Trifolium fragiferum														X															
Trifolium subterraneum														X															
Melilotus alba														X															
Melilotus indica														X															
Melilotus sativa														X															
Triticum vulgare														X															
Avena sativa														X															
Hordem vulgare														X															
Chloris gayana														X															
Lolium perenne														X															
Bautiloua rothrockii	Rothrock gramma grass	X																											
Eschscholtzia mexicana	California poppy	X																											
Holostium umbellatum						X																							
Viscaria alpina		X																											
Melandrium dioecum		X																											
Polycarpaie spirostylus	Carnation	X																											
Rosa woodsi	Wild rose	X																											
Equisetum variegatum	Horsetail	X							X																				
Dasiphara fruticosa	Bush cinquefoil	X																											
Cneoridium dumosum														X															
Carya sp.																													
Nyssa sylvatica																				X									
Clethra barbineris																				X									

Species		
Asplenium viride		
Thea sinesis		
Alyssum bertolonii		
Armeria maritima var halleri		
Minuartia verna	Creosote	
	Ironwood	
	Foothill palo-verde	
Acacia raddiana		
Acacia ehrenbergiana		
	White birch	
	Trembling aspen	
Dichapetalum cymosum		
Feijoa sellowiana		
Pseudotsuga menziesii		
Nothofagus fusca	Red beech	
Nothofagus menziesii	Silver beech	

Source. By NASA, 1968, taken from Carlisle and Cleveland, 1958; additions made here from Schutte, 1964; Bowen, 1966; Warren, Delavault and Barakso, 1968; Whitehead and Brooks, 1969a, 1969b; Huff, 1969; Timperly, Brooks and Peterson, 1970; Chaffee and Hessin, 1971; El Shazly et. al. 1971; and Hornbrook, 1971.

ever be the type of indicator plant, its presence provides evidence that a specific class of rock or mineral may be found in the subsurface. Whenever possible, the geologist–geochemist in the field should work with more than one indicator in order to have more confidence in preliminary or pilot studies. As is the case with geobotany, the geochemical exploration project will have more chance of success if a botanist who knows the study area is included as much in the planning of the program as in the field work.

Finally, the botanogeochemical method most used is based on the chemical analysis of the dry organs of plants or of organ ash. The analysis is not made on all vegetation found in a study zone, but generally on a plant or some plants that are known accumulators of one or more metals. The presence of metals in concentrations that would be toxic for other species apparently does not affect the so-called accumulators in any way. Carlisle and Cleveland (1958) published an excellent compilation of accumulator plants and the metal or metals that they concentrate; Table 6-4 gives this listing with some additions from other publications. Compilations such as Table 6-4 (and others given in this book) are guides and should not be considered as infallible, especially if they are derived from somewhat older publications. This was well illustrated by Cannon, Shacklette, and Bastron (1968) in their study of metal absorption by Equisetum (Horsetail). Equisetum had been cited as an accumulator of gold, but in an evaluation of 28 elements from a variety of mineralized and nonmineralized areas of continental United States and Alaska, the authors found this not to be true. According to Cannon, Shacklette, and Bastron, it is possible that earlier reports of high concentrations of gold in Equisetum may be erroneous because with the analytical procedures used, other metals may have been measured as well and may have been reported as amounts of gold. Equisetum was found to be an accumulator of zinc.

It is always convenient to compare the geochemical data obtained from a study area with what may be considered as an approximation of the average global value in order to determine if the data represents a metallogenic area of significant interest. The geochemical data from plants is most often reported in one of two ways: (a) the content of the chemical element in plant ash or (b) the content of the chemical element in the dry plant material. Table 6-5 gives representative sets of such values together with comparative data (from the compilations by Vinogradov, 1959 and Bowen, 1966) on the general composition of the lithosphere and soils. Of more use with respect to the flora are the values in Table 6-6 that provide us with excellent numbers on the average content of some metals in the ash of vegetation that grows in nonmineralized terrain. This table is a contribution of Cannon (1960a) and permits comparisons of data with more detailed information than that given in Table 6-5, thus giving more credence to the evaluations and interpretations of the biogeochemical analyses.

Table 6-5 Chemical Element Contents of the Lithosphere, Soils, and Plants (ppm) (compiled from Vinogradov, 1959; Taylor, 1964; Bear, 1964; and, Bowen, 1966)

Element	Lithosphere	Soil (with range)	Plants (in ash)	Plants (in dry weight)
Li	20	30 (7-200)	11	0.1
Be	2.8	6 (0.1-40)	2	<0.1
B	10	10 (2-100)	400	50
F	625	200 (30-300)	10	0.5-40
Na	26,150	6,300 (750-7,500)	20,000	1,200
Mg	21,950	5,000 (600-6,000)	70,000	3,200
Al	81,650	71,000 (10,000-300,000)	14,000	500 (0.5-4,000)
Si	279,600	330,000 (250,000-350,000)	150,000	200-5,000
P	1,050	650	70,000	2,300
S	260	700 (30-900)	50,000	3,400
Cl	130	100	n00	2,000
K	23,950	14,000 (400-30,000)	30,000	14,000
Ca	38,650	13,700 (7,000-500,000)	30,000	18,000
Ti	5,050	5,000 (1,000-10,000)	1,000	1
V	135	100 (20-500)	61	1.6
Cr	100	100 (5-3,000)	250	0.23
Mn	950	850 (100-4,000)	7,500	630
Fe	53,000	38,000 (7,000-550,000)	10,000	140
Co	25	8 (1-40)	15	0.5
Ni	75	40 (10-1,000)	50	3
Cu	55	20 (2-100)	200	14
Zn	70	50 (10-300)	900	100
As	1.8	6 (0.1-40)	30	0.2
Se	0.05	0.2 (0.01-2)		0.2
Br	2.5	5 (1-10)	150	15

Table 6-5 (Continued)

Element	Lithosphere	Soil (with range)	Plants (in ash)	Plants (in dry weight)
Rb	90	100 (20-600)	100	20
Sr	375	300 (50-1,000)	300	26
Zr	165	300 (60-2,000)		0.64
Mo	1.5	2 (0.2-5)	20	0.9
Ag	0.07	0.1 (0.01-5)	1	0.06
Cd	0.2	0.06 (0.01-0.7)	0.01	0.6
Sn	2	10 (2-200)	5	0.3
I	0.5	5	50	0.42
Cs	3	50	n	0.2
Ba	425	500 (100-3,000)	n00	14
Au	0.004		1	< 0.0005-0.002
Hg	0.08	0.03 (0.01-0.3)	0.001	0.015
Pb	12.7	10 (2-200)	11	2.7
Ra	0.000001	8×10^{-7} ($3 - 20 \times 10^{-7}$)	0.0000002	
Th	7.2	5 (0.1-12)		
U	1.8	1 (0.9-9)	0.5	
Bi	0.2			
C	200	20,000		454,000
Dy	4.1			0.02
Eu	1.2			0.021
Ga	15	30 (0.4-300)		0.06
H	1,400			55,000
Hf	3	6		0.01?
Ir	0.001			< 0.02
La	30	30 (1-5,000)		0.085
N	20	1,000 (200-2,500)		30,000

Nb	20		0.3
O	465,000	490,000	410,000
Ru	0.01		0.005
Sb	0.2	(2-10?)	0.06
Sc	22	7 (10-25)	0.008
Sm	6.6		0.0055
Tb	1.1		< 0.0015
Te	0.01		2-25
Tm	0.35		0.0015
W	1.5		0.07
Y	33	50 (25-250)	< 0.06
Yb	3.2	0.1	< 0.0015
Tl	0.47		
Ge	1.5		
Rh	0.005		
Pd	0.01		
In	0.1		
Ce	63		
Pr	7.3		
Nd	28		
Gd	6.3		
Ho	1.3		
Er	2.8		
Lu	0.6		
Ta	2		
Re	0.001		
Os	0.005		
Pt	0.01		

Table 6-6 Average Metal Content of the Ash of Vegetation that Grows in Nonmineralized Terrain (ppm) (from Cannon, 1960a)

Element	Average of More Than 100 Plant Species	Grasses (above ground)	Herbs (above ground)	Deciduous		Conifers (needles)
				Shrubs (leaves)	Trees (leaves)	
Al	8610					
Fe	6740					
Mn	4815					
Zn	1400	850	666	2303	1585	1127
B	700					
Cu	183	119	118	249	223	133
Pb	70	33	44	54	85	75
Ni	65	54	33	87	91	57
V	22	25	23	16	25	21
Mo	13	34	19	7	15	5
Cr	9	19	10	5	14	8
Co	9	10	11	5	10	7
Sn	5					
Be	2					
Ag	1					
U	0.6					
Au	0.007					

When we considered the significance of the chemical analysis in geochemistry, emphasis was placed on the constancy of operation that must be followed in order to obtain reliable comparative data. In biogeochemistry, certain norms should be followed in the collection of samples for a study. For example, it is known that the different organs of a plant specimen may accumulate different quantities of chemical elements; thus, only comparisons between the same organs should be made. The general sequence of metal accumulation by different floral organs in increasing concentrations has been proposed as follows:

For trees: bark < roots < wood < cones < twigs < leaves–needles
For smaller plants: stems < roots < leaves

These sequences provide only a general concept and in each biogeochemical program it must be decided (from a pilot study) which will be the best organ to collect. For example, in an orientation study of some silver vein deposits in the area of Cobalt, Ontario, Canada, Hornbrook (1971) investigated the contents of Ag, Co, Ni, Mn, Cu, Pb, and Zn in the bark, the leaves, the spurs, and the twigs of trembling aspen and white birch trees and the metal contents of the B and $(A_0 + A_1)$ soil horizons in order to establish the best sample(s) for a general geochemical exploration of the zone. In this pilot study area, the principal veins are in the Cobalt sediments and have dimensions of 15 to 205 meters in length, 15 to 60 meters of vertical extension, and some 16 cm in width; overlying the sediments is a mantle of Nipissing diabase about 50 to 60 meters thick and the sediments themselves overlie Keewatin greenstone. An area of 150 by 600 meters perpendicular to the strike of the parallel ore veins was selected for the program and the various types of orientation samples were taken on a grid of about 7.5 meters along five traverses 30 meters apart. The evaluation of geochemical results showed that the soil samples of the combined horizons $(A_0 + A_1)$ sized to -10 to $+80$ mesh indicated the locations of both the general anomalies and the individual vein systems; of equal value were the data from the geochemical analyses of the spurs of white birch. The data from the twigs of trembling aspen were less definitive in locating anomalies. Analysis for seven elements showed only the concentrations of Ag, Co, and Ni were indicative pathfinders to the mineralization. Very important also is the fact that on the basis of the pilot study, a grid of 30 by 60 meters was established as the most efficient sampling pattern for finding similar mineralization.

In their research on the content of arsenic in the tree *Pseudotsuga menziesii* (Douglas fir) as a guide to some economically viable deposits of gold, silver, and the base metals, Warren, Delavault, and Barakso (1968) noted that the twigs contained more arsenic than their attached needles. The authors found that *Pseudotsuga menziesii* has a great affinity for arsenic and that it contains

Table 6-7. Arsenic Content of Different Organs of Different Ages from Individual Trees (Pseudotsuga menziesii— Douglas fir) Growing in Various Localities (from Warren, Delavault, and Barakso, 1968)

Location	Sample		Organ	As (ppm)
H. B. Mine	64	3	First year twigs	510
			First year needles	120
			Second year twigs	70
			Second year needles	25
H. B. Mine	64	5	First year twigs	780
			First year needles	450
			Second year twigs	280
			Second year needles	60
King Vein of the Bralorne Mine	64	2	First year twigs	2110
			First year needles	1060
			Second year twigs	1390
			Second year needles	180

from 10 to 100 times as much arsenic as any of the neighboring trees that grow in the mineralized areas being studied. Because arsenic is recognized as being a pathfinder element for many base-metal and precious-metal deposits, the application of this technique based on a tree's affinity for arsenic would appear to have an excellent future in the search for relatively small but rich silver veins. In addition to the influence of the plant organ used for the analysis, the relative age of the floristic part used in the biogeochemical prospecting may affect the results of a program since older and younger specimens of the same organs may give significantly different data. Warren, Delavault, and Barakso used the twigs and needles of the most recent growing season because these growths contained more arsenic than their older counterparts. This influence and the effect of the organ used for the analysis are quite clearly illustrated by the comparative data presented in Table 6-7. One drawback in the use of arsenic in the Douglas fir in geochemical exploration is that the mineralization must contain a determined quantity of arsenic for the chemical analyses of the tree to be useful; in the case cited here, the authors consider that the ore must have a tenor of approximately 1% of arsenic. In the western part of North America, where Warren, Delavault, and Barakso took samples of the Douglas fir, the number of areas with ores that contain this limiting tenor of arsenic is large. In addition to being capable of absorbing and retaining more arsenic (as reflected in its ash) than the soil in which it is growing, *Pseudotsuga menziesii* is important in biogeochemical prospecting because of the species' extremely wide geographic extension in Western North America. On the basis of analyses already made, Warren, Delavault, and Barakso established background and anomaly values that can serve as guides for other geochemists (when the ash contents of the needles and twigs varied between 1.5 and 3.5% of the dry vegetation weight):

1. In zones where there is no known mineralization with arsenic, the twigs of the most recent growing season should contain < 100 ppm of arsenic.

2. In areas of glacial "drift" that have been contaminated by modest quantities of arsenic, the twigs should contain between 100 and 500 ppm of arsenic.

3. Within approximately 60 meters of manifestations of mineralization with arsenic, the twigs should contain > 1000 ppm of arsenic.

4. The value of the so-called anomaly will be between 500 and 1000 ppm of arsenic.

Thus, the geochemist who is to initiate an exploration program in Western North America for Cu, Pb, or Zn in fairly widespread deposits or for veins of Ag or Au of little extension should consider using this particular biogeochemical method as a fundamental part of his pilot project. For this author there is no doubt that there are other examples of flora in other areas of the world that act in a manner similar to *Pseudotsuga menziesii* and because of this could be used with success in geochemical prospecting for Cu, Pb, Zn, Ag, and Au. It is essential, however, to carry out proper orientation studies in collaboration with botanists in order to find the accumulator guides.

In 1966 Warren, Delavault and Barakso published a paper on geochemical prospecting with mercury. They showed that the different floral species and plant organs vary a great deal in their capacity to accumulate mercury. Some exploration norms based on a limited sampling were presented and the authors wrote that if the ash of the last 2 years growth contains between 1 and 10 ppm of Hg, this can be considered normal for mineralized areas. If the ash contains between 10 and 20 ppm of Hg, this would indicate a possible anomaly, and when the ash has more than 20 ppm of Hg, this would be considered a probable anomaly. Of all the trees studied in the project, the Lodgepole pine gave better results in the prospecting for mercury or for deposits for which the mercury could be a pathfinder element.

Warren (1966) established similar norms for lead in biogeochemical prospecting. He determined that the most recent vegetation growth should not be used for Pb analysis (although such a procedure is common in analysis for other metals) because the content of Pb varies significantly in the new growths. This variation results from lead input by man-generated pollution. According to Warren, wherever it is feasible, the growth representative of the season previous to the present field season should be analyzed. For Pb the emphasis should be put on twigs of deciduous trees and on the needles of conifers. Commonly, the twigs will contain between 25 and 50 ppm of Pb, with variation of up to 50% being possible. When the trees are growing in zones where the soil has formed from rocks containing abnormal amounts of Pb (but not present as ore), the quantity of Pb in the biogeochemical specimens would be between 50 and 250 ppm. In areas of economic or potentially economic mineralization of Pb beneath the soil cover, great variations in the quantity of Pb in the vegetation are obtained; the ash of twigs from trees at four Pb-mining areas in Western Canada gave analytical results of between 200 and 2000 ppm of Pb. However, Warren feels that for Pb, it would be more practical to investigate well-developed residual soils than plants, but notes also that where a residual soil does not exist or where poorly developed residual soils are found, the biogeochemical technique can be very useful and sometimes more practical than the soil geochemistry technique.

The general concept that leaves of a tree contain higher metalliferous concentrations than the twigs or small branches to which they are attached is often correct, but the geochemist must assure himself of this by means of his own analyses of specimens from a new exploration area. In a biogeochemical pilot study, Timperly, Brooks, and Peterson (1970) made a statistical evaluation of the contents of Cu, Ni, and Zn in four plant species native to the Basic Riwaka Complex, New Zealand, and their related soils. The researchers used leaves and twigs of *Nothofagus fusca, Nothofagus menziesii, Quintinia acutifolia,* and *Weinmannia racemosa,* collected on a grid of 30 by 60 meters in an area of approximately 400 by 600 meters; also, additional samples were taken at intermediate locations within the grid and from an area adjacent to the gridded zone. At all the sampling sites, the scientists tried to follow a very constant methodology and selected trees of equal size for each species (for sampling), taking about 120 grams each of leaves, twigs, and soil; for the biogeochemical specimens, a sample representative of each tree was taken. When the authors reviewed their results, it was clear that all the species gave correlation coefficients significant at the probability level of 0.1% for Ni; for Cu, only the species *Nothofagus fusca, Nothofagus menziesii,* and *Quintinia acutifolia* had equal significance with respect to the relation between the concentration in plants and the concentration in soils. It was concluded that it was very feasible to use native trees in this zone of New Zealand in biogeochemical prospecting for Cu and Ni (in the Basic Riwaka Complex), and the best sample would be the ash of leaves of *Nothofagus fusca* and *Nothofagus menziesii;* statistically, these two trees can be treated as one for biogeochemical sampling in this area. Biogeochemical prospecting would be especially useful in the Basic Riwaka Complex because there is a dense vegetation and difficult topography and because it requires much more work to take a soil sample than a sample of leaves. In addition, the biogeochemical sample has the advantage of weighing less than a good soil, stream sediment, or rock sample and its preparation in the laboratory may require less true time than the rock or detrital samples. Whitehead and Brooks (1969a) made a similar study in New Zealand on the relative usefulness of soils and some plant species in the search for uranium. They showed that the fluorometric analysis of either plants or soils gave good indications of the presence of uranium in the area of Buller Gorge. However, as in the Basic Riwaka Complex area, the biogeochemical prospecting approach would be the most efficient method for geochemical exploration because of the rugged topography and dense cover of vegetation in Buller Gorge.

Botanogeochemistry also may have a practical application for prospecting in regions where there are thick layers of alluvium concealing deposits in underlying rocks. In 1971 Chaffee and Hessin made an evaluation of various classes of geochemical samples that could be used in the search for hidden copper-molybdenum porphyry mineralization in the pediments of Southern Arizona (Basin and Range Province). The pediments are covered by as much as 90 meters of alluvium. Chaffee and Hessin collected 2400 samples of soil (they used the −60 mesh size fraction of material 6 to 15 cm, deep for analysis), mother rock (where it was available), and the leaves and twigs of three species of plants that have deep roots and are common to the region (creosote, ironwood, and foothill palo-verde). The samples were analyzed for Zn, Cu, Mo, and Mn. A plot of the values for Zn, Cu, and Mo in the ash of the organs of the three species used clearly focused on the anomalous areas up to 240 meters beyond the mineralized outcrops in places where the postmineralization alluvium overlies the hidden deposit. There were no corresponding anomalies indicated by the geochemistry of the soils. Since it was known that the roots of the three plant species used for the prospecting could penetrate up to 30 meters into the alluvium, Chaffee and Hessin thought that the anomalous metal contents in the leaves and twigs directly indicated the location of the mineralized deposit by tapping either the mineralized rock itself or groundwater enriched by mineralization. They proposed, therefore, that the sampling of plants with deeply penetrating roots could be very useful in the exploration for hidden ores in areas similar to the Basin and Range Province.

The White Sand Formation covers much of the potential mineralized terrain of Western Guyana so that usual geochemical methods do not serve for exploration. However, Gibbs (1970) recognized that the Baramalli tree is very common in the soil developed from the weathering of the White Sand formation and that it had a large and long taproot. In three regions (Tamour, Barabara, and Groete Creek), samples were taken of the wood from the heart of the Baramalli tree and analyzed via emission spectrography. Significant detectable quantities of the elements Co, Ni, Pb, Mo, Cr, and Cu were found. In the Groete Creek mineralized zone, residual soils with unknown metal contents and others with known trace metal concentrations, both from the White Sand formation, were sampled. Anomalies for Cu and Mo (from the tree samples) were established and were from 5 to 10 times as great as background values; these corresponded to known soil anomalies and hence proved that the mineralized zone continued beneath the covering of the White Sand formation. Gibbs did not come to a firm conclusion on the usefulness of the Baramalli tree as an accumulator of metals so that it could be used as a botanogeochemical tool because factors that influenced the observed metal accumulations, such as the thickness of the sand, the presence of sedimentary clays beneath the sand, and the preferential accumulation of certain metals by the tree and its effect on the dispersion of the metals, were then being evaluated. More recently, Banerjee, Gedeon, and Inasi (1972) have considered the problems of mineral exploration in the White Sand formation covered Precambrian shield (more than 40,000 km^2 or 21% of Guyana's land area) and reemphasized the use of the Baramalli tree (*Catostemma commune, C. fragrams, C. altsoni*) as first reported by Gibbs (1970). The authors believe that for a successful exploration program in the White-Sand-covered area of Guyana, especially where there is a relatively thin cover of sand, biogeochemical sampling could play an important role, together with the ground geophysical survey and the extrapolation of geological information from adjacent areas, especially that related to groundwater movement.

Shallow-rooted plants may also be used in future geochemical studies of arid or semiarid regions, not only to indicate the possible presence of mineralized particles within the zone of their root systems, but also to assist in defining the extension of these mineralized fragments and the rocks from which they derive. In a preliminary study, the writer analyzed samples of the cactus *Cardon sp.* from within and outside of a zone of copper porphyry mineralization in a desert environment near the location called Farallon Negro, Catamarca Province, Northwest Argentina. The limited sampling of cactus (interior pulp) and their related thin soils gave geochemical data that showed that the samples from within the known mineralized zone contained about twice as much Zn and Pb as the samples taken from outside the known limits of the mineralization. Copper contents of the samples were about the same from both sample sets.

Where do we stand then with respect to biogeochemistry or to be more precise, botanogeochemistry, because it is this branch of biogeochemistry that has immediate practical application in the search for ore (and as a guide in groundwater surveys and geologic mapping; Cannon, 1971). First, geobotany, universal and local indicator plants, and botanogeochemistry are interesting research topics not to be left for the future, but rather to be used at present. Second, we are dealing not with a branch of the geosciences that has a potential value for application, but with one with a field of "now" value, currently in use, because the bases for its use have already been established, although research into unusual accumulations of elements by plants (and animals) under varying conditions must be continued (Peterson, 1971). Third, the majority of advances of this part of exploration geochemistry will come from the collaboration of a team of geologists, chemists, and geochemists working with botanists or geobotanists. Finally, there is an ample field of work open for innovative basic research such as the type of investigation made by Warren and his colleagues on trout livers or that carried out by Finnish researchers on the canine olefates. One may consider the use of animals employed for food and having limited and known geographic ranges for prospecting, as perhaps the sheep of Patagonia or the cattle from the mountainous and difficultly accessible areas of Colombia, by the analysis of some anatomical part before or after the animal enters the slaughterhouse. For a fuller treatment of biogeochemistry based on botanical specimens, the reader is referred to the informative works of Malyuga (1964) and Brooks (1972).

REFERENCES

Banerjee, A. K., Gedeon, A. Z., and Inasi, J. C., 1972, Problems of mineral exploration in the white sand covered Precambrian Shield of Guyana, South America, Abstracts, 4th International Geochemical Exploration Symposium, The Institute of Mining and Metallurgy, London, pp. 14–15.

Bear, F. E. (editor), 1964, Chemistry of the Soil, 2nd edition, Reinhold Publishing Corp., New York, 515 pp.

Bowen, H. J. M., 1966, *Trace Elements in Biochemistry,* Academic Press, London, 241 pp.

Boyle, R. W., and Garrett, R. G., 1970, Geochemical prospecting – a review of its status and future, *Earth-Sci. Rev.,* **6,** 51–75.

Brooks, R. R., 1972, Geobotany and Biogeochemistry in Mineral Exploration: Harper & Row, Publishers, New York, 290 pp.

Canney, F. C., 1971, Remote detection of geochemical anomalies – a preliminary feasibility study, in *Geochemical Exploration,* (Boyle, tech. ed.), special volume No. 11, Canadian Institute of Mining and Metallurgy, p. 479.

Cannon, H. L., 1960a, Botanical prospecting for ore deposits, *Science,* **132,** pp. 591–598.

Cannon, H. L., 1960b, The development of botanical methods of prospecting for uranium on the Colorado Plateau, *U.S. Geol. Survey Bull.* **1085-A,** 1–50.

Cannon, H. L., 1971, The use of plant indicators in ground water surveys, geologic mapping, and mineral prospecting, *Taxon,* 20(2/3), pp. 227–256.

Cannon, H. L., Shacklette, H. T., and Bastron, H., 1968, Metal absorption by *Equisetum* (Horsetail), *U.S. Geol. Surv. Bull.,* **1278-A,** 21 pp.

Carlisle, D., and Cleveland, G. B., 1958, Plants as a guide to mineralization, State of California, Division of Mines, Special Report 50, 31 pp.

Chaffee, M. A., and Hessin, T. D., 1971, An evaluation of some geochemical sampling in the search for concealed "porphyry" copper–molybdenum deposits on pediments in

Southern Arizona, U.S.A., in *Geochemical Exploration* (Boyle, tech. ed.), Special Volume No. 11, Canadian Institute of Mining and Metallurgy, pp. 401–409.

Cole, M. M., 1971, The importance of environment in biogeographical-geobotanical and biogeochemical investigations, in *Geochemical Exploration* (Boyle, tech. ed.) Special Volume No. 11. Canadian Institute of Mining and Metallurgy, pp. 414–425.

El Shazly, E. M., Barakat, N., Eissa, E. A., Emara, H. H., Ali, I. S., Shaltout, S., and Sharaf, F. S., 1971, The use of acacia trees in geogeochemical prospecting, in *Geochemical Exploration,* (Boyle, tech. ed.), Special Volume No. 11, Canadian Institute of Mining and Metallurgy, pp. 426–434.

Fortescue, J. A. C., and Hornbrook, E. H. W., 1967, A brief survey of progress made in biogeochemical prospecting research at the Geological Survey of Canada 1962-1965, in *Proceedings, Symposium on Geochemical Prospecting* (Cameron, Ed.), Geological Survey of Canada, Department of Energy, Mines and Resources, Paper 66-54, pp. 111-133.

Fortescue, J. A. C. and Hornbrook, E. H. W., 1969, Progress report on biogeochemical research at the Geological Survey of Canada 1963-1966, Geological Survey of Canada, Department of Energy, Mines and Resources, Paper 67-23, 143 pp.

Gibbs, A. K., 1970, Baramalli sampling in base metals exploration: a promising new technique, in Annual Report 1969, Geological Survey Department, Ministry of Agriculture and Natural Resources, Georgetown, Guyana.

Hornbrook, E. H. W., 1971, Effectiveness of geochemical and biochemical exploration methods in the Cobalt Area, Ontario, in *Geochemical Exploration,* (Boyle, tech. ed.), Special Volume No. 11, Canadian Institute of Mining and Metallurgy, pp. 435–443.

Huff, L. C., 1969, Accumulation of ore metals by mesquite plants, *Quart. Colo. School Mines,* **64,** 510.

Keller, W. D. and Frederickson, A. F., 1952, Role of plants and colloidal acids in the mechanism of weathering, *Amer. Jour. Sci.,* **250,** 594–608.

Malyuga, D. P., 1964, *Biogeochemical Methods of Prospecting,* Consultants Bureau, New York, 205 pp.

NASA, 1968, Application of Biogeochemistry to Mineral Prospecting, National Aeronautics and Space Administration Special Publication 5056, 134 pp.

Peterson, P. J., 1971, Unusual accumulations of elements by plants and animals, *Sci. Progr., Oxford,* **59** 505–526.

Scharrer, K., 1955, *Biochemie der Spurenelemente,* Parey, Berlin, 404 pp.

Schutte, K. H., 1964, *The Biology of the Trace Elements,* Crosby Lockwood, London, 228 pp.

Sutcliffe, J. F., 1962, Mineral Salt Absorption Plants, Pergamon, New York, 192 pp.

Taylor, S. R., 1964, Abundance of chemical elements in the continental crust: a new table, *Geochim. Cosmochim. Acta,* **28,** 1273-1285.

Timperly, M. H., Brooks, R. R., and Peterson, P. J., 1970, Prospecting for copper and nickel in New Zealand by statistical analysis of biogeochemical data, *Econ. Geol.,* **65,** 505–510.

Vinogradov, A. P., 1959, Geochemistry of Rare and Dispersed Chemical Elements in Soils [translation from Russian], Consultants Bureau, New York, 209 pp.

Warren, H. V., 1966, Some aspects of lead pollution in perspective, *J. Coll. Gen. Practitioners,* **11,** 135–142.

Warren, H. V., and Delavault, R. E., 1960, Observations on the biogeochemistry of lead in Canada, *Trans. Royal Soc. Can., Ser. of III*, **LIV**, 11–20.

Warren, H. V., and Delavault, R. E., 1967, A geologist looks at pollution – mineral variety, *Western Miner*, **40**, 22–32.

Warren, H. V., Delavault, R. E., and Barakso, J., 1966, Some observations on the geochemistry of mercury as applied to prospecting, *Econ. Geol.*, **61**, 1010–1028.

Warren, H. V., Delavault, R. E., and Barakso, J., 1968, The arsenic content of Douglas Fir as a guide to some gold, silver, and base metal deposits, *Can. Mining Met. Bull.*, **61**, (675), 860–866

Warren, H. V., Delavault, R. E., and Cross, C. H., 1965, Mineral contamination in soil and vegetation and its possible relation to public health, Department of Geology, University of British Columbia, Background Paper A3-3, 11 pp.

Warren, H. V., Delavault, R. E., and Cross, C. H., 1966, Geochemistry in mineral exploration, *Western Miner*, **39**, 22–26, and 101–102.

Warren, H. V., Delavault, R. E., Peterson, G. R., and Fletcher, K., 1971, The copper and zinc content of trout livers as an aid in the search for favorable areas to prospect, in *Geochemical Exploration* (Boyle, tech. ed.), Special Volume No. 11, Canadian Institute of Mining and Metallurgy, pp. 444–450.

Whitehead, N. E., and Brooks, R. R., 1969a, A comparative evaluation of scintillometric, geochemical and biogeochemical methods of prospecting for uranium, *Econ. Geol.*, **64**, 50–56.

Whitehead, N. E., and Brooks, R. R., 1969b, Radioecological observations on plants of the Lower Buller Gorge region of New Zealand and their significance for biogeochemical prospecting, *J. Appl. Ecol.*, **6**, 301–310.

7

CONCEPTUAL PATTERNS OF ANOMALOUS GEOCHEMICAL DISTRIBUTIONS AND THEIR RELATIONS TO ACTUAL PATTERNS

Geochemical anomaly patterns present significant variations dependent on the physical, chemical, and biological factors (or parameters) that have been directly or indirectly considered in other chapters of this book. However, because of the variety of patterns that have been found and described in relation to the geology, it is evident that the most important factor influencing the observed anomaly (especially in detailed prospecting based on residual soil samples and/or the unweathered or weathered mother rock) is the shape (and type) of the mineral deposit being sought and its strike and dip. Among the other principal factors influencing anomaly patterns are the regional and local geological structure, the nature of the host rock, sampling density (and, of course, the sample being used), the topography, the hydrological regime, and the mobility of the element or elements being studied in the physical, chemical, and biological conditions dominant in the geological–geochemical environment. All these parameters are generally interrelated as regards the total anomaly pattern found for a given deposit in a given area.

Geochemical dispersions and their related areal patterns may be classified as syngenetic and epigenetic (Ginzburg, 1960; Hawkes and Webb, 1962). Syngenetic patterns result from the dispersion of chemical elements that took place at the same time that the host rock was being formed; epigenetic patterns are those that result from chemical elements being introduced into the host rock after its formation or those that have been separated from the host rock after its original formation. Thus, the environment in which the primary geochemical dispersion develops can be characterized by conditions of relatively high temperature and pressure, that is, conditions existent at depth; using this con-

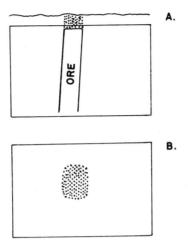

Figure 7-1. Idealized situation in which the geochemical anomaly develops in the residual soil directly above the ore.

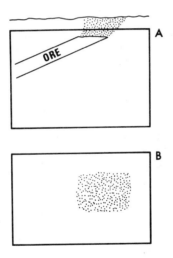

Figure 7-2. Idealized situation in which the geochemical anomaly develops in the residual soil with greater dispersion with respect to the real position of the vein because of the gentle dip of the ore-bearing vein.

cept, the formation of syngenetic dispersions is limited to rocks of igneous, metamorphic, or hydrothermal origin that are now close to the Earth's surface or exposed at it. The existence of secondary or epigenetic dispersions in a rock is related to hydrothermal, metamorphic, pneumatolitic, infiltrational, or assimilatory processes that affect the preexisting rock.

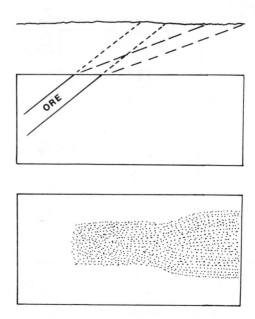

Figure 7-3. Idealized situation in which the geochemical anomaly develops with relatively greater dispersion with respect to the vein (compared to Figure 7-2, for example) because of the dip of the vein and a subsequent compaction of the residual soil.

A semantic problem now exists as to whether a dispersion should be considered a primary or secondary effect when weathering and the subsequent physical, chemical, and biological movement of detrital elements are treated. This problem has been considered by James (1967) who felt that the concept proposed by Hawkes (1957) relating dispersion with the two principal phases of the geochemical cycle (given below) should be set aside:

1. The metamorphic phase or the phase of igneous differentiation that takes place at great depths and results in a dispersion designated as primary.
2. The weathering, erosion, transport, and sedimentation phases that act at or close to the surface of the planet and result in a dispersion designated as secondary.

At present, this type of classification for geochemical patterns or dispersions is a problem because the patterns in a host rock for ore bodies formed by syngenetic sedimentary processes are not patterns of primary dispersions in the strict interpretative sense, but really represent secondary fossil patterns. Therefore, because many important mineral deposits are of sedimentary origin, a reevaluation of the concepts of primary and secondary dispersions is necessary.

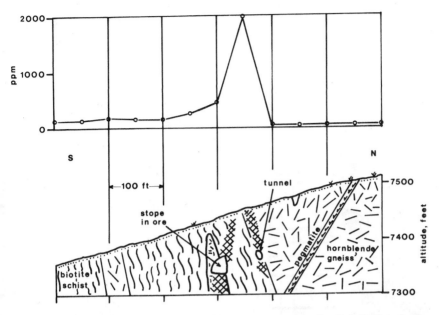

Figure 7-4. Actual situation in which the geochemical anomaly (for Cu) develops in the residual soil directly above the ore (chalcopyrite and pyrrhotite in a mineralized fracture zone), Malachite Mine, Jefferson County, Colorado. In the same area, local zones have abnormal Cu concentrations up to 900 feet downslope from the mineralized fracture. (From Huff, 1960.)

James (1967) suggested that the term "pattern of primary dispersion" should be used to describe the distribution of the chemical elements in unweathered rock situated around an ore body without attributing importance to the method of formation (or genesis) of that body. The term "pattern of secondary dispersion" can then be used for those patterns developed from the redistribution of the elements in the primary dispersion (or in the primary pattern) in the zone of surface oxidation and weathering. If the researcher feels the need to use a concept based on a genetic association, the adjectives "hypogenetic" and "supergenetic" can be used to describe the deep and surface environments, or the presently preferred designations "endogenetic" and exogenetic" can be used to refer to the two dominantly important geochemical environments.

As regards the geochemical patterns related to mineral deposits, one may initially conceive of a simple geological situation in which the ore is present as a vein or fissure filling in the host rock and in which weathering in an area of gently undulating topography and good drainage has resulted in the development of a residual soil, the sample type used in geochemical prospecting programs to detail the position of a deposit. In this case the geochemical anomaly

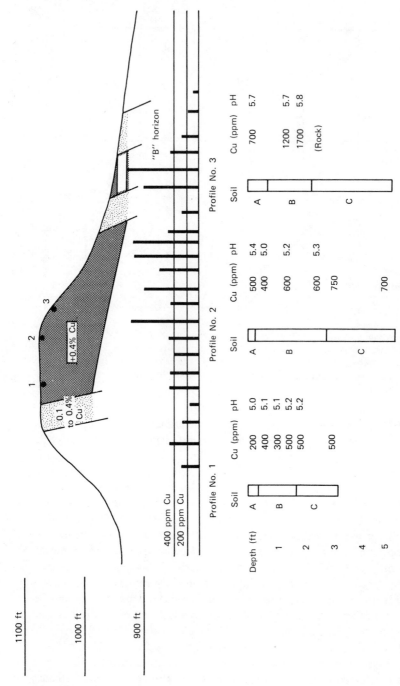

Figure 7-5. Actual situation in a laterite zone in which the topography and perhaps soil pH influence the position of an anomaly. The strongest anomaly in an area of plateaus and valleys occurs on the slopes of the mineralized zones instead of on the plateaus, but maintains a good spatial relation with the underlying copper porphyry ore. It is possible that the pH of soils on the plateaus allows more Cu to be mobilized, whereas there is less mobilization downslope where the soils have notably higher pH values. (Reprinted from Clarke, Hatchet Creek prospect, West Clay County, Alabama, U.S.A., 1971, from *Geochemical Exploration*, (Boyle, tech. ed.) Special Volume No. 11, The Canadian Institute of Mining and Metallurgy.)

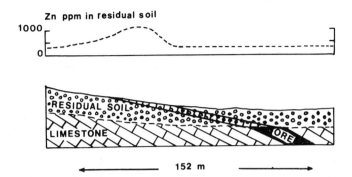

Figure 7-6. Actual situation in which there is a displacement and amplification of the anomaly (for Zn) in a residual soil caused by both the compaction of the soil and the relatively gentle dip of the sphalerite-bearing deposit replacing limestone, Tennessee. (From Hawkes and Lakin, 1949, *Econ. Geol.,* **44,** p. 289.)

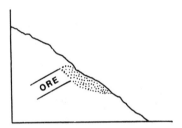

Figure 7-7. Idealized situation in which there is a dispersion of the anomaly downslope and a lateral dispersion that may be due to the gradual creep of soil downwards.

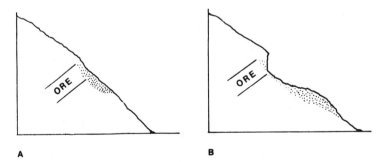

Figure 7-8. Idealized situation in which there was a slide or movement of a unit of material downwards, thus displacing the anomaly from the true position of the ore.

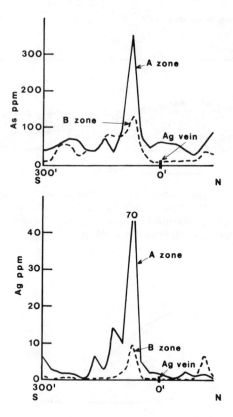

Figure 7-9. Actual situation in which there was a displacement downwards of an anomaly, but in a gently dipping zone (across the O'Brien silver vein no. 6, Cobalt, Ontario, Canada). (Reproduced from Boyle and Dass, 1967, *Econ. Geol.*, **62**, 275.)

associated with the vein or fissure filling would appear almost directly above the position (in space) of the mineralization (see Figure 7-1) if other factors have not enhanced the dispersion of the element(s) analyzed. In a field situation where the mineralized vein has a gentle dip instead of that shown in Figure 7-1, the geochemical anomaly may show greater dispersion (see Figure 7-2) depending on the degree of weathering (with depth) of the underlying rock. Another ideal situation can be considered in which the weathering has been deep and where there has been a compacting of the residual soil above the vein, producing in this way a geochemical anomaly pattern wider than would be expected without the compaction (see Figure 7-3). These hypothetical geological–geochemical situations have real analogies as shown in Figures 7-4, 7-5, and 7-6.

Now, if prospecting is being carried out in an environment where the topog-

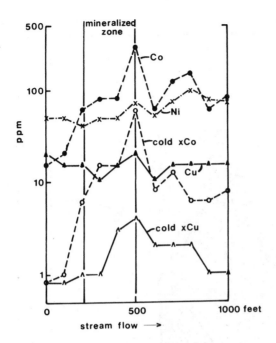

Figure 7-10. Distribution of Co, Ni, and Cu in stream sediments crossing a mineralized norite-gabbro, Somerset County, Maine. (Reproduced from Canney and Wing, 1966, *Econ. Geol.*, **61**, 198.)

raphy is relatively steep, soil creep by gravity towards the lower areas, influenced by soil type, slope, presence or absence of vegetation, and availability of water to serve as a lubricant, can disperse the anomaly being studied and give a longer pattern and a somewhat wider pattern laterally (see Figure 7-7). Obviously, if there are slides, flows, or falls of the study material, there is most probably not going to be a very direct relation between the anomalies detected in the fallen material and the real position of the mineralized vein (see Figure 7-8). Of course, the dispersion and displacement of an anomaly can also be found in gently sloping terrain (see Figure 7-9).

The anomalies and dispersions of the materials just described for the most part represent a dispersion or a development of the geochemical anomaly of an almost completely mechanical and relatively *in situ* character. However, geochemical anomalies can show up at a significantly great distance from a mineral deposit. The sample that has the greatest areal dispersion (linearly) farthest from a mineralized zone is that of the stream sediment. Running water moves material by traction, saltation, and in suspension and in solution, fairly large distances from the area of weathering and erosion of an ore. One

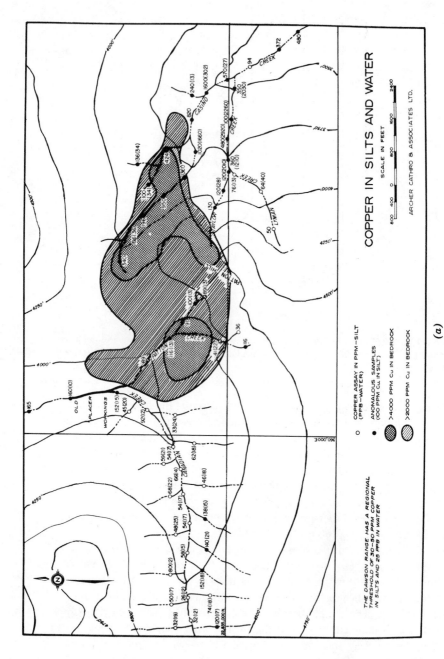

COPPER IN SILTS AND WATER

○ COPPER ASSAY IN PPM—SILT
    (PPB—WATER)

● ANOMALOUS SAMPLES
    (IOO PPM Cu IN SILT)

▓▓ >4000 PPM Cu IN BEDROCK

▒▒ >2000 PPM Cu IN BEDROCK

THE DAWSON RANGE HAS A REGIONAL
THRESHOLD OF 30-50 PPM COPPER
IN SILTS AND 25 PPB IN WATER

SCALE IN FEET

ARCHER CATHRO & ASSOCIATES LTD.

(a)

Figure 7-11 (a).

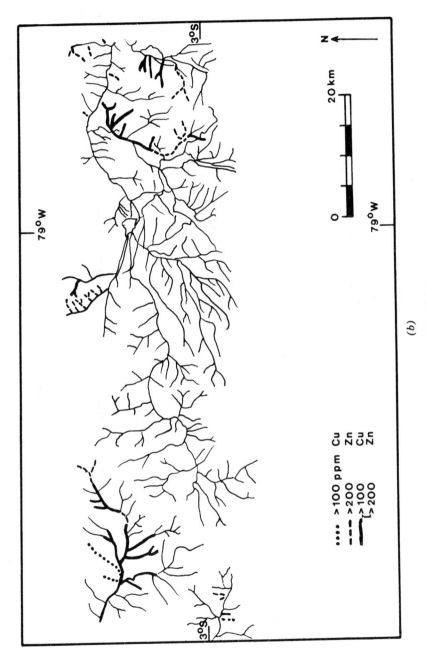

Figure 7-11 (b).

161

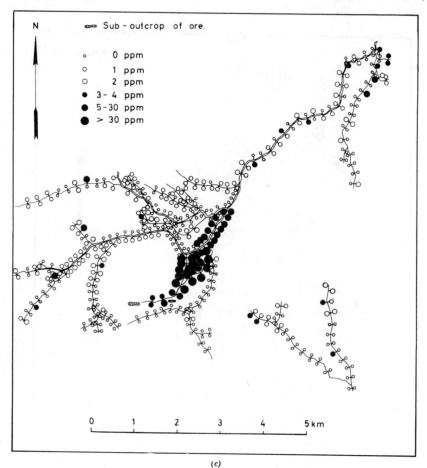

(c)

Figure 7-11 (c).

may conceive of a situation in which a slight anomaly downstream will increase in intensity (and contrast) upstream (towards the mineralization) until reaching an area where it starts decreasing significantly or almost completely disappears; a real situation ot this type is illustrated in Figure 7-10. Prospecting may then be intensified where the largest anomaly exists or upstream from this anomaly, depending on what the program director projects as being most logical on the basis of his evaluation of the geochemical-geological-hydrogeo-logical-geomorphological system. Sometimes this detailed or intensified study may initially require a more dense sampling of the stream sediments or may necessitate using the soils or vegetation of the zone of interest in an attempt to localize and delimit the true mineralized area. The pattern of the anomaly

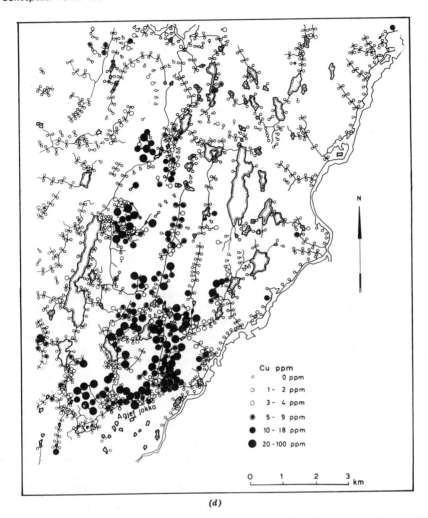

(d)

Figure 7-11. Actual examples of anomaly patterns that are found in stream sediments: (a) Cu in silts and waters draining the Cu–Mo porphyry deposit, Casino, Yukon, Canada. (Reprinted from Archer and Main, 1971, *Geochemical Exploration* (Boyle, tech. ed.), Special Volume No. 11, The Canadian Institute of Mining and Metallurgy); (b) Cu and Zn in stream sediments from a zone of Central Ecuador (reproduced from Kroon and de Grys, 1970, *Econ. Geol.,* 1970, **65,** 558); (c) cold extractable Cu in stream sediments from Hjerkinn, central Norway; and (d) cold extractable Cu in stream sediments from Agjetjokka, northern Norway (c and d reproduced from Bolviken, 1967, *Geochemical Prospecting in Fennoscandia* (Kvalheim, ed.), by permission of John Wiley & Sons, Inc.) Copyright © 1967, by John Wiley & Sons, Inc.

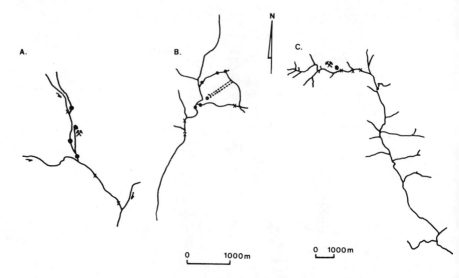

Figure 7-12. Actual examples of anomaly patterns found in waters: (A) fluoride distribution in stream waters from the drainage area of the Gottesehre fluorite mine, St. Blasien, Black Forest, Germany; (B) fluoride distribution in stream waters from the drainage area of the Clara fluorite–barite mine, Wolfach, Black Forest, Germany; (C) fluoride distribution in stream waters of the drainage area of the Menzenschwand uranium mine, near St. Blasien, Black Forest, Germany. (Reprinted Friedrich and Pluger, 1971, *Geochemical Exploration* (Boyle, tech. ed.), Special Volume No. 11, The Canadian Institute of Mining and Metallurgy.

that is found in stream sediments assumes the shape of the river or of the rivers and streams that comprise the drainage network, for example, linear, sinuous, or dendritic (see Figure 7-11). Similar patterns are found in the waters of mineralized zones (see Figure 7-12). It should also be noted that the position of a geochemical anomaly in stream sediments can vary with the type of analysis made on the stream sediment samples. For example, Figure 7-13 shows that the distribution of copper extracted from stream sediments with hot HNO_3 has the highest values about 80 meters downstream from known mineralization for about 200 meters; however, the distribution of copper extracted from these same stream sediments with cold HCl is fairly constant in the entire zone represented by the map in Figure 7-14. When the maximum values of an anomaly found in stream sediments is not very close to the mineralogically interesting zone, several factors may be interrelated in explaining the displacement; among these are the gradient of the river, the shape of the channel, the amount of water carried continuously or seasonally, the shape of the river (and drainage basin) in space, the type of weathering in the area, the size and weight of the material that gives the anomalous values, and those

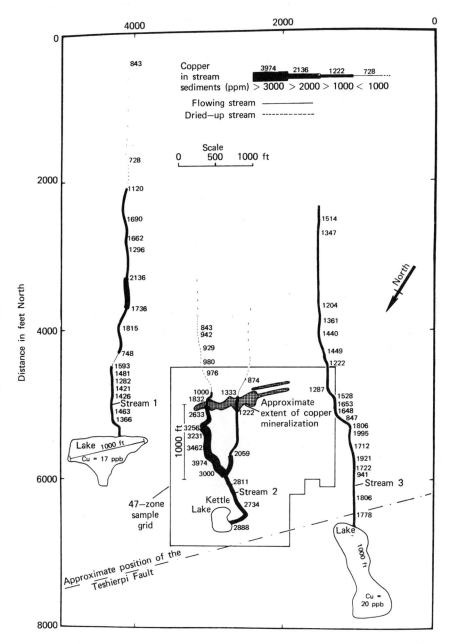

Figure 7-13. Concentration of Cu extracted with hot HNO_3 from stream sediments in the 47-zone area, Coppermine Basalt Belt, N.W.T., Canada. (Reprinted from Allan and Hornbrook, 1971, *Geochemical Exploration,* (Boyle, tech. ed.), Special Volume No. 11, The Canadian Institute of Mining and Metallurgy.)

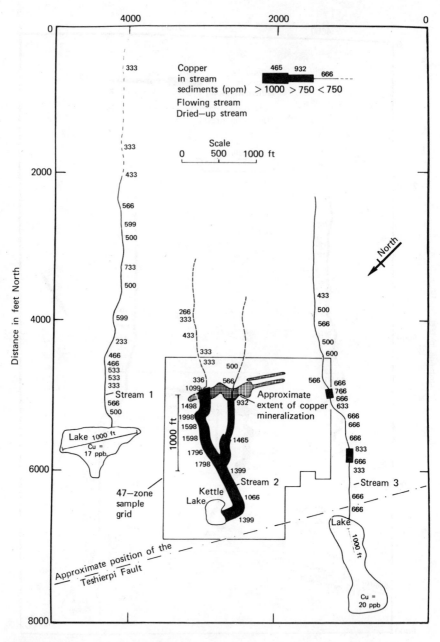

Figure 7-14. Concentrations of Cu, cold extracted with HCl from stream sediments in the 47-zone area, Coppermine Basalt Belt, N.W.T., Canada. (Reprinted from Allan and Hornbrook, 1971, *Geochemical Exploration* (Boyle, tech. ed.), Special Volume No. 11, The Canadian Institute of Mining and Metallurgy.)

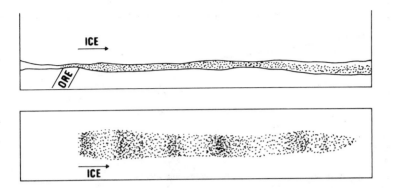

Figure 7-15. Idealized situation in which the anomaly related to mineralization has been dispersed linearly by glacial activity in the direction of glacier movement.

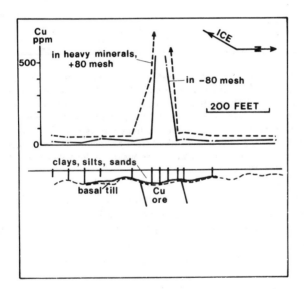

Figure 7-16. Actual situation in which the dispersion often associated with glaciated areas is not evident, thus indicating that the anomaly in the basal till of the Louvem Cu deposit, Quebec, Canada, developed after the last glacial activity. (Reprinted from Garrett, 1971, *Geochemical Exploration* (Boyle, tech. ed.), Special Volume No. 11, The Canadian Institute of Mining and Metallurgy.)

chemical and biological factors that influence the total system being studied. For example, Jones (1970, personal communication) reported that in the Department of Tolima, Colombia, the strongest anomalies (associated with known

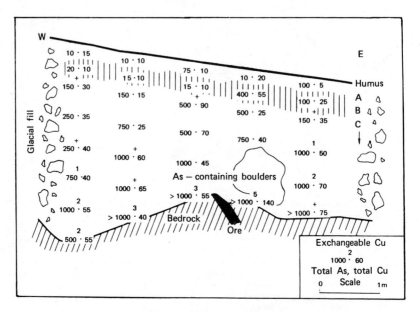

Figure 7-17. Boulder trains of glacial material at Kotanen, central Finland, influenced geochemists to dig trenches during exploration. One of these trenches uncovered a tillite rich in As at a depth of 100 cm; below, in the granitic gneiss host rock, a vein of arseno-pyrite 5 to 10 cm thick was found. This, and the presence of other materials containing As and Cu at the surface, 700 meters from the trench, served as a basis for systematic geochemical studies of the tills. (Reproduced from Kauranne, 1967b, *Geochemical Prospecting in Fennoscandia*, (Kvalheim, ed.) by permission of John Wiley & Sons, Inc.) Copyright © 1967, by John Wiley & Sons, Inc.

mineralization upstream) appeared in a perimeter from 1½ to 3 km from known mines, and this displacement could be understood in terms of the first four factors cited above. Several examples of stream sediment anomalies and their relations to mineralization are presented in Figures 7-10, 7-11, 7-13, and 7-14. It should be emphasized here that the length of stream-sediment-type anomalies is a function of the statistical norms that are established for a region being investigated; the higher the elemental concentration fixed to delimit the anomaly from the nonanomaly, the shorter may be the length of the anomalous zone.

Dispersions and developments of mechanical (physical) geochemical anomalies are also found in glacial and eolian environments. Let us first consider the dispersions related to glacial activity. When a glacier passes over mineralized terrain that may or may not have been physically, chemically, and/or biologically weathered, the mass of ice moves along picking up and dragging with it all the classes of materials in its path, material that is deposited when the ice

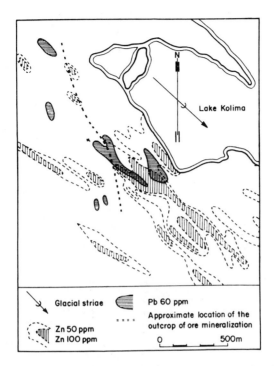

Figure 7-18. Distribution of Zn and Pb in the B layer of podsol profile, 0.05 mm fraction of the glacial till of Kolima, central Finland. Glacial erratics rich in Zn (3.6 to 20%) and Pb (1.5 to 6%) are numerous in the area. (Reproduced from Kauranne, 1967a, *Geochemical Prospecting in Fennoscandia,* (Kvalheim, ed.) by permission of John Wiley & Sons, Inc.) Copyright © 1967, by John Wiley & Sons, Inc.

melts. In such a situation it is possible to find a broad dispersion of chemical elements with rather average values in the samples of till generally used in sampling glaciated areas. Also, due to the predominant linear movement of the ice, the geochemical concentration pattern very often associated with glaciated areas may be strung out, with a final form dependent on the distance of the deposit from which the chemical elements were derived, and it may then be used to establish the existence of an anomaly (see Figure 7-15). With an anomaly determined on this basis, the exploration can be intensified in the area in the direction of last movement of the ice. Nonetheless, epigenetic dispersions in till from underlying veins may appear in the basal till directly above the positions of the veins, a situation illustrated in the Louvem deposit, Quebec, Canada (see Figure 7-16). Figure 7-17 shows a similar relation, but one associated with boulder trains. Many dominantly physical geochemical dispersion patterns have been found in glacial or glaciated areas; some representations of these are

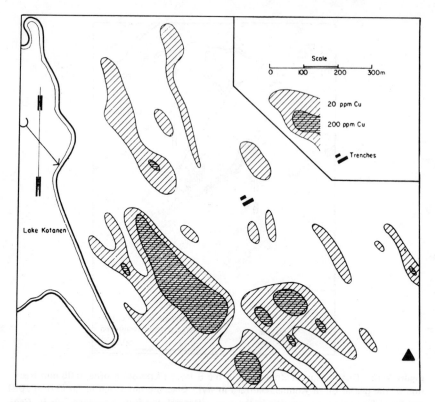

Figure 7-19. Distribution of Cu in the C horizon, 0.05-mm fraction, at a depth of 60 cm in the glacial till. (Reproduced from Kauranne, 1967b, *Geochemical Prospecting in Fenno-scandia* (Kvalheim, ed.) by permission of John Wiley & Sons, Inc.) Copyright © 1967, by John Wiley & Sons, Inc.

given in Figures 7-18, 7-19, and 7-20. As is evident from Figure 7-21, the geo-chemical anomalies developed in glaciated areas can be rather discontinuous due to the action designated as "override."

In areas of much wind and especially of rather directional wind, it would seem that the dispersion of the detrital weathering products of ore deposits would be found in somewhat extended shape in the direction of the dominant winds (see Figure 7-22). If a zone of anomalies is found in an environment known for the directional nature of its winds, such knowledge can be used by prospecting in an upwind area. The author does not recall any case in which a mineralized zone has been found by a combination of meteorology and the study of detrital eolian materials, and the discovery of an anomaly in these materials; nevertheless, the possibility of finding such anomalies cannot be dis-counted. Approximately 20% of the Earth's continental surface is of a desert

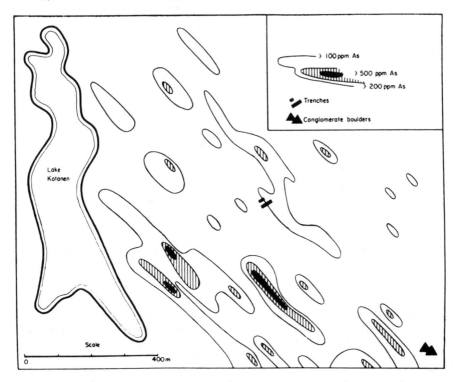

Figure 7-20. Distribution of As in the same material of Figure 7-19. The elongated forms of the anomalies in the direction of glacier movement as indicated by glacial striations are very marked. (Reproduced from Kauranne, 1967b, *Geochemical Prospecting in Fennoscandia* (Kvalheim, ed.) by permission of John Wiley & Sons, Inc.) Copyright © 1967, by John Wiley & Sons, Inc.

or semidesert nature so that conditions could be propitious for prospecting by studies of eolian transported detritals. To a certain degree the work of Weiss (1971) provides an example of prospecting using the forementioned study materials. He collected aerosols during the dry and sometimes windy seasons in South Africa and Southwest Africa. After analyzing the aerosols for their size distributions and their contents of various elements, and after applying corrections for wind direction and velocity, he found that the larger particles with their contents of indicator metals were in the air both directly above an area of mineralization and all around the center of mineralization.

Physical alterations to a system subsequent to those related to the initial development of a physical anomaly can hide that anomaly and it is essential that the director of a geochemical exploration program be alert to this possibility. For example, a slide of material from a steep area above an ore and its corresponding anomaly can cover the anomaly formed in the residual soil that

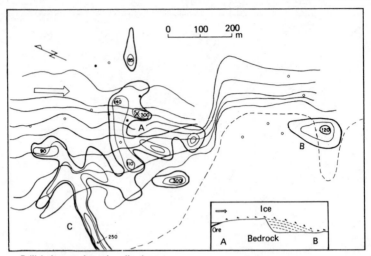

• Drill hole exposing mineralization
◦ Drill hole exposing no mineralization
– – – Limit of peat bog
⟹ Direction of glacial transport

Figure 7-21. Actual pattern of geochemical anomalies contoured on 30, 50, and 100 ppm of Pb. The topographic contours are at 6 meters. Mineralization is exposed in the trench at X and in some boreholes. The insert provides an explanation for the lack of continuity between the anomalies at *A* and *B* due to the presence of a stationary obstruction of ice. The position of the anomaly at *C* is due to the drainage effect as indicated from its shape, its termination in a bog, and its position normal to the topographic contours. (From Brotzen, 1967, *Geochemical Prospecting in Fennoscandia* (Kvalheim ed.) by permission of John Wiley & Sons, Inc.) Copyright © 1967, by John Wiley & Sons, Inc.

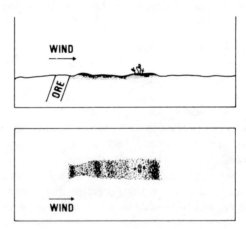

Figure 7-22. Idealized situation in which an anomaly developed from directional wind in a desert zone.

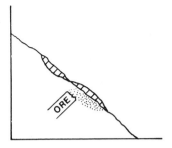

Figure 7-23. Idealized situation in which an anomaly is covered by material from a slide above a mineralized vein.

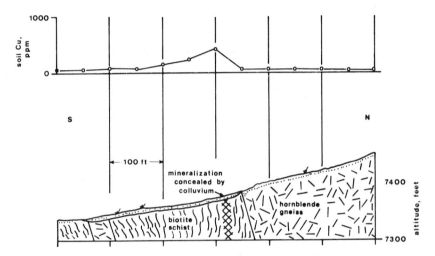

Figure 7-24. Actual situation in which the mineralized zone is hidden by a colluvium deposit in a part of the Malachite Mine, Jefferson County, Colorado. (From Huff, 1960.)

may have been dispersed somewhat mechanically by gravity-influenced creep in a previous cycle of development (see Figure 7-23). In the same way, a colluvial deposit can originate from the weathering and gravitational movement in an area and/or hide mineralization or anomalies buried below it (see Figure 7-24). The mineralization or its related anomaly can be hidden by other natural and predominantly physical events. For example, in areas where there has been recent volcanic activity or where volcanic activity with associated emission of ash to the atmosphere and its subsequent deposition is taking place, the ash or the soil developed on the ash may be covering what would be considered a paleosoil, which itself could have developed as a residual soil and thus be a carrier of geochemical anomalies (see Figure 7-25). Such a situation exists in the Central Cordillera of Colombia, Department of Antioquia; in the

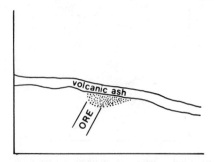

Figure 7-25. Idealized situation in which a fall of volcanic ash over an anomaly developed in a residual soil hides the anomaly.

area of the Mina Cristallina, a soil developed on an amphibolite into which there had been the injection of a chalcopyrite- and pyrrhotite-bearing quartz vein. This soil was buried by a volcanic ash fall from which a soil covering the paleosoil on the old amphibolite surface formed. Again, then, a point must be made of the fact that the geochemist in the field must be prepared to recognize the possible existence of the effects of both the ancient and existing geological processes.

In these cases of dominantly mechanical dispersion, we arbitrarily represented the ore as a vein or fissure filling, fairly localized areally. However, if we remember the many and varied classes of ore deposits and their very different areal extensions, such as those in the genetic classification proposed by Bateman (1950) or the classification presented by Vol'fson (1955) based on genesis and structure, we understand perfectly well how the size and extension of the geochemical anomaly pattern may also vary according to the type of mineralization (genetic and structural) and its own extension in three dimensions (see Tables 7-1 and 7-2).

Thus, the geochemical anomaly patterns due essentially to physical (mechanical) dispersions could be relatively limited in extent if one works with pegmatites developed from magmatic concentration and injection of residual fluids, or with cavity fillings originating from hydrothermal processes in fissures, shear zones, and volcanic breccias; conversely, such patterns may be relatively widespread in mineral segregation zones caused by early magmatic concentrations, fillings from hydrothermal processes in solution cavities and channels, disseminated replacements caused by hydrothermal activity, or from residual concentrations over extensive areas. Each general class of deposits has its own characteristics and extensions as does each deposit in each general class. Nonetheless, certain regularities of the geochemical anomaly that can exist after a mechanical dispersion can be projected into the plan of geochemical prospecting and the program must be properly coordinated so as not to lose a potentially valuable deposit.

Table 7-1 Simple Genetic Classification of Ore Deposits (from Bateman, 1950)

Process	Deposits	Examples
1. Magmatic concentration	Early magmatic: Disseminated crystallization	Diamond pipes
	Segregation	Chromite deposits
	Injection	Kiruna magnetite?
	Late magmatic: Residual liquid segregation	Taberg magnetite
	Residual liquid injection	Adirondack magnetite, pegmatites
	Immiscible liquid segregation	Insizwa sulfides
	Immiscible liquid injection	Vlackfontein, S. Africa
2. Sublimation	Sublimates	Sulfur
3. Contact metasomatism	Contact metasomatic: Iron, copper, gold, etc.	Cornwall magnetite, Morenci (old), etc.
4. Hydrothermal processes A. Cavity filling	Cavity filling (open space deposits): Fissure veins	Pachuca, Mexico
	Shear-zone deposits	Otago, New Zealand
	Stockworks	Quartz Hill, Colorado
	Ladder veins	Morning Star, Australia
	Saddle-reefs	Bendigo, Australia
	Tension-crack fillings (pitches and flats)	Wisconsin Pb and Zn
	Breccia fillings: Volcanic	Bassick pipe, Colorado
	Tectonic	Mascot, Tennessee, Zn
	Collapse	Bisbee, Arizona
	Solution-cavity fillings Caves and channels	Wisconsin-Illinois Pb and Zn
	Gash veins	Upper Mississippi Valley Pb and Zn

Table 7-1 (Continued)

Process	Deposits	Examples
	Pore-space fillings	"Red bed" copper
	Vesicular fillings	Lake Superior copper
B. Replacement	Replacement	
	Massive	Bisbee copper
	Lode fissure	Kirkland Lake gold
	Disseminated	"Porphyry" coppers
5. Sedimentation (exclusive of evaporation)	Sedimentary: iron, manganese, phosphate, etc.	Clinton iron ores
6. Evaporation	Evaporites: Marine	Gypsum, salt, potash
	Lake	Sodium carbonate, borates
	Groundwater	Chile nitrates
7. Residual and mechanical concentration A. Residual concentration	Residual deposits: iron, manganese, bauxite, etc.	Lake Superior iron ores, Gold Coast manganese, Arkansas bauxite.
B. Mechanical concentration	Placers: Stream	California placers
	Beach	Nome, Alaska, gold
	Eluvial	Dutch East Indies tin
	Eolian	Australian gold
8. Surficial oxidation and supergene enrichment	Oxidized, supergene sulfide	Chuquicamata, Chile Ray, Arizona, copper
9. Metamorphism	Metamorphosed deposits	Rammelsberg, Germany
	Metamorphic deposits	Graphite, asbestos, talc, soapstone, sillimanite group, garnet

Table 7-2 Metalliferous Deposit Structures Distinguished on Their Genesis (from Vol'fson, 1955)

1.	Magmatic deposits; magmatic proper or due to liquation, associated with alkaline intrusions; pockets or schlieren (tabular bodies); bottom deposits; steeply or gently dipping veins
2.	Pegmatitic deposits; veins; lenses; pipe-shaped bodies
3.	Greisen; stocks; veins; pipes; other forms
4.	Skarns; deeply or gently dipping; nestlike; pillarlike; layerlike; veins; pipelike; bodies of complex shapes
5.	Hydrothermal deposits; saddlelike; layerlike; ribbonlike bodies; simple lenses; complex fissure veins; pillarlike; stocks; ladderlike; others

Table 7-3 Supergene Mobility of Some Elements in a Siliceous Environment Free of Sulfides (from Hawkes and Webb, 1962)

Relative Mobility	Major Elements	Minor Elements
Very mobile	S, Cl	Br, I, Mo, B, Se
Moderately mobile	Ca, Na, Mg, K	Zn, Ba, U
Moderately immobile	Si, Mn	Ni, Co, Cu, As, Sb, Pb
Very immobile	Fe, Al, Ti	Cr, rare earths

Source. Reprinted from Geochemistry in Mineral Exploration, Harper & Row, 1962, p. 17, Table 2.2.

As has already been indicated in Chapter 4, the chemical elements respond differently with respect to their transport in aqueous media from a mineralized zone and they have their own mobilities that are dependent on a combination of determined factors. The multielement relations of the majority of the mineral-forming elements and the relative mobilities of the elements in an assemblage allow us to use various pathfinder or guide elements, such as Hg or As, in addition to the principal elements of a deposit in the general exploration and localization of sulfide deposits. The mobility of some elements in different geological environments can be appreciated from the data presented by Hawkes and Webb (1962) in Tables 7-3 and 7-4. It can be clearly seen that, in general, in siliceous weathering environments for sulfides, Zn is very mobile and Cu has an intermediate mobility, but in calcareous weathering environments, Zn has an intermediate mobility and Cu is essentially immobile; in both weathering environments Pb is rather immobile. Thus, it would be expected that the dispersions related to one environment or the other could be spatially different (other factors being the same). In siliceous environments, Zn would

Table 7-4 Mobility of Some Elements during the Weathering of Sulfide Deposits (from Hawkes and Webb, 1962)

Relative Mobility	Siliceous Environment	Calcareous Environment
Mobile	S, Mo, Zn, Ag	
Intermediate mobility	Cu, Co, Ni, Mo[a], As	S, Mo, Zn, Ag
Immobile	Fe, Pb, As[a]	Fe, Cu, Pb

Source. Reprinted from *Geochemistry in Mineral Exploration,* Harper & Row, 1962, p. 18, Table 2.3.
[a]Environment rich in Fe.

show an areally wider dispersion than Cu, which in turn would have a wider dispersion than Pb; Cu could have dispersion patterns similar to those of Pb in calcareous environments.

In a like manner, studies of textures and the paragenetic relations of ore minerals show a general order or sequence of deposition (crystallization) of the minerals with their principal (and/or) minor element or elements that is the result of changes in the physical–chemical conditions of the mineralizing fluids, which in turn are influenced by changes in the character of the host rock. The order of crystallization depends primarily on the solubilities of the minerals being considered and on the concentration of the elements in the solutions that are changing absolutely and relatively as deposition and the formation of minerals takes place. Among the principal causes for the deposition of ore minerals from solutions that carry the mineral-forming elements as complexes, for example, of sulfides, one may cite: cooling as the fluids follow a thermal gradient, mixing or dilution of the fluids with meteoric waters, adiabatic expansion, and reactions with the enclosing rock; finally, any reaction that decreases the activity (effective concentration) of the complexes also decreases their solubility and results in mineral deposition. For the sulfides, the mobility of ore metals in a fluid transporting medium generally follows a sequence in decreasing order of $Hg > Pb > Zn > Cu > Sn > Ni > Fe > Co$, and, according to the publication by Barnes and Czamanske (1967), when the complexes of these metals are formed in mineralizing fluids, their stabilities are very similar in order to that given above and show a decreasing order of $Hg > Cd > Pb > Cu > Zn > Sn > Ni > Fe > Co > Mn$. Therefore, when the zones sequences related to sulfide mineralization are studied, one may conceptualize that Hg will be found farthest away from the principal source of mineralization and it is this characteristic of Hg that allows us to use it as a pathfinder element in the search for sulfide deposits or other classes of deposits containing sulfides as nonprincipal minerals. In fact, Hg may be found in a relatively remote position in the sequential zonation of real deposits reported upon in many research projects. In addition, as Chan (1969) noted, any high-tempera-

Table 7-5 Lateral and Vertical Zoning of Endogenous Halos in Different Types of Deposits (from Ovchinnikov and Grigoryan, 1971)

Deposits	Lateral Zoning[a]	Vertical Zoning[b]
Lead–zinc deposits in skarns	Ba, Zn, Pb, As, Ag, Cu, Sb	Sb, Cu, As, Ba, Ag, Pb, Zn, Cu
Lead–zinc deposits in acid effusive rocks	Pb, Ba, Zn, Ag, Cu, As, Co	As, Ba, Ag, Pb, Zn, Cu, Co
Skarns – scheelite deposits	W, Mo, Cu, Ba, Zn, Pb	Ba, Pb, Zn, Cu, W, Mo
Gold–quartz deposits	Au, As, Bi, Ag, Pb, Sb, Cu, Be, Mo, Co, Zn	Sb, As, Ag, Pb, Zn, Cu, Bi, Mo, Au, Co, Be
Copper–gold deposits	Au, Cu, Mo, Ag, As, Sb	Sb, As, Ag, Cu, Mo, Au
Copper–bismuth deposits	Cu, Bi, Pb, Ag, As, Ba, Zn, Co	Ba, Ag, Pb, Zn, Cu, Bi, Co
Uranium–molybdenum deposits	U, Mo, Pb, Cu, Zn, Ag	Ag, Pb, Zn, Cu, Mo, U
Mercury deposits	Hg, As, Ba, Cu, Pb, Zn, Ni, Ag, Co	Ba, Hg(?), Sb, As, Ag, Pb, Zn, Cu, Ni, Co
Sulfide–cassiterite deposits	Ag, Zn, Pb, Sn, Cu, Mo	Ag, Pb, Zn, Cu, Mo, Sn
Stratiform lead–zinc deposits	Ag, Pb, Cu, As, Ba, Co, Zn, Ni	As, Ba, Ag, Cu, Pb, Zn, Co

Source. Reprinted from *Geochemical Exploration* (Boyle, tech. ed.), Special Vol. No. 11, The Canadian Institute of Mining and Metallurgy.

[a] Elements are given in decreasing order of the width of their halos in cross section; i.e., in lead–zinc deposits in skarns, Sb has the narrowest, halo, whereas Ba has the most extensive.

[b] Reading from left to right, the indicators of the supraore parts grade or pass downward to the indicators of the subore parts of an ore zone.

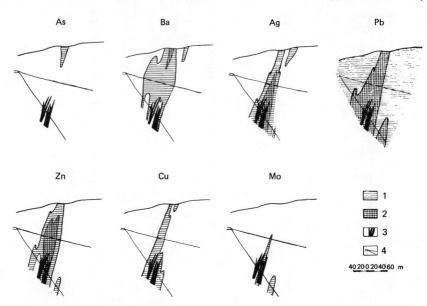

Figure 7-26. Endogenous geochemical halos around a lead–zinc ore in skarn. The contents of the elements in the dispersion halos are: (*1*) 0.0015 to 0.003% As, 0.01 to 0.1% Ba, 0.00003 to 0.0001% Ag, 0.003 to 0.01% Pb, 0.005 to 0.01% Zn, 0.003 to 0.05% Cu, 0.003 to 0.001% Mo and (*2*) 0.1 to 0.2% Ba, 0.001 to 0.008% Ag, 0.01 to 0.1% Pb, and 0.01 to 1.0% Zn. (*3*) ore and (*4*) prospecting boreholes. (Reprinted from Ochinnikov and Grigoryan, 1971, *Geochemical Exploration* (Boyle, tech. ed.), Special Volume No. 11, The Canadian Institute of Mining and Metallurgy.)

ture mineralization subsequent to the formation of an ore deposit could cause a movement of Hg farther away from its originating point.

Ovchinnikov and Grigoryan (1971) studied the primary geochemical dispersions associated with hydrothermal sulfide deposits of the plutogenetic telethermal classes and of the skarn group of superimposed ores and found that two characteristics of these dispersions permit their application in geochemical exploration:

1. The dimensions of the primary geochemical dispersions of several elements are much wider than those of the deposits around which the dispersions are developed.
2. The vertical dimension of the primary dispersions is especially great above steeply dipping ore bodies.

Table 7-5 indicates the lateral and vertical zonation of endogenous halos in different types of deposits and Figures 7-26 and 7-27 give examples of how the dispersions for some elements in the two general groups of deposits appear.

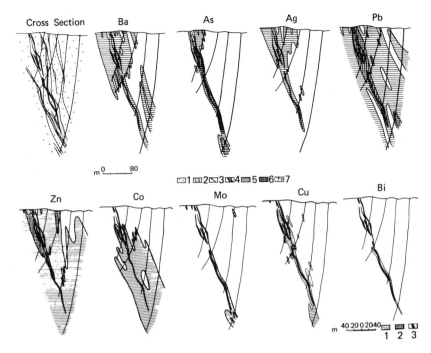

Figure 7-27. (upper) primary halos around ore bodies of the *V.* Kanimansur Deposit: (*1*) quartziferous porphyry (welded tuff); (*2*) quartz porphyry; (*3*) diabase porphyry (dikes); (*4*) ore; (*5*) 0.05 to 0.5% Ba, 0.005 to 0.01% As, 0.00002 to 0.0001% Ag, 0.005 to 0.1% Pb; (*6*) 0.01 to 0.1% As and 0.1 to 1.0% Pb; (*7*) sampling interval. (Lower) at same locations as above. Contents: (*1*) 0.005 to 0.01% Zn, 0.005 to 0.01% Cu, 0.00008 to 0.001% Bi, 0.003 to 0.02% Sn, 0.0002 to 0.006% Co, 0.0001 to 0.008% Mo, 0.0002 to 0.0008% U; (*2*) 0.01 to 0.8% Zn, 0.01 to 0.03% Cu; (*3*) ore. (Reprinted from Ochinnikov and Grigoryan, 1971, *Geochemical Exploration* (Boyle, tech. ed.), from Special Volume No. 11, The Canadian Institute of Mining and Metallurgy.)

In areas where the mother rock is hidden by a thin cover, the secondary dispersion patterns formed by the disintegration of primary material can be used directly in prospecting. However, Ovchinnikov and Grigoryan noted that the breakup of the primary dispersion patterns causes numerous changes in the distribution of the indicator elements cited in Table 7-5, and because of this, special studies (pilot studies) are needed to understand the relations between the primary and secondary dispersions of each region that is being investigated, advice given several times in this text.

In the most common exogenetic geochemical exploration environments, hydromorphic, chemical, and biogeochemical actions are superimposed on physical (or mechanical) actions. Differences exist only of intensity and contribution (or control) of each active process in the study environment. Thus, the

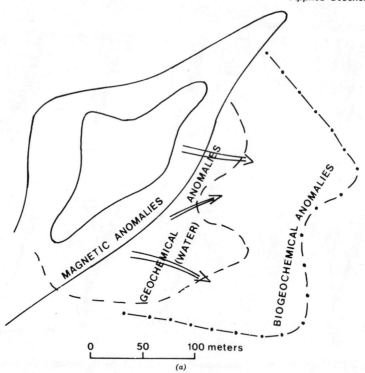

0   50   100 meters

(a)

Figure 7-28a. Relation between biogeochemical, geochemical, and magnetic anomalies in an area of the Nokia, Finland, region. The arrows indicate direction of groundwater flow. (From Marmo, 1958.)

interpretations and evaluations of the element distribution patterns associated with geochemical anomalies, and made with the forementioned considerations in mind, can be significantly different from those made in regions where the physical processes are very dominant. In general, the hydromorphic, chemical, and biogeochemical factors serve to increase the recognizable dispersion areas (and/or show them to be displaced) by causing a separation of dominantly multielement anomalous areas into mono- and bielement zones and by developing false anomalies. These effects are directly related to the geochemical reactions of the chemical elements in water and these reactivities are tied to the different solubilities of the elements under the physical–chemical conditions dominant in an environment and to the presence or absence of reaction surfaces (on colloids, for example) that may markedly concentrate selected elements.

As a general example of the separatory processes just mentioned, the relations between the magnetic, hydrogeochemical, and biogeochemical anomalies in an area of the Nokia region, Finland, described by Marmo (1958) can be

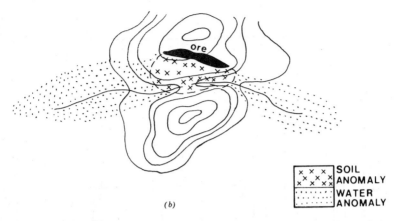

(b)

x x x
x x x x   SOIL
x x x x   ANOMALY
· · · · · WATER
· · · · · ANOMALY

Figure 7-28b. Distribution of cations, derived from an ore, in soils and waters influenced by the topographic characteristics of the area. The ore is in the upper slope of a wind gap. In soils the anomalies are generally found between the ore body and the gap, but in waters the anomalies extend in the direction of river flow on both sides of the gap. (From Marmo, 1958.)

considered. In Figure 7-28 the influence of the flow direction of subsurface waters in amplifying the anomaly zone indicated by geomagnetic studies is clearly presented. Here, the geochemical (or hydrogeochemical) anomalies as found in the waters extend from the magnetic anomaly zone in the direction of subsurface water flow for a distance of about 100 meters. However, the area of the biogeochemical anomalies extends still another 100 meters in the direction of subsurface water movement; the subsurface waters provided the cations absorbed and concentrated by the biogeochemical samples, although these same cations analyzed in the waters did not indicate the extent of the anomalous zone.

The relations between biogeochemical anomalies and other geochemical anomalies are sometimes very excellent areally, but may vary with the capacity of the sample type considered to absorb and retain one or several elements and with the availability of the nutrient elements for the live sample forms. The availability depends on several factors and in Figure 7-29 we examine one of these, the effect of pH. The widths of the bands in Figure 7-29 indicate the favorability of the influence of the reaction with respect to the presence of adequate quantities of nutrients in easily available forms, but do not necessarily indicate the quantities of the elements present. Figure 7-30 shows the relations between the average concentrations of trace elements in leaves and twigs of trees and shrubs or bushes that grow in areas of siliceous, ultrabasic, and calcareous rocks and compares these concentrations with the corresponding values for rocks and soils. For example, there is a truly excellent positive

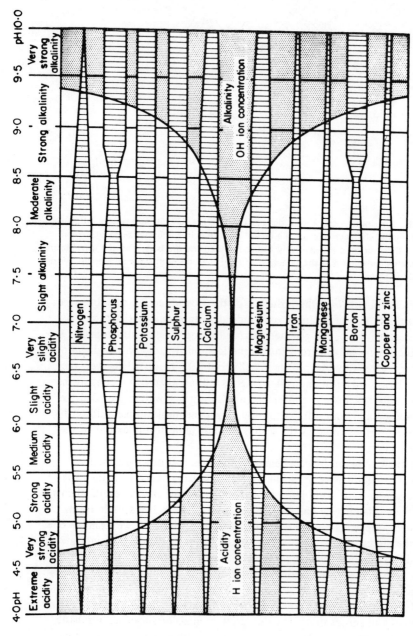

Figure 7-29. General tendency of the pH of reaction to the availability of nutrients to plants. The width of the bands indicate the favorability of the pH of reaction to the presence of adequate quantities of nutrients in easily available forms and not necessarily to the quantities present; these quantities are influenced by various other factors. (From Troog, 1948, as reproduced in Lag, 1967, *Geochemical Prospecting in Fenno-scandia* (Kvalheim, ed.) by permission of John Wiley & Sons, Inc. Copyright © 1967, by John Wiley & Sons, Inc.)

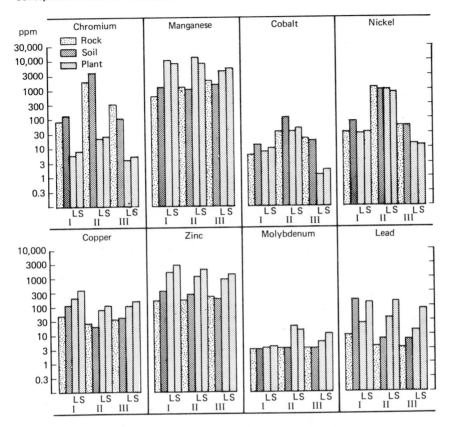

Figure 7-30. Average concentration of trace elements in leaves (*L*) and twigs (*S*) of trees and bushes that grow over silicate rocks (*I*), ultrabasic rocks (*II*) and calcareous rocks (*III*) compared with corresponding values for rocks and soils (ppm). (From Lounamaa, 1967, *Geochemical Prospecting in Fennoscandia,* (Kvalheim, ed.) by permission of John Wiley & Sons, Inc. Copyright © 1967, by John Wiley & Sons, Inc.)

correlation between the concentration of Ni in soils derived from ultrabasic rocks and its concentration in leaves from trees and shrubs or bushes that grow in these soils. As an adjunct to the generalities given in Figure 7-30, one may cite the results of a study on the relation between the anomalies for Ni (and other metals) in residual soils and some species of trees in the Riwaka Basic Complex, New Zealand, in which there was a positive correlation between the leaf ash and the soil for two species (correlation coefficient estimates of +0.82 and +0.75) and Ni. Also, from the figures presented in the report on this complex (see, for example, Figure 7-31), it may be seen that, in general, the anomalous zones indicated from the leaf ash correspond well with the anomalous zones indicated by the soils, although it is clear that for the species considered

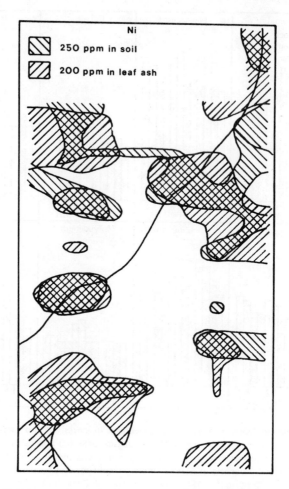

Figure 7-31. Nickel anomalies in the Riwaka Basic Complex, New Zealand, indicated by Ni concentrations in soil and leaf ash from *Nothofagus fusca* and *Nothofagus menziesii* (treated as a single species). Reproduced from Timperly et al. 1970, *Econ. Geol.,* 1970, **65,** 507.)

in the figures of the original paper, the biogeochemically anomalous zones are greater than those determined pedogeochemically. However, in Figure 7-31 and others in the original study, we see that there are anomalous areas indicated by geochemical analyses of the biogeochemical samples that do not correspond to soil anomalies and vice versa. The authors of the investigation (Timperly, Brooks, and Peterson, 1970) believe that the reason why plants give anomalies in certain places where soils do not give them may be the presence of a relatively large concentration of soluble (available) nickel upslope or deep

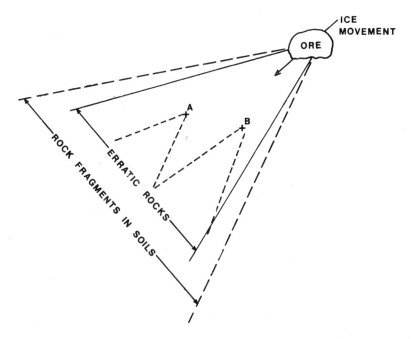

Figure 7-32. Distribution of ore fragments by glacial transport. *A* and *B* are large erratics that have produced new secondary anomaly fans because of more advanced disintegration. (From Marmo, 1958.)

mineralization that does not show up in the soil analyses. These geological controls could result in elevated concentrations of nickel in plants and not in soils. Where the soils gave anomalies and the plants did not, there may be a very low concentration of available nickel, although a total anomalous concentration exists; an efficient drainage system that continually carries away the greatest part of the weathering products is considered by the authors to be the most reasonable explanation for this low concentration of available nickel.

Previously, the anomaly patterns associated with glaciated areas were considered and we observed that under conditions of physical (mechanical) dispersion, such anomaly patterns are often strung out linearly depending on the distance between the deposits from which sediments were derived that gave chemical element data for establishing the presence of anomalies. However, in geological-topographical conditions similar except for differences in the flow direction of subsurface water after the glacial activity terminated, we must modify our concepts regarding patterns of anomalies related to glacial processes. For example, Figure 7-32 shows that the movement of subsurface water in the same direction as the last movement of a glacier amplifies the dispersion of the elements studied well beyond the actual limit of the glacial material. The method

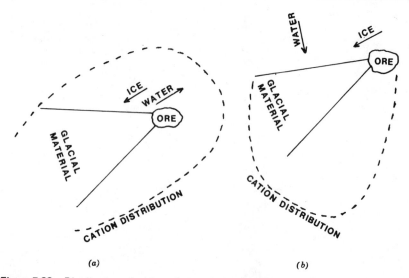

Figure 7-33. Distribution of cations due to glacial transport of moraine material and subsequent subsurface water activity (or surface water activity). (*a*) The direction of movement of the water is opposite to the direction of movement of the ice; (*b*) the direction of movement of the water is subnormal to the last direction of movement of the ice. (From Marmo, 1958.)

used to find the ore from which the anomaly was derived would be to intensify the prospecting in the direction from which the ice came and the water comes. However, if the subsurface waters of a study zone were moving in a direction opposite (see Figure 7-33 *a*) or intermediate (see Figure 7-33 *b*) to the movement of the preexisting glacial ice, the resulting anomalies would have to be investigated in a different manner, basing the prospecting plans for localizing an ore on a knowledge of the hydrological system in the area of interest. This type of reasoning must be applied in any program in which a mechanical dispersion of material from a mineral deposit is related to aqueous activity.

Several of the chemical effects that influence the geochemical anomaly patterns have already been considered in the discussions on the responses and activities of the elements during the crystallization of magmas and during weathering and the development of soils. However, once elements in solution or in suspension in the form of fine particles or colloids enter into the hydrological system outside the immediate zone of mineralization and weathering, they can be deposited or precipitated close to or far from that zone in strong or very slight concentrations. This activity can sometimes provide us with true or false anomalies depending on what is seen and interpreted from an evaluation of the conditions at the sampling areas and their surroundings.

Some of these effects can be examined in swampy environments where there are bogs in which peat is forming. Many authors have evaluated the concentration of elements in samples from these environments as a function of factors such as the pH, the degree of humification, and the presence of iron and/or manganese oxides. These evaluations demonstrate the complexity resulting from the partial controls on the presence and distribution of some elements. For example, in the Tynagh, Ireland, area, Sutherland-Brown (1967) found that the concentrations of Cu, Hg, and Zn in peats were much greater than their concentrations in adjacent areas having free drainage; however, Pb has similar concentrations in both environments. In the same way, the analyses of materials from peat bogs in mineralized areas of Finland showed greater concentrations of Cu, Zn, and Ni than those from peat bogs in nonmineralized zones. Gleeson and Coope (1967) noted that although Zn is very soluble (especially in environments with relatively low pH conditions) and can be carried far by water, it was found in great concentrations in the peats of Eastern Canada.

In addition to the concentration of certain elements in peat bogs or swampy environments, notable but certainly not "standard" distributions of elements have been found in the thicknesses of peat sections studied. In one area Fe, Mn, Mo, and Pb are found concentrated in the upper parts of the peat section and Ni, Ti, V, and Cd are concentrated in the lower parts of the section. In another area, Pb, Zn, and Mo are concentrated in both the upper and lower sections of peat bog sequences. Gleeson and Coope (1967) related the observed metal distribution to the degree of humification because there were increases in metal concentrations with an increase in humification and because experiments showed that the base exchange capacity of humified peat at a pH = 6, is 8 times as great as that of clay due to the formation of organic complexes that have a strong affinity for metals. The degree of humification increases with pH; with decomposition of organics, the pH increases because of the liberation of bases and the decomposition of organic acids. The relations of equilibrium concentrations of metals in solutions of purified humic acid have been investigated by Chowdhury and Bose (1971); for some +2 valence ions, this equilibrium followed the sequence of Pb < Cu < Zn < Ni < Co, although the stability of the chelates is the inverse with Pb > Cu > Zn > Ni > Co. On the basis of these experimental data, the authors concluded that the adsorption of metals by reaction surfaces or by chemical compounds during the weathering process is a function of atomic weight and valence and that the relative adsorption of elements of the same valence is proportional to their atomic weights.

In addition to these factors, there are others that influence the distribution of elements in peat bog areas and hence the geochemical anomaly patterns. The real or false anomalies can be derived by evaporation and transpiration of large quantities of humidity from the peat bog surfaces. Also, the vegetation

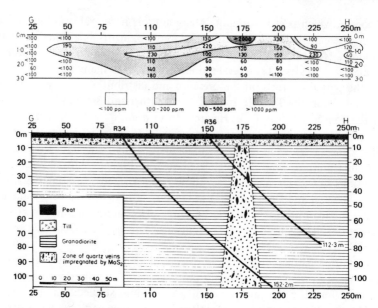

Figure 7-34. The Susineva bog, Finland, Traverse G-H in Figure 7-35: (below) geologic profile as determined from boreholes; (above) distribution of Mo in the upper part of the bog. (From Salmi, 1959.)

that is part of the bog environment can concentrate elements and contribute them to the system being studied when each life cycle is completed. This last-mentioned potential contributor to the development of anomalies in peat zones that have been covered by glaciers has been effectively demonstrated by Salmi (1959, 1967). In the Susineva (Finland) peat bog area, there are quartz veins with lenses and impregnations of MoS_2 intruded into a granodiorite, the granodiorite is covered by till, and the till is covered by peat (see Figure 7-34). The geochemical analyses of the clays, sands, and gravels comprising the till did not show that there was much capacity of the till to concentrate the ions that were being filtered from the underlying mineral deposit and, as a result, these ions could enter the peat and concentrate in the peat phase of the section. Dissolved in water, the elements move in the material overlying the ore-bearing rocks; the elements may also move by diffusion. Natural peat contains more than 90% water. Mineral soils beneath the bog are saturated with water so that a layer of soil extends from the mother rock towards the surface of the bog providing an environment in which the solutions can be moved easily. Abundant evaporation from the peat bog surface and transpiration from the bog plants, both in excess of precipitation during the summer, act jointly with processes of diffusion and root suction, and are the factors that control the transport of material dissolved from a mother rock across mineral soils towards the peat layer and the bog plants. Due to the poor absorption capability of the till and clay, their trace element contents are small. Nevertheless, these elements can penetrate the peat layer and concentrate there in the leaves and branches as bog vegetation. The trace elements in the peat and plants are very much enriched (relatively) in their ash, and their concentrations are very different from those of the mineral soil from the same place. Iron, Mo, and Pb are more concentrated in the upper part of the peat, whereas Ti and V are abundant at the bottom of the bog. This may be caused by the fact that Ti and V form compounds somewhat soluble in the anaerobic conditions at the bottom of the bog and are absorbed by the peat layer, while Fe, Mo, and Pb migrate across the peat layer towards the surface and are oxidized there. At Susineva, the peat gave geochemical anomalies for Mo, especially in its upper parts, an observation that can be studied and then serve as a good example of how secondary events are related to the interpretations of geochemical data from areas of scientific interest. In this region there was a preglacial event during which the MoS_2-bearing quartz vein was injected into the granodiorite. Glacial activity stripped away whatever soil could have been developed by the weathering of the granodiorite and its included quartz vein; this event was followed by a postglacial stage with the deposition of till. Because of the nature of the glacial till, an imperfect drainage was developed and the vegetation of the area rhythmically died and accumulated to form a peat bog and, subsequently, the peat. Later, the anomaly was formed in the peat in the way already described, and instead of a linear-type anomaly dispersion often associated with glaciated regions, a very localized anomaly was found, tied in spatially with the underlying mineralized vein, the presence of which was proven by drilling (Figure 7-34). Also, Figure 7-35 demonstrates that there is a lessening of the anomaly to the north and to the east in the direction of lateral movement of swampy water that was following the dip of the peat bog surface.

Earlier we cited the work of Nigrini (1971) in which it was demonstrated that subsurface water charged with CO_2 and with a low pH can carry several elements in solution, but that with its exposure to the atmosphere, its chemical characteristics changed because of a loss of CO_2 and a resulting increase of pH. This caused a precipitation of Fe and Mn oxides and a decrease in the concentrations of Cu, Pb, and Zn in the waters because of their capture by the precipitated oxides. This type of element accumulation by Fe or Mn oxides must be taken into consideration in the interpretation of geochemical data because a significant concentration of metals can be accumulated over a period of time and result in false anomalies. The capability of the cited oxides to adsorb and absorb chemical elements has been known for many years. Several papers have been published on the accumulation of metals in manganese nodules from marine environments (see Chapter 10) and from limnological environments (Lake Superior, for example). In his investigation on the

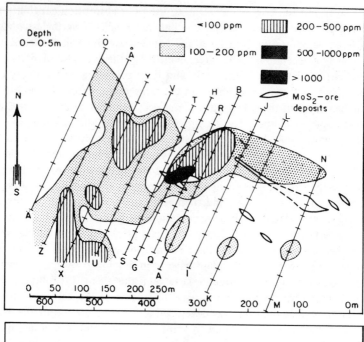

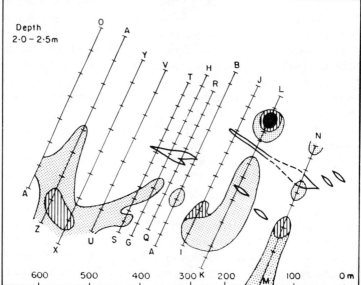

Figure 7-35. The Susineva bog, Finland: (above) distribution of Mo in the 0 to 0.5 meter horizon; (below) distribution of Mo in the 2.0 to 2.5-meter horizon. (From Salmi, 1959.)

localization of serpentinites via geochemistry of tropical soils, Kilpatrick (1969) found that Ni was dissipated in the soils during weathering of the serpentinites, whereas Co was enriched (relatively), probably due to the process of adsorption (sequestering) of the Co by Mn oxides. Nickel is more mobile than Co in environments rich in Mn. Thus, Co can be used as a pathfinder or guide element in prospecting for Ni, but the exploration geochemist must be aware of this enrichment process for Co in certain environments (in this case an environment rich in Mn) when he is making the geochemical interpretations.

In a like manner, the Fe and Mn oxides can cause metal anomalies in stream sediments. Horsnail, Nichol, and Webb (1969) made a comparison of geochemical data from rocks, soils, and stream sediments from drainage basins in zones of imperfect drainage and in zones of free drainage in an evaluation of the factors that influenced the secondary dispersion of the chemical elements. They found that in the −80 mesh fraction of stream sediments from a zone with gleys and peats, and podzols from environments with impeded drainage, there were elevated contents of Mn associated with very elevated quantities of Fe, Co, and As, and often with more than half of the available amounts of Ni, Pb, Zn, and Cu. The adjacent upstream areas where there was an environment of brown earth with free drainage had sediments in which the concentrations of the above metals were rather constant, but lower. In another area, the rocks from a heath with poor drainage had similar concentrations of Mn, Co, As, and Fe, but the stream sediments deriving from the area were rich in Mn (×30), Co (×13), As (×5), and Fe (×2), thus making for anomalous but possibly invalid anomalies as regards geochemical prospecting. These authors noted that the differences were due to the responses of the chemical elements to changes or differences in the pH–Eh environment and that the pH–Eh values between soils and stream sediments had greater differences in areas with imperfect drainage. Soils in areas of imperfect drainage have lower pH–Eh values than related stream sediments, and under these conditions, elements such as Co, Fe, Mn, Ni, and Zn (all of a relatively low ionic potential) enter into solution and are transported to drainage channels. Once they enter these channels, there are increases of Eh and pH with oxidation and precipitation of Fe and Mn oxides which themselves sequester metals such as Co, Ni, and Zn available from the solutions. In general, stream sediments show little variation in pH–Eh conditions once equilibrium is approximated in the physical–chemical environment of the surroundings. Therefore, in channels that receive detritus and suspended and dissolved materials from poorly drained areas, there is an active precipitation of Fe and Mn oxides, often as crusts on bank materials, gravels, and boulders, and the relative concentration of other elements such as those cited above. Elements with higher ionic potentials, such as Cr, Ti, and Ga, do not have their dispersion patterns affected by the presence of Fe and Mn oxides and generally show a "normal" clastic dispersion. Complications may arise if

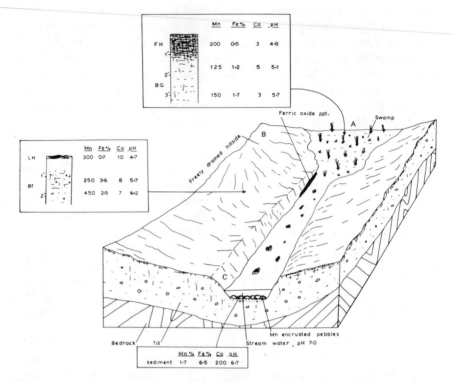

Figure 7-36. Precipitation of Mn and Fe oxides in a drainage channel and the relation to trace element distribution and pH. (Reprinted from Horsnail and Elliot, 1971, *Geochemical Exploration* (Boyle, tech. ed.), Special Volume No. 11, The Canadian Institute of Mining and Metallurgy.)

bank material collapses (because of undermining) with its contained oxide crusts and/or with soils from this environment into the stream or river. The result is that the oxides become part of the fluvial system and may be concentrated locally with fine-sized sediments thus causing high and real geochemical values, but values invalid with respect to prospecting if the dominant potential cause of these values (addition of previously precipitated materials with their concentrations of elements) is not taken into consideration. Horsnail and Elliot (1971) made a similar study and demonstrated that the factors that influenced element dispersion in such environments are ionic potential, pH, Eh, chelation, sequestering, and precipitation close to caliche beds, factors already cited in the section on the theoretical aspects of element dispersion. The effects of some of these factors are well illustrated in Figure 7-36 in which we see how the concentrations of some elements change in samples taken from areally close but environmentally different subenvironments. The lowest con-

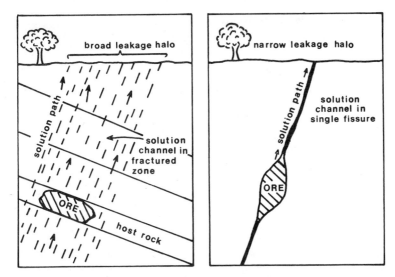

Figure 7-37. Conceptual diagrams that show the relations of solution channels of ores to the width of halos of the resulting dispersion. (From Hawkes and Webb, 1962, *Geochemistry in Mineral Exploration,* Harper & Row Figure 4.9, p. 59.

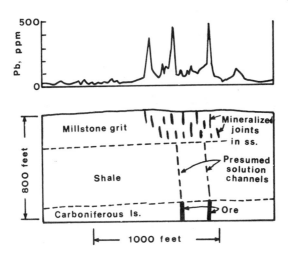

Figure 7-38. Actual situation in which the anomaly pattern is influenced by the disposition of joints in rock overlying ore and underlying residual soil, Gregory Mine, Derbyshire, England. The ore is at a depth of 700 feet and the geochemical data were obtained from the −80 mesh fraction of the residual soil. (From Webb, 1958.)

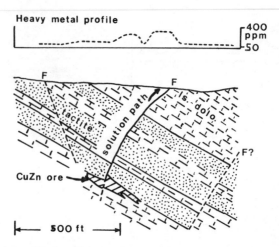

Figure 7-39. Actual situation in which the change in the direction of solution channels upward (in a fault zone) results in a displacement of the anomaly found in the rock exposed at the surface. Modified from Cooper and Huff, 1951, *Econ. Geol.,* 1951, **46**, 735, 742, 751.

centrations of Fe, Mn, and Co are found under swampy conditions (site A) in the system being studied; under oxidation conditions in areas of free drainage (Site B), there is the precipitation of Fe oxide on the channel banks and there are higher concentrations of Fe, Mn, and Co on the river banks and in the soils associated with the free drainage subenvironment. Finally, in the most oxidizing subenvironment of the same river (site C), there is precipitation of Mn oxide on gravels and pebbles and the resulting highest concentrations of the three elements analyzed. If the influence of Fe and Mn oxides or other chelation or sequestering agents is not appreciated, incorrect conclusions can be made in the interpretations of the geochemical data.

In concluding this section, let us consider a physical–structural situation that can influence the position or displacement of an anomaly pattern. Hawkes and Webb (1962) showed conceptually how the amplitude or areal extension of a secondary dispersion halo can be controlled by the structures associated with concentrations of ore or by the structures that provided pathways for the solutions coming through during the original mineralization or for the secondary solutions that moved towards the surface (see Figure 7-37). However, if the solution channels originating, for example, as fractures, joints, or faults, curve continuously or discontinuously as they project towards the surface, the anomalies developed from the ore may be displaced spatially from the ore in the subsurface with the possible result that vertical test drilling will not hit the target. Also, if the host rock containing discontinuous ore bodies has a relatively slight structural dip or if vein projections towards the surface are gentle

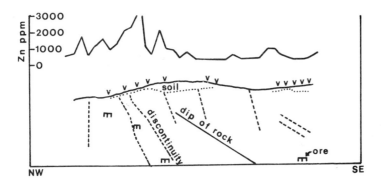

Figure 7-40. Geological–geochemical profiles that show the dependence of Zn concentrations in soils on the geologic structure, Austinville, Virginia. The *v* indicates the position of boreholes. (Fron Ginzburg, 1960.)

in their ascension, the anomalies that one could find in unconsolidated surface materials or in geochemical samples derived from these materials may be significantly displaced spatially in a direction horizontal to the ore body in the underlying rock. Examples of these geological situations can be found in the publications of Hawkes and Webb (1962) and Rostad (1971) and some illustrations from these studies are reproduced in Figures 7-37, 7-38, 7-39, 7-40, 7-41 and 7-42. Rostad (1971) worked on a mineralization assemblage of W, Ag, Cu, and Mo that was associated with a Tertiary stock (Lemhi County, Idaho). In the paragenetic sequence of the study zone, two epochs of mineralization were recognized: molybdenum mineralization was a first stage event with pyrite and less-important amounts of chalcopyrite and molybdenite in the granitic stock and the quartzite into which the stock was injected; the second stage of mineralization was represented by quartz veins bearing heubnerite and several other minerals. From his analytical results, Rostad observed that the concentration of Mo in soils directly over the stock was generally low (see Figure 7-41), but was greater in the outcrops or projections of outcrops crystallized before the weathering of the heubnerite-bearing quartz veins. Because of the up dip distance of the vein projections from the granitic stock to the surface, these projected anomaly-probable areas are greatly displaced from the area on the surface directly above the granite stock (see Figure 7-42). The knowledge of the geology of the study zone made the investigator aware of the fact that the displaced anomalies coincided with a molybdenite mineralization in the heubnerite-bearing quartz veins; this may have been due to the fractures in the veins being open when the molybdenite mineralization took place or to the possibility that there was a later remobilization of Mo. In this way, Rostad demonstrated that the Mo anomalies in the soils he studied do not indicate molybdenite mineralization directly below (vertically), and it becomes clearly

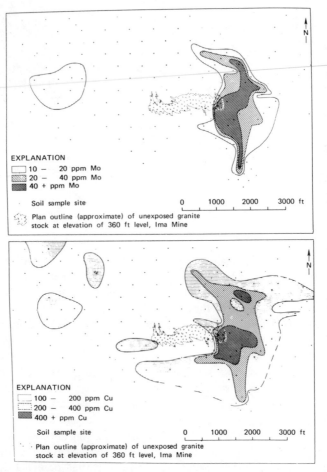

Figure 7-41. The distribution of Mo (upper) and Cu (lower) in soils related to a Tertiary stock, Lemhi County, Idaho. The anomalies are displaced to the east. Reprinted from Rostad, 1971, *Geochemical Exploration,* Special Volume No. 11, The Canadian Institute of Mining and Metallurgy.

understood that geological knowledge is necessary to properly and efficiently interpret geochemical results.

REFERENCES

Allan, R. J., and Hornbrook, E. H. W., 1971, Exploration geochemistry evaluation study in a region of continuous permafrost, Northwest Territories, Canada: *Geochemical Exploration* (Boyle, tech. ed.) Special Volume No. 11, Canadian Institute of Mining and Metallurgy, pp. 53–66.

Archer, A. R., and Main, C. A., 1971, Casino, Yukon – A geochemical discovery of an unglaciated Arizona-type porphyry: *Geochemical Exploration* (Boyle, tech. ed.), Special Volume No. 11, Canadian Institute of Mining and Metallurgy, pp. 67–77.

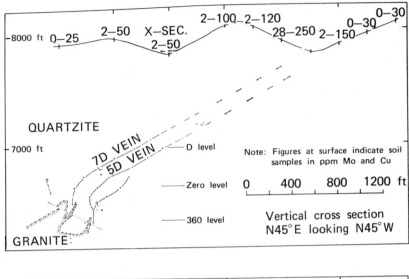

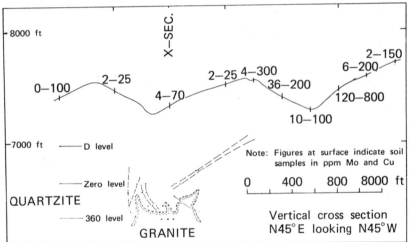

Figure 7-42. Explanation for the displacement seen in Figure 7-41. The strike and dip of the mineralized structures with relation to the stock influence the positions of the anomalies. (Reprinted from Rostad, 1971, *Geochemical Exploration* (Boyle, tech. ed.), Special Volume No. 11, The Canadian Institute of Mining and Metallurgy.)

Barnes, H. C., and Czamanske, G. K., 1967, Solubilities and transport of ore minerals, in *Geochemistry of Hydrothermal Ore Deposits* (Barnes, ed.), Holt, Rinehart and Winston, New York, pp. 334–381.

Bateman, A. M., 1950, *Economic Mineral Deposits,* 2nd edition, Wiley, New York, 916 pp.

Bolviken, B., 1967, Recent geochemical prospecting in Norway, in *Geochemical Prospecting in Fennoscandia* (Kvalheim, ed.), Interscience, New York, pp. 225–254.

Boyle, R. W., and Dass, A. S., 1967, Geochemical prospecting – use of the A horizon in soil surveys, *Econ. Geol.*, **62**, 274–276.

Brotzen, O., 1967, Geochemical prospecting in northern Sweden, in *Geochemical Prospecting in Fennoscandia* (Kvalheim, ed.), Interscience, New York, pp. 203–223.

Canney, F. C., and Wing, L. A., 1966, Cobalt: useful but neglected in geochemical prospecting, *Econ. Geol.*, **61**, 198–203.

Chan, S. S. M., 1969, Suggested guide for exploration from geochemical investigation of ore veins at the Galena Mine deposits, Shoshone County, Idaho, *Quart., Colo. School Mines*, **64**, No. 1, 139–168.

Chowdhury, A. N., and Bose, B. B., 1971, Role of "humus matter" in the formation of geochemical anomalies: *Geochemical Exploration*, (Boyle, tech. ed.), Special Vol. No. 11, Canadian Inst. Mining and Metallurgy, pp. 401–409.

Clarke, O. M., 1971, Geochemical prospecting in lateritic soils of Alabama, *Geochemical Exploration* (Boyle, tech. ed.), Special Volume No. 11, Canadian Institute of Mining and Metallurgy, pp. 122–125.

Cooper, J. R., and Huff, L. C., 1951, Geological investigations and geochemical prospecting experiment at Johnson, Arizona, *Econ. Geol.*, **46**, 731–756.

de Grys, A., 1970, Copper and zinc in alluvial magnetites from central Ecuador, *Econ. Geol.*, **65**, 714–717.

Friedrich, G. H., and Pluger, W. L., 1971, Geochemical prospecting for barite and fluorite deposits, *Geochemical Exploration* (Boyle, tech. ed.), Special Volume No. 11, Canadian Institute of Mining and Metallurgy, pp. 151–156.

Fulton, R. B., 1950, Prospecting for zinc using semiquantitative analyses of soils, *Econ. Geol.*, **45**, 654–670.

Garrett, R. G., 1971, The dispersion of copper and zinc in glacial overburden at the Louvem Deposit, Val d'or, Quebec, *Geochemical Exploration* (Boyle, tech. ed.), Special Vol. No. 11, Canadian Institute Of Mining and Metallurgy, pp. 157–158.

Ginzburg, I. I., 1960, *Principles of Geochemical Prospecting*, Pergammon, New York, 311 pp. (translated from the original work published in Russian, 1957).

Gleeson, C. F., and Coope, J. A., 1967, Some observation on the distribution of metals in swamps in eastern Canada, Geological Survey of Canada, Department of Energy, Mines, and Resources, Paper 66-54, pp. 145–166.

Hawkes, H. E., 1957, Principles of Geochemical Prospecting, *U. S. Geol. Surv. Bull.*, **1000-F**, 225–355.

Hawkes H. E., and Lakin, H. W., 1949, Vestigial zinc in surface residuum associated with primary zinc ore in east Tennessee, *Econ. Geol.*, **44**, 286–295.

Hawkes, H. E., and Webb, J. S., 1962, *Geochemistry in Mineral Exploration*, Harper & Row, New York, 415 pp.

Horsnail, R. F., and Elliot, I. L., 1971, Some environmental influences on the secondary dispersion of molybdenum and copper in western Canada, *Geochemical Exploration* (Boyle, tech. ed.), Special Vol. No. 11, Canadian Institute of Mining and Metallurgy, pp. 166–175.

Horsnail, R. F., Nichol, I., and Webb, J. S., 1969, Influence of variation in the surface environment on metal distribution in drainage sediments, *Quart., Colo. School Mines*, **64**, No. 1, 307–322.

Huff, L. C., 1960, Comparison of soil analysis with other prospecting methods at a small high-grade copper lode, *Intern. Geol. Congress, 22nd Symp. de Exploracion Geoquim., Ciudad de Mexico*, Tomo III, 487–500.

James, C. H., 1967, The use of the terms "primary" and "secondary" dispersion in geochemical prospecting, *Econ. Geol.*, **62**, 997–999.

Kauranne, L. K., 1967a, Trace element concentration in layers of glacial drift at Kolima, central Finland, in *Geochemical Prospecting in Fennoscandia* (Kvalheim, ed.), Interscience, New York, pp. 181–192.

Kauranne, L. K., 1967b, Prospecting for copper by geochemical and related methods at Kotanen, Viitasaari, central Finland, in *Geochemical Prospecting in Fennoscandia* (Kvalheim, ed.), Interscience, New York, pp. 255–260.

Kilpatrick, B. E., 1969, Nickel, chromium and cobalt in tropical soils over serpentinites, Northwest District, Guyana, *Quart. Colo. School Mines,* **64**, No. 1, 323–332.

Kroon, T. P. and de Grys, A., 1970, A geochemical drainage survey in central Ecuador, *Econ. Geol.,* **65**, 557–563.

Lag, J., 1967, Soils of Fennoscandia and some remarks on the interference of the soils in geochemical prospecting, in *Geochemical Prospecting in Fennoscandia* (Kvalheim, ed.), Interscience, New York, pp. 85–95.

Lounamaa, J., 1967, Trace elements in trees and shrubs growing on different rocks in Finland, in *Geochemical Prospecting in Fennoscandia* (Kvalheim, ed.), Interscience, New York, pp. 287–317.

Marmo, V., 1958, On biogeochemical prospecting of ores, *Intern. Geol. Cong., 22nd Symp. Exploracion Geoquim., Ciudad de Mexico,* Tomo I, 223–234.

Nigrini, A., 1971, Investigation into the transport and deposition of copper lead and zinc in the surficial environment (Abstr.), *Geochemical Exploration* (Boyle, tech. ed.), Special Vol. No. 11, Canadian Institute of Mining and Metallurgy, p. 235.

Ovchinnikov, L. N. and Grigoryan, S. V., 1971, Primary halos in prospecting for sulphide deposits, *Geochemical Exploration* (Boyle, tech. ed.), Special Vol. No. 11, Canadian Institute of Mining and Metallurgy, pp. 375–380.

Rostad, O. H., 1971, Offset geochemical anomalies at the Ima Mine, Lemhi County, Idaho, *Geochemical Exploration* (Boyle, tech. ed.), Special Volume No. 11, Canadian Institute of Mining and Metallurgy, pp. 241–246.

Salmi, M., 1959, On peat-chemical prospecting in Finland, *Interna. Geol. Congr. 22nd Symp. Exploracion Geoquim., Ciudad de Mexico,* Tomo II, 243–254.

Salmi, M., 1967, Peat in prospecting, Applications in Finland, in *Geochemical Prospecting in Fennoscandia* (Kvalheim, ed.), Interscience, New York, pp. 113–126.

Sutherland-Brown, A., 1967, Investigations of mercury dispersion haloes around mineral deposits in Central British Columbia, Geological Survey of Canada, Department of Energy, Mines and Resources Paper 66-54, pp. 73–83.

Timperly, M. H., Brooks, R. R., and Peterson, P. J., 1970, Prospecting for copper and nickel in New Zealand by statistical analysis of biogeochemical data, *Econ. Geol.,* **65**, 505–510.

Vol'fson, F. I., 1955, in Russian, Certain regularities in the distribution of endogene deposits of different genetic distribution (In Problems of Study of Structure of Ore Fields and Formations), *Tr. Inst. Geol. Nauk, Akad. Nauk SSSR,* **162**, 5–24.

Webb, J. S., 1958, Notes on geochemical prospecting for lead-zinc deposits in the British Isles, Technical aids to exploration, Symposium on the Future of Non-Ferrous Mining in Great Britain and Ireland, Institute of Mining Metallurgy, London, Paper 19, pp. 23–40.

Weiss, O., 1971, Airborne geochemical prospecting, *Geochemical Exploration* (Boyle, tech. ed.), Special Vol. No. 11, Canadian Institute of Mining and Metallurgy, pp. 520–514.

8

THE SELECTION AND
EVALUATION OF THE
DESIGNATED ANOMALY

The anomaly in geochemical prospecting is a deviation (generally positive) from the measured characteristics that are considered "normal" for a given area of geochemical–geological–geomorphological environment. The background value, the limits of local and regional fluctuations of the background value (thresholds), and the value of a measurement above which a geochemical concentration may be considered abnormal (the anomaly) can be determined by evaluating a series of numbers by means of simple statistics. Although statistical data are being used in geochemical prospecting with increasing enthusiasm (very often in areas already known for their economic exploitation), it should be emphasized that these should serve as guides only in conjunction with visual appraisals (by means of geochemical profiles, line densities, circle diameters, size of triangles, and other cartographic models) of geochemical maps in a qualitative assessment of available information. Nonetheless, most important for any geochemical exploration program is a knowledge of the geology and the mineralization of the prospect zone both from published and unpublished studies and from the observations and descriptions made in the field by the project geologist–geochemist who planned and directed the sampling phase.

The basis for the geochemical-statistical studies is the construction of the relative frequency histogram that is also known as the "density function" (frequency), or as will be used in this exposition, "the distribution." A normal distribution (Gaussian) has the graphic form of a symmetrical bell and is defined by the expression

$$y = \frac{1}{s\sqrt{2\pi}}\, e^{-(x-\bar{x})^2/2s^2}$$

in which y (with a linear scale) = the height of the curve at any point along
the x scale (also linear)

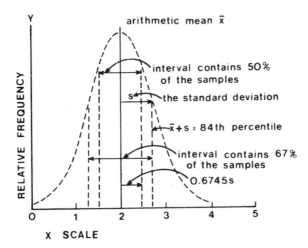

Figure 8-1. Normal (Gaussian) distribution showing the graphical positions of some statistical parameters. (From Krumbein and F. A. Graybill, 1965, *An Introduction to Statistical Models in Geology* with permission of McGraw-Hill Book Company. Copyright © 1965 by McGraw-Hill, Inc.)

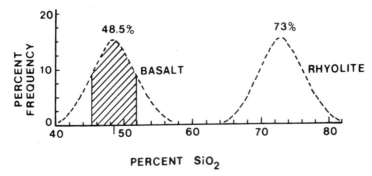

Figure 8-2. Frequency curves of silica distribution in basalts and rhyolites. Both distributions are bell-shaped symmetrical and are typical of normal distribution. The ruled area represents two thirds of the total area under the curve and corresponds to the limits $\bar{x} \pm s$. The arithmetic means are 48.5 and 73% for basalt and rhyolite, respectively. (From Ahrens, 1965, *Distribution of the Elements in Our Planet,* with permission of McGraw-Hill Book Company. Copyright © 1965 by McGraw-Hill, Inc.)

$\bar{x}$ = the average or the arithmetic mean
s = the standard deviation (see Figure 8-1)
However, in geochemistry, experience has demonstrated that there are relatively few cases of normal distribution, with two notable exceptions being the distribution of silica in basalts and rhyolites (see Figure 8-2). For a natural pop-

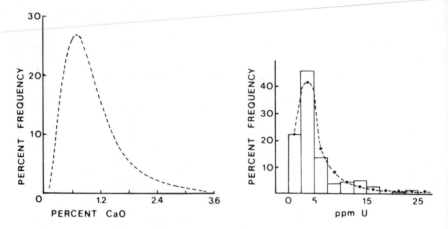

Figure 8-3. Frequency distribution curves showing positive asymmetry that is most common in geochemistry. (*A*) The distribution of CaO in 141 granite samples. (From Ahrens, 1965, *Distribution of the Elements in Our Planet* with permission of McGraw-Hill Book Company. Copyright © 1965 by McGraw-Hill, Inc.) (*B*) The distribution of U in 164 samples of granite from Spain; here we also see the histogram from which the frequency curve was drawn. (From Nicolli, 1969.)

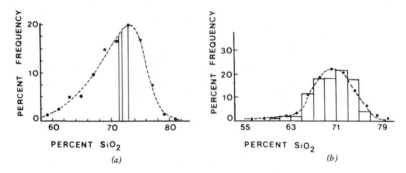

Figure 8-4. Frequency distribution curves for silica in granites, showing negative asymmetry that is not common in geochemistry. (*A*) The distribution of silica in 797 granite samples. (From Ahrens, 1965, *Distribution of the Elements in Our Planet* with permission of McGraw-Hill Book Company. Copyright © 1965 by McGraw-Hill, Inc.) (*B*) The distribution of silica in 164 granites from Spain; here we also see the histogram from which the frequency curve was drawn. (From Nicolli, 1969.)

ulation, the geochemical distribution of the chemical elements in rocks, sediments, soils, and waters, most often approximates a lognormal distribution (Ahrens, 1954, and others), although Hawkes and Webb (1962) examined studies of many authors and noted that the natural distribution in geochemistry is not necessarily normal or lognormal, but perhaps has its own independent form of distribution.

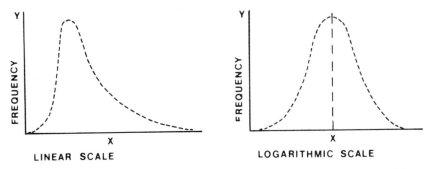

Figure 8-5. The logarithmic transformation. The distribution of *x* is lognormal, while the distribution of log *x* is normal; a curve of arithmetic distribution for a lognormally distributed population is asymmetrical, but the logarithmic distribution has a symmetrical bell shape.

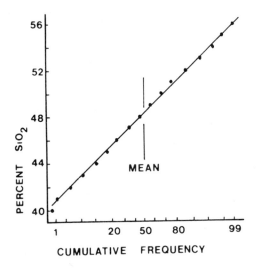

Figure 8-6. Normality proof for the distribution of silica in 401 basalt samples. A straight line accommodates the plotted points well when an arithmetic scale is used with probability paper; thus, the population can be considered normal. The cumulative frequency scale is on the abscissa, but it can also occupy the ordinate (see Figure 8-10). (From L. H. Ahrens, 1965, *Distribution of the Elements in Our Planet,* with permission of McGraw-Hill Book Company. Copyright © 1965 by McGraw-Hill, Inc.)

If a plot of the lognormal distribution of a natural population is made using linear (arithmetic) scales, an asymmetrical curve (skewed positive or negative) is obtained (see Figures 8-3 and 8-4), with the positive asymmetry being the most common in geochemistry (Ahrens, 1965). But if the relative frequency of the same population is plotted on a linear scale ordinate against the geo-

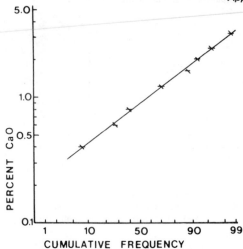

Figure 8-7. Lognormality proof for the distribution of CaO in granites (see Figure 8-3*A* for frequency distribution curve). A straight line accommodates the plotted points well when a log scale is used with probability paper. (From L. H. Ahrens, 1965, *Distribution of the Elements in Our Planet,* with permission of McGraw-Hill Book Company. Copyright © 1965 by McGraw-Hill, Inc.)

chemical value on a logarithmic scale abscissa, the resulting curve could approximate a symmetrical bell shape, symmetrical around a geometric mean instead of an arithmetic mean as in a normal distribution (see Figure 8-5).

Instead of preparing a histogram and tracing its corresponding curve to determine the normal or lognormal nature of a geochemical distribution, the normality, or in the common geochemical case, the lognormality of data from a population can be tested using probability paper plots. For a proof of normality, the probability paper used has an ordinate with a linear scale, and for the proof of lognormality, the ordinate has a logarithmic scale; the probability scale of the abscissa serves for the plotting of the cumulative frequencies. With a normal distribution, the data plot on probability paper with a linear scale will show nearly a straight-line relation (see Figure 8-6); in a lognormal distribution, the data plot will approximate a straight line when the probability paper has a logarithmic scale (see Figures 8-7 and 8-8). If there is a sampling of more than one, or mixed populations, the probability paper plot will show more than one nearly straight line (see Figures 8-9 and 8-10).

The most critical figure necessary for establishing values for the background, local and regional variation (thresholds), and possible and probable anomalies in a series of geochemical concentration values is the standard (or typical) deviation, which can be calculated by the equation

$$s = \pm \sqrt{\frac{\Sigma d^2}{n-1}}$$

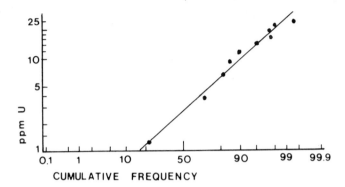

Figure 8-8. Lognormality proof for the distribution of U in 164 granites from Spain (see Figure 8-3*B* for the frequency distribution curve and the histogram). A straight line accommodates the plotted points well when a log scale is used with probability paper. From Nicolli (1969).

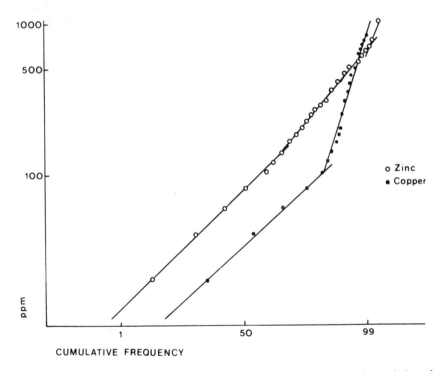

Figure 8-9. Example of lognormal distribution of more than one statistical population of geochemical data for Zn and Cu in stream sediments from more than 2100 sites in Central Ecuador. (Reproduced from Kroon and de Grys, 1970, *Econ. Geol.*, **65**, 560.)

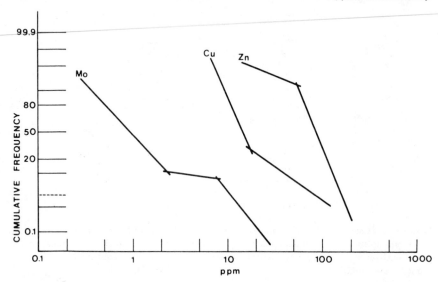

Figure 8-10. Proof of lognormality of two populations for Cu and Zn and of two populations plus a mixed population for Mo in Guatemalan stream sediments. (Reproduced from Lepeltier, 1969, *Econ. Geol.,* **64,** 545.)

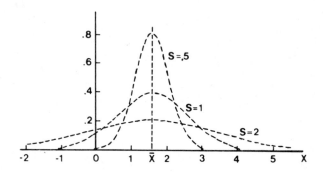

Figure 8-11. Examples of normal curves with the same mean and different values of the standard deviation *s.* (From Krumbein and Graybill, 1965, *An Introduction to Statistical Models in Geology,* with permission of McGraw-Hill Book Company. Copyright © 1965 by McGraw-Hill, Inc.)

in which d = the deviation from the average $\bar{x}$ of the value of each measurement in the population

  n = the number of observations

The standard deviation can be very different in populations that have the same mean (see Figure 8-11). According to classical statistics, in a normal distribution, 68.26% of the population will have values between the average $\bar{x}$ and $\pm$ s,

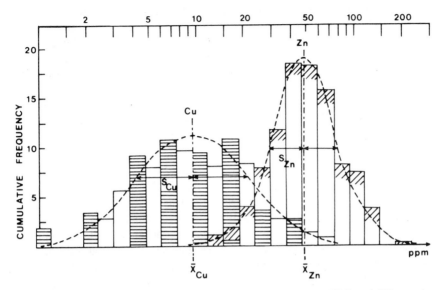

Figure 8-12. Proof of lognormality for Cu and Zn distributions in 2778 and 989 samples, respectively, of stream sediments from the Motagua drainage basin, Guatemala, by means of a logarithmic transformation; the resulting bell-shaped frequency curves are fairly well defined. (Reproduced from Lepeltier, 1969, *Econ. Geol.*, **64**, 543.)

95.44% of the population will have values between $\bar{x}$ and $\pm\, 2s$, and 99.74% of the population will have values between $\bar{x}$ and $\pm\, 3s$. This means that in a geo-chemical measuring in a population of 1000 samples, 682 samples will theoretically have values of a measured quantity between the average and plus and minus one standard deviation. Because we are most interested in positive values in geochemical prospecting, the above sentence can be restated such that 159 samples will have values greater than $\bar{x} + s$. Using this approach, 23 samples will have values greater than $\bar{x} + 2s$, and 1 sample will (theoretically) have a value greater than $\bar{x} + 3s$.

As indicated previously, the asymmetrical diagram of a lognormal distribution (plotted on linear scales) can be normalized to assume a symmetrical bell shape by means of a logarithmic transformation (using the logarithmic scale on the abscissa for the plotting of elemental concentrations (see Figures 8-5 and 8-12). The geochemical data can then be analyzed using the standard deviation, as in a normal distribution, which has already been discussed in the last paragraph.

In geochemical prospecting it is most common to select as the background value the figure that corresponds to $\bar{x} + s$, and as the limit of regional and local fluctuations (the threshold) the figure corresponding to $\bar{x} + 2s$. The measurements that exceed $\bar{x} + 2s$ can be considered as the geochemical anomalies; the areas where samples give anomalous values merit an intensified investigation. All exploration geochemists do not use the same method to select the geochemical anomaly. For example, Hawkes and Webb (1962) indicated that values greater than $\bar{x} + 2s$ but less than $\bar{x} + 3s$ can be classified as possible anomalous concentrations and that values greater than $\bar{x} + 3s$ can be considered as probable anomalous figures; when we remember that in a population of 1000 samples only one sample will theoretically have a value greater than $\bar{x} + 3s$, it would seem that the evaluation of Hawkes and Webb is too restricted.

In their study on the geology, mineral deposits, and geochemical and radiometric anomalies of the Serpentine Hot Springs area, Seward Peninsula, Alaska, Sainsbury et al. (1970) determined background values for various metals and then calculated numerical values for the designated anomaly by dividing the total content of a metal by its background; this procedure results in a ratio in which the background is represented by the number 1. These authors also give a total anomaly for each sampling site by adding the numbers that represent the concentration ratios and subtracting from them the sum of the backgrounds; thus, if one is working with 15 ppm of Mo (over a background of 3), 100 ppm of Sn (over a background of 15), and 15 ppm of Ag (over a background of 1), the concentration ratios are 3, 6.6, and 15, which total 24.6, and this figure minus 3 (the sum of the backgrounds of the concentration ratios) gives the magnitude of the total anomaly, 21.6, for Mo, Sn, and Ag at the sampling site.

At an earlier date, DeGeoffroy, Wu, and Heins (1968) described a method that they used to select geochemical targets in the metallogenic zinc area of Southwest Wisconsin. The method consisted of several steps:

1. Selection of areal reference units.

2. Selection of significant data based on the calculation of a trend of the anomalies and of the residual components.

3. Evaluation of significant data by means of an index.

4. Selection and plotting of geochemical targets for test drilling.

The authors separated the nonsystematic local changes in the concentration of zinc from the continual regional changes at any geographic point P_i by the equation:

$$O_i = T_i + A_i + R_i$$

in which O_i = the observed value

T_i = the regional component that varies in a continual manner in the area

A_i = the discontinuous anomaly

R_i = the residual component

When P_i is changing, the other values are changing with the development of a trend surface and anomaly surfaces that are combined then to give an approximate representation of the Zn surface.

The data designated as significant are the positive deviations over the anomalies.

In the above study, the trend component is the arithmetic mean of Zn in a section, the anomaly component is the arithmetic mean of the deviations over the tendency in a cell, and the residual component is the positive deviation over the anomaly. The authors detailed how calculations are made for each step in the selection of a target and observed that the method can be used in metallogenic zinc areas adjacent to the zone already studied and in areas of similar geology and mineralization.

Bolviken (1971) considered techniques being used to establish positions and levels of anomaly values and wrote that because the concepts of background and anomaly can be considered as statistical distributions, both should plot as straight lines on probability paper. In his paper he proposed that background and anomaly for a specific geochemical element in a specific zone be defined as follows:

1. Background is the distribution to which the majority of observations in a set of random samples from the zone belongs.

2. The anomaly is any distribution differing from the background.

Thus, according to the definitions set up by Bolviken, the concept of threshold loses meaning, and he suggests that the concept of anomaly significance or contrast can be developed from some statistical test such as that being tested at the Geological Survey Norway. According to the technique, the empirical population is cumulated and plotted on probability paper, a precalculated combination of distributions that can be best fitted to the empirical population is obtained from an atlas of nomograms, and the background and anomaly are estimated as the two single distributions of the nomogram from which there may also be an indication of the relative portions of background and anomaly in the joint distribution. Although the method indicates how the sample set can be divided between background and anomaly, the problem of determining which samples of the set belong to the background and which belong to the anomaly has not yet been resolved and is still being investigated.

If we accept the concept that the major part of significant geochemical targets will be indicated if a limit of $\bar{x} + 2s$ is established for the anomaly, we can use the graphical method proposed by Lepeltier (1969) to determine the values for $\bar{x}$, $\bar{x} + s$, $\bar{x} + 2s$, and $\bar{x} + 3s$. Lepeltier prepared a graph (see Figure 8-13) by plotting the chemical element concentrations on the logarithmic scale

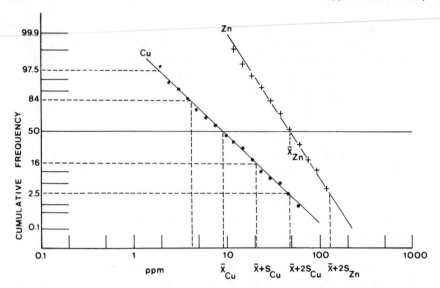

Figure 8-13. Probability paper lognormality proof of the same data in Figure 8-12. The two straight lines accommodate the plotted points well so that the population can be considered lognormal. (Reproduced from Lepeltier, 1969, *Econ. Geol., 64,* 541.)

of probability paper against the cumulative frequencies (on the probability scale) after selecting the logarithmic interval for the classes (of element concentration) that were to be plotted. The selection is of utmost importance in the preparation of a histogram or a plot of cumulative frequencies against intervals of measured values. The use of too few intervals can hide important characteristics in the distribution of a population and the use of too many intervals can obscure details in a very detailed scheme. Because lognormal distribution is very common in geochemistry, Shaw (1964) proposed the rule that the width of an interval, expressed in logarithmic form, should be equal to or less than 0.5 of the standard deviation. Lepeltier (1969) suggested that it is convenient to work with a minimum of 9 to a maximum of 19 intervals on the line of cumulative frequencies. In his work with Guatemalan stream sediments, he chose the logarithmic limits of the intervals using the relation:

$$\text{Logarithmic interval} = \frac{\text{logarithm } R}{n}$$

in which R = the ratio between the maximum and minimum value of the
         population
      n = the number of classes with which one works
Since the average experimental values of R generally vary from 6 to 300 when

n varies between 10 and 20 classes and the logarithm of R is between 0.78 and 2.48, the limiting values for the logarithmic intervals have been calculated as:

$$\text{Logarithmic interval} = \frac{0.78}{20} = 0.039$$

$$\text{Logarithmic interval} = \frac{2.48}{10} = 0.25$$

For the Guatemalan study, Lepeltier used a logarithmic interval of 0.10 because this gave good definition of the distribution curve and served as an excellent framework for the majority of the distributions studied. With the graph illustrated in Figure 8-13, therefore, one can rapidly find the background, threshold, and other values mentioned previously. The figure corresponding to 50% of the cumulative frequencies is equal to the geometric mean $\bar{x}$, that corresponding to 84% is equal to the background value $\bar{x} + s$, and that corresponding to 97.5% is equal to the limit of regional and local variation of the background $\bar{x} + 2s$; above this limit, a concentration may be considered a geochemical anomaly (see Figure 8-13).

Recently Polikarpochkin (1971) published an equation by which the geochemical anomalies of drainage basins can be differentiated into two groups. One group is possibly related to mineral deposits and requires more field investigation, and the other is related to other factors and can be dismissed from consideration. In this equation,

$$P_n = k'P'(L_{2n} - L_{1n}) \frac{1}{e^{-\sigma(X-L_{2n})} - e^{-\sigma(X-L_{1n})}} \frac{\phi(X)}{f(X)}$$

$$- \sum_{i=1}^{n-1} P_i \frac{L_{2n} - L_{1n}}{L_{2i} - L_{1i}} \frac{e^{-\sigma(X-L_{2i})} - e^{-\sigma(X-L_{1i})}}{e^{-\sigma(X-L_{2n})} - e^{-\sigma(X-L_{1n})}}$$

$$f(X) = \frac{1}{2} X^2; \qquad \phi(X) = X - \frac{1}{\sigma}(1 - e^{-\sigma X})$$

$$f(X) = X; \qquad \phi(X) = 1 - e^{-\sigma X}$$

in which P' = the productivity of the dispersion train

P = the productivity of a mineral deposit for a particular element

k' = the proportionality coefficient

x = the distance of each sampling site from the head of a basin

L_1 and L_2 = the distances of the upper and lower limits of an ore-bearing section measured from the head of a basin

n = the number of sections with ore, counting from the head of a valley

σ = the dynamic parameter

e = the base of natural logarithms

According to Polikarpochkin, if the values of P', L_1, L_2, and σ are known and the reserves determined for at least one mineral deposit, the values of k' can be calculated in absolute figures from the sampling of the dispersion trains. If k' cannot be calculated, the productivity of a mineral deposit can be presented in conditional units of P/k'; this is considered a valid method for a comparison of data, but not for obtaining absolute figures for the productivity potential of a zone.

Complementing the basic statistical methods used in geochemical prospecting are the so-called sophisticated statistical methods that aid us in locating the subtle geochemical anomalies and elemental trends not easily recognizable in any other way. In the group of sophisticated methods, the two most used today are trend-surface analysis and factor–vector analysis, both of which require a computer to handle the data because of the complicated mathematical calculations involved. The two techniques have their best application in the interpretation of regional geochemical data.

Nichol, Garrett, and Webb (1969) described trend surface analysis as a procedure by which the general trend of an area (the background) is separated from random local variations in a series of data in the form of a trend surface and some residuals. The precise form of the calculated surface is that in which the sums of the squares of the residuals (or the differences between the observed and computed values) have minimum values. Thus, the trend surface contains the systematic or regional component of the variations of the geochemical data, and the residuals include the local (or erratic) geochemical values not related to the principal elemental distribution. In this way the chemical signals that may be caused by ore deposits are separated from other chemical signals and random "noise." As Cameron and Hobbs (1971) wrote, the trend surface eliminates the geochemical variation which is attributed to such factors as regional lithologic changes, climate, vegetation, and topography.

This is then a polynomial surface fitting, that is, a type of regression analysis by which a level of confidence related to each surface by variance analysis can be established. The designated residuals (or the deviations) of the polynomial surfaces are derived from two components (Nichol, Garrett and Webb, 1969):

1. The error due to the sampling and the chemical analysis (already cited in the section on significance of the chemical analysis).

2. A real variation between the determined value and the computed surface at the sampling site.

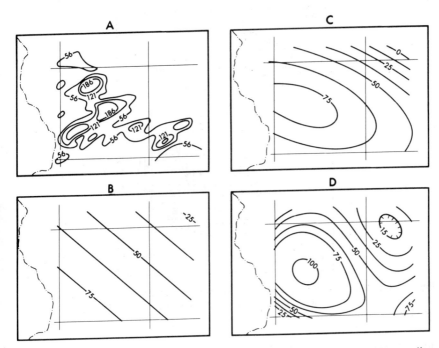

Figure 8-14. Geochemical maps of the Park City area, Utah: (A) geochemical anomalies for Cu; (B) linear trend surface for Cu; (C) quadratic trend surface for Cu; (D) the cubic trend surface for Cu. (Reproduced from Nackowski et al., 1967, *Econ. Geol.*, **62**, 1080, 1084.)

In this way, a limiting value layer that can be attributed to the error in sampling and in the chemical analysis can be established for each side (±) of the computed surface; the residuals outside the limits of the layers are those that have the greatest significance. In many recent investigations, these limits are $\bar{x} \pm 2s$, and, of course, the economic geochemist is most interested in the geochemical anomalies greater than $\bar{x} + 2s$.

In 1967, Nackowski, Mardirosian, and Botbol beautifully illustrated the utility of trend surfaces in geochemical prospecting. They compared geochemical maps prepared on the basis of trend surfaces with geochemical maps of isochemical contours indicating background areas, threshold areas, and areas with anomaly values, in the area of Park City, Utah. The elements studied were Cu, Pb, Zn, and Mn. Figure 8-14 presents a comparison of the different maps for Cu. It is clearly seen that although the isochemical contour map with the Cu anomalies (related to the regional structural geology) reflects mineralized structures with a northeast strike, the mineralized structures apparently have an en echelon form towards the southeast. It can also be seen that the gener-

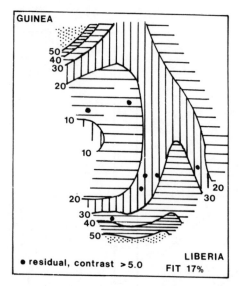

Figure 8-15. Cubic trend surface for Ni in stream sediments from Sierra Leone showing the residuals with a contrast >5.0. (Reproduced from Nichol et al., 1969, *Econ. Geol.,* **64,** 207.)

alized linear, quadratic, and cubic trend surfaces indicate that the strike of the Cu anomalies more correctly has an orientation towards the southeast.

Nichol, Garrett, and Webb (1969) used trend-surface analysis (and other statistical methods) to focus attention on local areas in which the geochemical data deviated significantly from the general trend in the Sierra Leone Basement Complex (an area of more than 40,000 km^2). In the case of Ni in stream sediments, the regional components of local variations that are not easily discernible visually in the graphical representation of the element concentration data via symbols are separated out by the third-order surface, thus highlighting significant residuals (see Figure 8-15). The authors feel that there is a possibility that the division of data into groups represented by symbols hid the apparent tendencies in the trend-surface maps, which themselves were prepared from the original chemical concentration information. The third-order surface computed from the grouped data is essentially equal to the map computed from the original data, thus indicating that the grouping of data before plotting the map does not eliminate any important areal trend. Nichol, Garrett, and Webb (1969) believe that the fact that the regional trend is not obvious at a visual inspection reflects the difficulty of the human brain to assimilate the original data and does not reflect on the form of presentation of these data.

Although statistical analysis by means of the confection of a trend surface reduced the random "noise" and reveals zones of consistent local variation,

Cameron and Hobbs (1971) note that the selection of the most appropriate form of the trend surface and of the parameters of the method for making geochemical contours is very subjective and requires a comparison of other relatively efficient methods for the fitting of surfaces. They suggested the potential use of a technique such as the Markov scheme, but also accept the use of other simpler methods to differentiate between chemical signals. If, for example, the rocks in a drainage basin can be divided into a few relatively simple homogeneous lithologies, the concentration of the elements in each sample can be normalized using the mean and the variance of all samples that lie over that particular type of rock.

Another basic calculation that is used in geochemical prospecting to establish the relation between a pair of factors (or two groups of values) is the correlation coefficient. This is determined by the equation:

$$\rho = \frac{\Sigma XY - \dfrac{(\Sigma X \cdot \Sigma Y)}{N}}{\left[\sqrt{\Sigma X^2 - \dfrac{(\Sigma X)^2}{N}} \right]\left[\sqrt{\Sigma Y^2 - \dfrac{(\Sigma Y)^2}{N}} \right]}$$

in which ρ = an estimation of the correlation coefficient

N = the number of samples of X and Y

ΣXY = the sum of the values of each X multiplied by the corresponding Y

$\Sigma X \cdot \Sigma Y$ = the sum of all the values of X multiplied by the sum of all the values of Y

ΣX^2 = the square of the sum of each value X

$(\Sigma X)^2$ = the square of the sum of all the values X

ΣY^2 = the square of the sum of each value Y

$(\Sigma Y)^2$ = the square of the sum of all the values Y

The resulting number, designated ρ, the estimation of the correlation coefficient, will have a value between 0 and ± 1.0. A correlation coefficient of 0 implies a random distribution without any relation between the factors being compared. A value of +1.0 implies a perfect correlation, whereas −1.0 indicates a perfectly invariant relation between the factors. If the correlation coefficient is squared and multiplied by 100, the resulting percentage is suggestive of the variation of the data and is called the variance percentage.

Lepeltier (1969) proposed a graphical method to make an estimation of the correlation coefficient, and although it is less precise than the statistical calculation, it is more rapid and certainly merits consideration here. A "correlation cloud" is constructed on logarithmic coordinates by plotting the values of each sample of the population being studied; the plot that results appears as a cloud of dots. Next, the axes that pass through the point on the graph representative

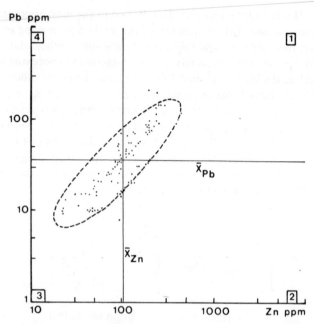

Figure 8-16. Correlation cloud diagram for the Pb–Zn couple for Guatemalan stream sediments. A rapid, acceptable estimate of the correlation coefficient can be made for the elements from such a plot. (Reproduced from Lepeltier, 1969, *Econ. Geol.,* **64,** 549.)

of the mean values of the two factors being analyzed are marked in parallel to the ordinate and abscissa, thus dividing the field of the plotted data into four quadrants (see Figure 8-16). The number of points falling in each quadrant are counted and later summed as follows:

N_1 = the number of points in the first and third quadrants
N_2 = the number of points in the second and fourth quadrants.

The correlation coefficient estimate is then calculated by the equation:

$$\rho = \text{sine} \left[\frac{\pi}{2} \frac{N_1 - N_2}{N_1 + N_2} \right]$$

For the example given in Figure 8-16, the correlation coefficient for Pb–Zn in 99 Guatemalan stream sediment samples is + 0.87, implying a very good correlation between the two elements from a polymetallic mineralization zone.

The methods described thus far are limited because they can be used for data available on a single chemical element or other variable in order to focus on an interesting geochemical concentration due to mineralization instead of any one of a number of possible sources of variation.

In recently published studies, multivariate techniques are being used and are based on the relations between concentrations of several chemical elements and other measurements that can be made on a collection of samples. Very much in use are factor–vector analysis and multiple regression analysis.

Rose, Dahlberg, and Keith (1970) approached the problem of different lithologies in the drainage basin of a geochemical prospecting area using multiple regression analysis. In addition to considering the mixing of all upstream lithologies of which stream sediments may be comprised, they also studied the effects of absorbent Fe and Mn oxides and of high-ion-exchange-capacity clay-size components in the sediments, as well as mentioning the probable effect of an organic material contribution as a metal carrier, but without making an estimation of the importance of such a contribution. The effect of each lithology was quantified according to its percentual areal presence in the drainage basin and Zn, Cu, Ni, Co, Cr, and V were expressed in regression equations as functions of these lithologic factors. In this way, 20 to 70% of the total variance of each element was explained. The content of Fe is very notable because of its influence on the elemental variation and thus may be a very important factor to be considered when using stream sediments as samples in geochemical prospecting. In multiple regression analysis, the background changes with the geological conditions and one works with a corrected background value for each sampling site. Therefore, instead of making a two-dimensional geometric space analysis, as is the case in trend surface analysis, we work with n-dimensional lithological–chemical space. After quantifying the lithologic effects in the study zone (of about 19,000 km^2), Rose, Dahlberg, and Keith determined the concentrations of Cr, Co, Cu, Fe, Mn, Ni, and V in the -100 mesh size fraction of 2800 samples and divided the total study area into cells for which a relief factor was determined by calculating an average of the maximum relief plus the minimum relief for each cell. Because the elements had distributions that approximated lognormal, the arithmetic concentrations were transformed to logarithmic values for the multiple regression analysis. Twenty-nine variables entered the analysis (see Table 8-1). The selection of the positive limit of the regional and local fluctuations (thresholds) of the background is established by the following equation:

$$V_m = \frac{\Sigma N(X_{im} - X_{im}')^2}{N - t - 1}$$

in which V_m = the variance of the elemental concentration m around the regression

X_{im} = the observed content in the cell

Table 8-1 The 29 Variables Used in the Regression to Correct Background Values in Stream Sediments from Pennsylvania (from Rose, Dahlberg, and Keith, 1970)

Variable Number	Description	Variable Number	Description
1	Zn (ppm)	17	Cambrian sandstone and shale
2	Cu (ppm)		
3	Ni (ppm)	18	Cambrian–Ordovician carbonates
4	Co (ppm)		
5	Cr (ppm)	19	Ordovician shale
6	V (ppm)	20	Ordovician–Silurian sandstone and shale
7	Mn (ppm)	21	Silurian red beds
8	Fe (%)	22	Silurian–Devonian brown to gray clastics
9	Graphitic gneiss and granite gneiss		
10	Gabbroic rocks	23	Devonian red beds
11	Ultramafic rocks	24	Devonian–Mississippian shale and sandstone
12	Granitic rocks		
13	Metabasalt	25	Mississippian red beds
14	Metarhyolite	26	Pennsylvanian sediments
15	Wissahickon schist and related rocks	27	Triassic sediments
		28	Triassic diabase
16	Wissahickon volcanics	29	Relief (difference between highest and lowest elevation in cell)

Source. Reprinted from *Economic Geology*, **65**, 162 (1970).

$X_{im'}$ = the calculated content for the cell based on the observed values of the independent variables
N = the total number of cells
t = the number of statistically significant coefficients in the regression equation

From this calculation, the standard error of the estimation, a factor analogous to standard deviation, is determined:

$$\sqrt{V_m} = S_m$$

It was arbitrarily established that when $(X_{im} - X_{im'}) > 2S_m$, there is an anomalous number, that is, the so-called anomaly. The authors of this method indicated that the selection of variables and the form in which they enter into the regression constitutes a great problem and that the geology of the study zone must be well known before using the method.

In accordance with CEGS (1968), factor and vector analysis has the purpose of finding the minimum number of variables that are necessary to explain the maximum percentage of variation observed in the data, identifying those variables of major importance, and determining the relative contribution of each one in terms of the variables that control the variation. The variables can be represented by vectors in a rectangular coordinate system in which the length of a vector is equal to the magnitude of the variable. The correlation between one variable and another is given by the cosine of the angle that relates the two factors being considered, and the total process has as an objective the expression of the relations of one group of variables by a minimum number of causal or control variables that are called factors. There are two types of factor analysis and these have been designated as "*R*" and "*Q*." The *R* factor analysis results in a comparison of relationships among the variable parameters in terms of the samples and thus is used to identify and highlight the dominant ion associations in the data. The *Q* analysis (very much used in geological studies) consists of making a comparison of sample pairs in terms of the variables, and from this comparison, maps can be made from geologic data (Krumbein and Graybill, 1965) which include for each sample its geographic coordinates and a determined composition so that the numerical value of each factor at each location can be plotted on a base map and subsequently contoured. Factor maps are distinct from maps of individual components because they project which are the assemblages of chemical elements in rocks (with hypothetical compositions) that can be combined to give deposits characteristic of the composite environmental effects instead of individual environmental effects.

In order that this type of analysis be used properly, some distribution schemes of the chemical elements (major, minor, or trace) in several assemblages of mother rocks (hypothetical) must be established for a principal environment and for a secondary environment (or the environment that represents the mineralization). Then the natural factors that influence the chemical composition of geological samples in a geochemical prospecting area are sequentially extracted. Each factor extracted represents the variation that is due to a single geological-geochemical process, remembering that chemical concentration of an element more commonly represents the sum of several geological and geochemical processes. As Nichol, Garrett, and Webb (1969) have pointed out, the contribution of each factor to the total composition in mixtures of several populations, each with its own background value (one factor), can be determined in this way and the content of the minor or trace elements that is due to some "abnormal" factor such as mineralization can be established. Generally, many factors are necessary to explain 100% of a multielemental distribution, but a few can project forward as important and indicate the minor role of other factors being considered in the problem. The areas where a composition may be found that may be due to an abnormal factor should be selected for more detailed geochemical prospecting.

Cameron and Hobbs (1971) wrote that the influence of each factor in each sample can be computed numerically and the resulting data analyzed by means of a trend surface, contoured and plotted in the same way as the original data comprised of element concentrations are contoured and plotted. It might sometimes be useful to apply this same type of analysis to the residuals of a trend surface because the trend surface may be efficient in extracting some "factors" and thus facilitate the work of the analyses by the multiple-variable processes. However, these same authors note that one cannot always depend on factor analysis for the interpretation of data from a geochemical prospecting program because factor-analysis methods are designed to extract an assemblage of factors that influence a single statistical population. Thus, if the area studied is heterogeneous both in geology and in the types of mineral deposits included, there are going to be a great number of different processes or different factors that will act on the same variables; this may be beyond the capability of factor analysis to resolve. An interpretation made on the basis of factor analysis in a case of heterogeneity of geology and type of mineral deposits can be very erroneous. Cameron and Hobbs (1971) insisted that it is necessary to determine, in this case, the quality of the interpretation by the evaluation of the factors that influence element concentration in the subareas of the major zone and determine if the matrices of the factors used to explain the element variations in the entire area are relatively constant with those of the subareas. They noted that another difficulty exists in using factor analysis in geochemical reconnaissance studies: the determination of only six or eight chemical elements may be inadequate for significantly resolving the factor structure because this is maintained constant and does not depend on the number of elements that are determined. Factor analysis has maximum applicability and utility when more factors are used.

Although it seems obvious, it must be noted that the best results of surface analysis are obtained when the areal distribution of the samples is evenly spaced and when the sample grid is optimized with respect to the probable size of the target and the resulting areal anomaly, and, of course, with respect to the economic limitations of the program. Cameron and Hobbs (1971) suggested that the major, minor, and trace elements should be analyzed simultaneously (perhaps with an emission spectrograph or a direct reader emission spectrometer) because when such data are subjected to statistical study, the effect of variation of the indicator (pathfinder) elements in trace amounts due to variable amounts of dominant minerals in subsamples is considerably reduced. We know, for example, that some elements with statistically strong or very strong dispersions in igneous rocks (trace elements) are, in general, very concentrated in accessory minerals, as Nicolli (1969) demonstrated very well for U in zircon.

One of the better applications of factor and vector analysis found in the geochemical prospecting literature is that by Nichol, Garrett, and Webb (1969) on a pilot area of approximately 75 km^2 in a mineralized band of schists from Sierra Leone where there are different types of rocks present: metamorphosed ultrabasic sediments and metasediments surrounded by granitized gneisses. In the investigation the authors determined 13 variables (As, Co, Cr, Cu, Ga, Mn, Mo, Ni, Pb, Sn, Ti, V, and Zn). The metal distribution in the stream sediments analyzed is notable because of the enrichment of Co, Ni, Cr, Mn, Cu, and V in sediments derived from the ultrabasic schists relative to the concentration of these elements in sediments derived from the metasediments which in turn have more elevated contents of these elements than the sediments derived from granites. The data were subjected to the Q type factor–vector analysis. Thirteen factors were required to explain 100% of the multielement distributions, but 3 factors alone explained 89.3% of the variation, thus indicating the less important role of the other 10 factors. Because the factors represent rocks with hypothetical chemical compositions, the samples that most approximated these compositions were selected as vectors, and then the variation of the minor elements in all the samples was computed as a function of various mixtures of the three vectors. The two-factor model accounts for 84.2% of the variation, and of this, 76.2% is explained by the first factor, which has a granitic composition; the second factor represents another 8.0% of the variation and has an ultrabasic composition. In the three-factor model, 89.2% of the variation of the minor elements is explained, with the first two factors corresponding to the granitic and ultrabasic compositions, but the hypothetical rocks selected to serve as vectors show a more extreme total composition when considered in terms of distribution frequencies of the metal contents. This was necessary to explain a greater part of the variation of the minor elements than in the two-factor model. The ultrabasic vector contained greater concentrations of Cu, Co, Ni, Cr, V, Mn, and Zn than in the two-factor model (more basic) and the first vector (granitic) had greater contents of Cu, Co, Ni, Cr, V, Mn, and Zn than in the two-factor model; however, the third vector was characterized by very low contents of these elements and could be representative of a granite poor in the metals analyzed. Nichol, Garrett, and Webb (1969) demonstrated the soundness of the interpretation of data from the entire area by making Q factor analyses of two subareas whose two- and three-vector models gave interpretations similar to that one proposed for the entire area. Type R factor analysis was also used with the data and explained 67.5% of the variation of the minor elements with three factors (associations of metals): (1) high Ni with Co, Cu, V, Zn, Cr, Mn, and Ti; (2) Ga with Pb, Ti, and Sn; and (3) Mo and As with Cr and V. For this region of Sierra Leone, Nichol, Garret, and Webb demonstrated the way in which multielement data can be studied with factor analysis by depending on a limited number of factors representative of the trace-element content of individual samples. These geochemists believed that the most important factors are normally related to the type of mother rock, to a secondary environment, or perhaps to mineralization in a very mineralized area. In this way variations in the total trace-element content can be expressed by a geochemical affinity that can assist indirectly for geological mapping where there are few outcrops. In addition, they observed that if multivariate techniques are used with the more common statistics (described in previous paragraphs), very significant contributions for the interpretation of multielement data from regional geochemical reconnaissance programs can be made.

Another good example of the use of factor and vector analysis in geochemical prospecting is given in the investigation by Armour-Brown and Nichol (1970) for regional geochemical reconnaissance and the localization of metallogenic provinces in approximately 200,000 km² of eastern Zambia. Instead of following their usual technique of collecting 1 to 5 samples per 2.5 km² (1 square mile), a sampling density of 1 sample for each 125 to 250-km² unit (50 to 100 square miles) was used; this was deemed allowable because Garrett and Nichol (1967) showed that this low density of sampling can serve to define variations in backgrounds of metal contents with widely spaced stream-sediment sampling. They identified the dominant characteristics of metal associations that contribute to the composition of the stream sediments and redescribed the multielement data in terms of components of these metal associations. With factor analysis there was a marked simplification in working with the data. For example, in the case of the drainage basin samples, the distribution of 13 minor elements (Co, Cr, Cu, Fe, Ga, Mn, Ni, Pb, Sn, Sr, Ti, V, and Zn) was described as a function of eight associations that represented the different mother rocks and the secondary environment that controlled the distribution of the minor elements. With the data presented in this form, there was an impressive improvement in the clarity of the relation of the data with the causal factors. The eight-factor model explained a total of 88.4% of the variation of the data; this model was used because the metal associations appeared to be related to the various types of rocks, the secondary environment, and the mineralization. Armour-Brown and Nichol (1970) did not present any information on the analysis of subareas to demonstrate the soundness of their results for the entire area.

Some geochemists are reluctant to use the computer approach to analyzing prospecting data because of the investment of time and money that such use initially requires; perhaps it is also because of the possibility that they do not understand clearly how the computer functions and thus do not have confidence in the preliminary results and interpretations. This difficulty may be overcome by having the geologist(s) and/or geochemist(s) attend one of the many seminars on computers that are often being given in different locations. Later, a mathematician can be incorporated into the geochemical program and can help not only in the common statistics operations, but also in determining which of the many "canned" computer programs can serve or be adapted to the geochemical program being considered. As important as communication is between the geologist and the chemist in planning and carrying out of a geochemical prospecting program, the communication that should be established between the mathematician and other members of a geochemical team is also fundamental in order to present the best or the better interpretations of the basic data.

Whichever statistical method is used to evaluate geochemical prospecting data, it is subject to the geoscientist's geochemical and geological experience and the observations made in the field. As Hawkes and Webb (1962) have written, "It should be emphasized that although statistics may help in presenting and analyzing geochemical data, it cannot provide the interpretation. A reliable interpretation of anomalies in terms of ore requires a combination of complex human experience and a capacity to recognize significant geometrical correlations. Pure mathematical analysis, therefore, is not likely to replace the subjective interpretative talents of the exploration geologist for some time to come."

REFERENCES

Ahrens, L. H., 1954a, The lognormal distribution of the elements, *Geochim. Cosmochim. Acta,* **5,** 49–73.

Ahrens, L. H., 1954b, The lognormal distribution of the elements, *Geochim. Cosmochim. Acta,* **6,** 121–131.

Ahrens, L. H., 1965, Distribution of the Elements in Our Planet, McGraw-Hill, New York, 110 pp.

Armour-Brown, A. and Nichol, I., 1970, Regional geochemical reconnaissance and the location of metallogenic provinces, *Econ. Geol.,* **65,** 312–330.

Bolviken, B., 1971, A statistical approach to the problem of interpretation in geochemical prospecting, *Geochemical Exploration* (Boyle, tech. ed.), Special Vol. No. 11, Canadian Institute of Mining and Metallurgy, pp. 564–567.

Cameron, E. M., and Hobbs, D., 1971, Computerized methods for interpreting reconnaissance geochemical surveys, *Geochemical Exploration* (Boyle, tech. ed.), Special Vol. No. 11, Canadian Institute of Mining and Metallurgy, p. 569.

CEGS (Council on Education in the Geological Sciences), 1968, *Bibliography of Statistical Applications in Geology,* CEGS Programs Publ. No. 2, Washington, D.C., 24 pp.

DeGeoffroy, J., and Wu, S. M., 1970, Design of a sampling plan for regional geochemical surveys, *Econ. Geol.,* 65, 340–347.

DeGeoffroy, J., Wu, S. M., and Heins, R. W., 1968, Selection of drilling targets from geochemical data in the Southwest Wisconsin Zinc Area, *Econ. Geol.,* 63, 787–795.

Garrett, R. G., and Nichol, I., 1967, Regional geochemical reconnaissance in Eastern Sierra Leone, *Trans. Inst. Min. Met., Sect. B., Appl. Earth Sci.,* 76, B97–B112.

Garrett, R. G., and Nichol, I., 1969, Factor analysis as an aid in the interpretation of regional geochemical stream sediment data: *Quart. Colo. School Mines,* 64, 245–264.

Hawkes, H. E., and Webb, J. S., 1962, *Geochemistry in Mineral Exploration,* Harper & Row, New York, 415 pp.

Kroon, T. P., and de Grys, A., 1970, A Geochemical drainage survey in Central Ecuador, *Econ. Geol.,* 65, 557–563.

Krumbein, W. C., and Graybill, F. A., 1965, *An Introduction to Statistical Models in Geology,* McGraw-Hill, New York, 475 pp.

Lepeltier, C., 1969, A simplified statistical treatment of geochemical data by graphical representation, *Econ. Geol.,* 64, 538–550.

Miller, R. L., and Kahn, J. S., 1962, *Statistical Analysis in the Geological Sciences,* Wiley, New York, 483 pp.

Nackowski, M. P., Mardirosian, C. A., and Botbol, J. M., 1967, Trend surface analysis of trace element data, Park City District, Utah, *Econ. Geol.,* 62, 1072–1087.

Nichol, I., Garrett, R. G., and Webb, J. S., 1966, Automatic data plotting and mathematical and statistical interpretation of geochemical data, *Proceedings of Symposium on Geochemical Prospecting,* (Cameron, ed.), Geological Survey of Canada, Department of Energy, Mines and Resources, Paper 66-54, pp. 195–210.

Nichol, I. Garrett, R. G., and Webb, J. S., 1969, The role of some statistical and mathematical methods in the interpretation of regional geochemical data, *Econ. Geol.,* 64, 204–220.

Nichol, I. and Webb, J. S., 1967, The application of computerized mathematical and statistical procedures to the interpretation of geochemical data, *Proc. Geol. Soc. London,* 1642, 186–198.

Nicolli, H. B., 1969, Distribucion estadistica de elementos en rocas graniticas, *Rev. Asoc. Geol. Arg.,* 24, 139–157.

Polikarpochkin, V. V., 1971, The quantitative estimation of ore-bearing areas by sample data of the drainage system, *Geochemical Exploration,* (Boyle, tech. ed.), Spec. Vol. No. 11, Canadian Institute of Mining and Metallurgy, pp. 585–586.

Rose, A. W., Dahlberg, E. C., and Keith, M. L., 1970, A multiple regression technique for adjusting background values in stream sediment geochemistry, *Econ. Geol.,* 65, 156–165.

Sainsbury, C. L., Hudson, T., Kachadoorian, R., and Richards, D., 1970, Geology, mineral deposits, and geochemical and radiometric anomalies, Serpentine Hot Springs Area, Seward Peninsula, Alaska, *U.S. Geol. Surv. Bull.,* 1312-H, H-1–H-19.

Shaw, D. M., 1964, Interprétation Geochimique Des Eléments En Traces Dans Les Roches Cristallines, Masson, Paris, 237 pp.

Sinclair, A. J., 1967, Trend surface analysis of minor elements in sulfides of the Slocan Mining Camp, British Columbia, Canada, *Econ. Geol.,* **62,** 1095–1101.

Vistelius, A. B., 1967, *Studies in Mathematical Geology,* Consultants Bureau, New York, 294 pp.

9

GEOCHEMICAL PROSPECTING
FOR HYDROCARBONS

The search for hydrocarbon deposits by means of geochemical prospecting is in reality a search directed first towards the localization of zones of maximum interest in areas of petroliferous (or natural gas) potential, although of no less importance is the actual find itself of the petroleum or natural gas area or of producing well sites. A region may be categorized as favorable for exploration for hydrocarbons after reviewing its geology (at the surface and perhaps in the subsurface), relating this geology to that of other producing regions, and evaluating direct indications of the existence of hydrocarbons such as filtrations of petroleum or natural gas. In general, the exploration phase of major initial importance in a hydrocarbon search project, after the geology is known, is the location of structures that are considered favorable for the trapping and accumulation of hydrocarbons; this step is carried out using geophysical techniques, especially seismology. Seismic projects are slow and costly; a seismic team can study or "make" some 2.5 to 6 linear kilometers per day at a cost that varies according to the conditions in the study zone and the country in which the work is being done, but one can estimate that the cost would be of the order of \$35,000 monthly (offshore seismic surveys may run as high as \$150,000 monthly). Because of such costs, it is important to consider any method, geochemical prospecting for example, that can delimit zones of most probability for hydrocarbons in a region of great areal extension during the exploration stage previous to the seismic study.

As is the case in seismic investigations, the cost of a geochemical exploration program is somewhat variable. In the United States the cost per month (22 working days) can be expected to be about \$20,000; this sum includes the collecting of samples by a trained team, the analyses for the factor being used in the exploration, evaluation of the data, preparation of the corresponding maps, and the indications of sites for intensifying the study of an area by means of seismology or drilling, that is, a complete interpretation. For an area of 300,000 acres (~ 764 km^2), and with a sample density of four samples per 2.6

km^2 ($\sim$ 1 mi^2), targets of 20,000 acres ($\sim$ 50 km^2) to 30,000 acres ($\sim$ 76 km^2) are sought, although in some areas very favorable zones of 5000 acres ($\sim$ 12 km^2) to 10,000 acres ($\sim$ 25 km^2) have been selected as targets. An area of 300,000 acres may be properly covered by 1180 samples, representing 26 days of field sampling, or an expense of $23,700 if a company is awarding the entire project to a geochemical exploration firm. However, the company can bring in a specialist to train its own prospecting team in the field itself and later pay only the costs of the geochemical analyses, the evaluation of the data with corresponding maps, and the indications of subareas where exploration should be intensified by means of drilling or seismology (or other geophysical method). The specialist represents an investment of approximately $2,000 per month, but if one calculates that at $15 per sample, the total cost for 1180 samples (which represent an area of $\sim$ 764 km^2) would be $17,000, it would seem worthwhile to contract an expert for a training period and subsequently save investment and provide more employment in a company, be it private or governmental.

Thus, if by means of geochemical prospecting, large areas of a prospect zone can be relegated to secondary exploration importance, financial investment and time during the exploration are saved since it is not necessary to do a complete seismic study. In areas for which geochemistry indicates greater petroleum or natural gas presence probabilities, seismology serves to prove the existence, below the surface, of faults and structures with structural relief greater than 65 to 70 meters, a figure that represents the resolution capability of seismic methods. However, for hydrocarbon traps with structural relief of less than 65 meters or for stratigraphic traps, both of which may not be detectable by geophysical methods, geochemical exploration (supplementing the geology) remains as the only technique that can improve a program's chances to localize such traps. The probability of finding a hydrocarbon-bearing trap has been considered and in many texts it is noted that the "success" ratio in the search for petroleum and/or natural gas based on all the geologic and seismic data available, is 1:9, that is, only 1 well of each 10 drilled will give petroleum and/or natural gas in exploitable quantities. Nonetheless, it must be mentioned that a company may have a "success" ratio of only 3 exploitable wells for each 100 drilled and make a significant profit (Levin, 1969). Therefore, a ratio of producing wells to nonproducing wells of 1:10 or 10:100 can obviously give very important profits. One company doing only service work has presented data on its past 25 years of activity showing that by its analyses in geochemical exploration (and all the geological data available), the company has the so-called "success" ratio of 1:4 in all areas studied and 1:5 in areas of stratigraphic traps. Davidson (1967) wrote that an evaluation of available information indicated that new field wildcats drilled on prospects where geochemistry played a vital role in exploration have yielded from 2 to 7 times as many producers as the national average where geochemistry was not used. In specific areas the Russian efficiency in gasometric reconnaissance studies to find producing zones in areas of propitious geological and geophysical conditions was 70 to 80% in the region of Central Volga and ~ 90% over the platform structures of the Komi region, both in the Soviet Union (Sokolov et al., 1971).

In general, petroleum companies spend 63.3% of their exploration budgets for seismology, 26.7% in geophysical reconnaissance by aircraft, and only 10% in programs related to research including geochemistry as a research field; for other mineral exploration programs, the methods used in prospecting and their corresponding percentages of exploration budgets include induced polarization (31.4%), electromagnetism (18%), geochemistry (11.8%), magnetism (8.5%), gravity (8.1%), resistivity (6.4%), test-drilling logs (5.5%), seismology (4.9%), radioactivity (3.7%), and others (1.7%) (Levin, 1969). This apparent inhibition towards investing more funds in geochemical prospecting studies for hydrocarbons might be due in part to the conservative nature of the petroleum investors and their lack of intimacy with a relatively little known and propagandized method. It might also be due to the fact that some companies that have specialized in geochemistry and petroleum and natural gas exploration do not reveal the factor(s) being measured and later used to prepare geochemical maps; that which is secretive is often subject to suspicion. In addition, the concept exists for many that with the classical methods of exploration there are great profits, so why invest funds and time in a method that for some has not been sufficiently proven? Nonetheless, in some countries and for several petroleum companies, geochemical prospecting for hydrocarbons already plays an important role in the exploration phases of programs related to the search for petroleum and natural gas (most notably in the Soviet Union). As should be obvious, if geochemical prospecting for hydrocarbons, seismology (which together with other geophysical methods is used in the detailing phase of exploration), and test drilling cost only a small fraction of the anticipated value of a deposit, their use is economically advisable.

What can be considered then as a geochemical factor that can be used in the search for hydrocarbons? And, on what theoretical base has the concept of using such factors been developed?

First, there is the identification of the gases present at or near the surface (2 to 10 meters) or at depth in test wells (to 600 meters) and the quantitative measurement of the contributions of each gas to a total gas assemblage. Also, the relationships between deposits of hydrocarbons and concentrations of various metals, salts, CO_2, H_2S, and electrical conductance and/or resistivity have been used or are being used and investigated in order to locate and delineate zones of maximum potential with respect to the trapping and accumulation of petroleum and natural gas. As is the case in prospecting for metals and non-metals, the geological and geochemical indications of the presence of a deposit are only indications. Although a possible deposit can be perfectly located, the proof of its exploitability derives from test drillings from which the quantity of a commodity present can be calculated. The economic prospects of the particular situation can be determined from such a calculation and adjunct factors. In the final analysis, it is the economic factor that dictates whether a deposit is exploitable or not; that which is not exploitable at a given time may be exploitable in the future depending on the necessities of a country or a region, advances in technology, and the change in prices for a commodity (supply and demand).

Let us now examine the techniques of geochemical prospecting for hydrocarbons, beginning with the gases (and vapors) associated with them.

Theoretically, the component gases of hydrocarbon accumulations may escape from them and move vertically (upwards) by diffusion and effusion due to the pressure gradient that exists and their much lower densities as compared to those of water, overlying rocks, and petroleum; the velocities of transfer upwards depends on this same pressure gradient, the viscosity of the gas, the molecular size, and of course, the degree of impermeability or permeability of the enclosing rocks. If these gases move in a more or less coherent mass, penetrating intermolecular spaces, and with no excessive lateral dispersion due to water movement (juvenile or groundwater) in the rocks across which the gases must pass, there could exist at or near the terrestrial surface (above the hydrocarbon accumulations) signs or clues that indicate the presence of such accumulations at depth. Also, deep rock samples over petroliferous deposits should contain greater amounts of gas than those rocks outside the zone of influence of such deposits. These signs, or perhaps it would be better to call them "pathfinders," can be weak concentrations of the same gases or notable filtrations (seepages) of petroleum or natural gas coming from the petroliferous deposits, modifications of material (soils) that is in contact with the gases, and various other things already mentioned above.

That the gases rise vertically with relatively limited lateral dispersion is a premise accepted by many geochemists doing prospecting for hydrocarbons. Experiments in the laboratory and in the field have indicated that the movement of gases from the lower to the upper part of a system generally follows Fick's law, a spherical function for diffusion, with the flow of gas in capillary openings following Poiselle's law. In addition, these same experiments have shown that the light gases (methane, CH_4; ethane, C_2H_6; propane, C_3H_8; and butane, C_4H_{10}) and perhaps vapors of the light liquids (C_5-C_8) do not have great difficulty in passing through rocks impervious to the passage of liquids. If the enclosing rocks are faulted and/or fractured, the process of vertical movement would be more rapid if the rock openings are not sealed by secondary deposition (for example, of carbonates). As for the influence of a lateral migration of waters trapped in the rocks overlying the petroleum accumulations on the dispersion of rising gases, such lateral migration is probably slow with respect to the velocity of vertical migration of the gases. However, what appears to be more important with respect to the ascent of gases is the probability that they move along films of water held by the surficial attraction of detrital material, films that are not significantly tied in with the waters that migrate laterally; also, the gases move by means of diffusion across static water in impermeable strata. Certainly, if there are routes open leading upward, the hydrocarbons, both gases and liquids, are going to rise until their ascent is stopped by an impermeable horizon that later may be crossed only by the light gases and perhaps vapors of light liquids. Accepting then that gases from petroleum deposits (exploitable and nonexploitable) arrive at the terrestrial surface possibly with a more or less regular flow (with respect to geologic time), investigators initially began to search for and categorize gaseous manifestations at or near the Earth's surface and subsequently detailed their studies by analyses of relatively deep materials and by the identification and quantitative determinations of all the gases that comprise the sample (analysis was aided greatly by advances in gas chromatography). Horvitz (1969) has reviewed the extraction apparatus and latest advances in analytical techniques. The gases found in natural sources are dominated by methane, but as is seen in Table 9-1, the compositions are very variable. In general, one may state that the most common concentrations are less than 100 ppb ($10^{-5}\%$). A value of 1000 ppb is considered high and worth checking out.

In the first investigations, a sample of soil air was obtained from a well drilled to about 60 cm; the well was capped for 1 or 2 days (depending on the planning of the research program) so that the gas in the well equilibrated with the environment; a volume of the soil air was then taken for analysis of methane (Laubmeyer, 1933). It was learned almost immediately that several factors influenced the quantities of gases measured in the soil airs from such a relatively shallow depth (60 cm), as for example, barometric pressure (which causes flow modifications due to irregular "breathing" in and out), wind, rain, heat, humidity, and the amount of methane produced by decomposition of organic matter and microbial activity in the soil. Because these difficulties and interferences affected the shallow samples, Russian scientists studied soil air samples from depths of 2 to 10 meters and observed a marked increase in the quantity of gases present (methane and a grouping of ethane and other heavier gases), plus a sharpness and constancy of results (Sokolov, 1934). The Russian scientists also worked to further develop their field sampling technique and bettered their results still more. Research on the gasometric method and its usefulness in the search for hydrocarbons continued, but instead of analyzing the soil air, the soil itself was analyzed for its content of free and sorbed gases (Rosaire, 1938; Horvitz, 1939). When these North American scientists compared their results on data obtained from soil air with those obtained by degassing of the

Table 9-1 The Composition of Some Natural Gases (after Sokolov, 1971)

Field, region	Composition of gas, %							
	CH_4	C_2H_6	C_3H_8	C_4H_{10}	C_5H_{12}	CO_2	N_2	H_2S
Kala, Aserbaijan (petroleum, gas)	96.1	2.1	0.2	0.4	1.0			
Gasli, Middle Asia (gas, gas condensate)	95.0	1.7	0.4	0.14	0.33	0.61	1.8	0.02
Romashkino, Volga-Ural (petroleum)	40.0	19.5	18.0	17.5	4.9	0.1	10.0	
Tujmasi, Volga-Ural (petroleum)	39.5	22.4	17.4	6.7	3.4	0.1	10.6	

Source. Reprinted from *Geochemical Exploration* (Boyle, tech. ed.), Special Volume No. 11, Canadian Institute of Mining and Metallurgy.

233

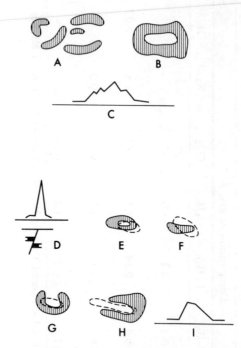

Figure 9-1. (Top) schematic shapes of some gas anomalies over oil and gas fields: (*A*) Mastagy-Buzovny; (*B*) Kogemjakino; (*C*) Usinskaja. (Reprinted from Sokolov et al. (1971), *Geochemical Exploration,* (Boyle, tech. ed.), Special Volume No. 11, The Canadian Institute of Mining and Metallurgy.) (Bottom) classes of (schematic) gas anomalies: (*A*) anomaly over a fault (Sachalin); (*B*) Joukovskaja; (*C*) Kluchevskaja; (*D*) Kochanskaja; (*E*) Muchanovskaja; (*F*) Iskogorinskaja. (Reprinted from Sokolov et al. 1971, *Geochemical Exploration,* (Boyle, tech. ed.), Special Volume No. 11, The Canadian Institute of Mining and Metallurgy.)

soils themselves, they found much greater quantities of gases present and detected heavier saturated hydrocarbons in the organic matter extracted from the soil. The study of organic materials extracted from solids by degassing and other techniques is the basis today for the gasometric method used in geochemical prospecting for hydrocarbons.

However, there were some problems at the beginning of the use of soils because of the biological activity that continued in the soils stored in their sample containers during the time in the field, the transport to the laboratory, and the time of storage in the laboratory before the degassing and subsequent analysis; this problem was alleviated by the vacuum packing of the samples in hermetically sealed jars. As is the case in the search for metals by geochemical exploration, the size fraction of the sample analyzed had its influence on the results. A fine fraction (clay, for example) was used because fine material

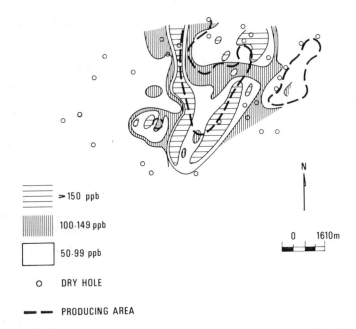

> 150 ppb

100-149 ppb

50-99 ppb

o   DRY HOLE

▬ ▬   PRODUCING AREA

N

0   1610m

Figure 9-2. Distribution of ethane and heavier hydrocarbons in the area of the Little Beaver oil field, Denver Basin, Colorado. The heavy dashed line indicates the vertical projection of the subsurface production. (From Horvitz, 1959.)

should have a greater capacity of sorption of gases than the coarser fractions, and researchers have continued using the fine fraction. Nonetheless, McCrossan and Ball (1971) reported that after making a geochemical study relating the distribution of hydrocarbons (C_1-nC_4) in the material sampled at a depth of approximately 1.5 meters in glacial deposits to known petroleum deposits (Caroline Area, Alberta, Canada) and fracture patterns (lineals of aerial photographs), they studied the distribution of gases in seven size fractions of a single sample, and on the basis of the results may now have to change their methods. In the fractions analyzed (between 0.00145 and 16 mm), there was a positive correlation between the quantity of gas and the fraction examined and also between the quantity of gas and proportion of carbonate in the sample. The clay-size fraction had a lesser content and the 16 mm gravel had the highest content of all the gaseous hydrocarbons determined. McCrossan and Ball theorize that much of the gas may be allocthonous, brought in and deposited with the glacial material in the form of carbonates, and are investigating this possibility. This particular report demonstrates again the importance of making a pilot study before initiating a regional program of geochemical prospecting.

In historical works and others that followed, one of the first observations

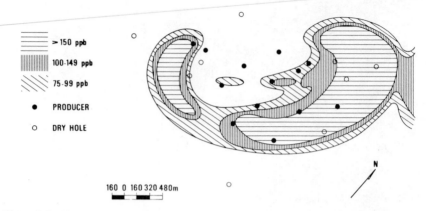

> 150 ppb
100-149 ppb
75-99 ppb
● PRODUCER
○ DRY HOLE

160 0 160 320 480m

Figure 9-3. Hydrocarbon anomaly over the Bonney oil field, Brazoria County, Texas. Samples were taken at a depth of about 3.6 meters. (From Horvitz, 1959.)

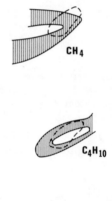

CH$_4$

C$_4$H$_{10}$

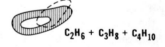

C$_2$H$_6$ + C$_3$H$_8$ + C$_4$H$_{10}$

Figure 9-4. Gas anomalies over the Sakar-Chaga (gas condensate) field. (Reprinted from Sokolov et al. 1971, *Geochemical Exploration,* (Boyle, tech. ed.), Special Volume No. 11, The Canadian Institute of Mining and Metallurgy.)

made in areas of known geology and production of hydrocarbons, as well as in unknown areas, was that the pattern of the gasometric manifestations (anomalies) did not generally have a full form, but rather appeared as a complete or incomplete halo (see Figure 9-1), with the greater concentrations of gases in the zone peripheral to hydrocarbon production (see Figures 9-2 and 9-3). The

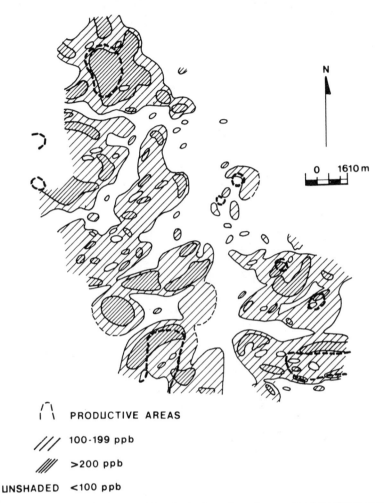

PRODUCTIVE AREAS

/// 100-199 ppb

/// >200 ppb

UNSHADED <100 ppb

Figure 9-5. Distribution of ethane and heavier hydrocarbons in an area of the Texas coast close to the Gulf of Mexico; the heavy dashed line indicates the vertical projection of the subsurface production. (From Horvitz 1959.)

halo patterns obtained for the distinct components of a group of gases are similar, but in some cases the heavier gases are more closely situated at the surface to the actual position of a deposit of hydrocarbons (see Figure 9-4). The anomaly patterns plotted can be fairly simple, as we have already seen, or they can be in complex forms, which, depending on the geology and surface conditions of the area studied and on the density of sampling, may or may not be spatially closely associated with hydrocarbon accumulations in the subsurface (see Figures 9-5 and 9-6). The observation by Horvitz (1969) on the perman-

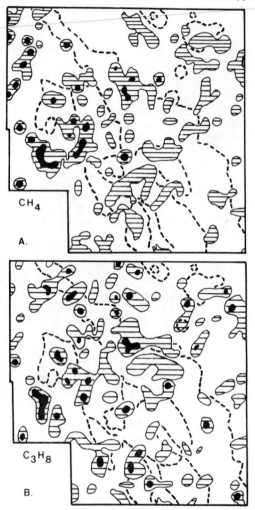

Figure 9-6. (*A*) Methane anomalies over oil and gas fields; (*B*) propane anomalies over oil and gas fields. The highest contrast anomalies are represented in solid black; the next highest anomaly level is shown by horizontal rulings. Heavy dashed lines indicate the vertical projection of the subsurface oil and gas production. (Reprinted from McCrossan and Ball, *Geochemical Exploration,* (Boyle, tech. ed.), Special Volume No. 11, The Canadian Institute of Mining and Metallurgy.)

ency of the anomaly patterns is very important. On the basis of his work on the Hastings oil field, Brazoria County, Texas, in 1946 and again in 1968, he found that there is a weakening and ultimate disappearance of an anomaly as the underlying hydrocarbon accumulation is withdrawn from its reservoir. As important is the fact that an anomaly remains in the soil, and its presence can be confirmed even though a period of many years has elapsed since it was first recognized if the hydrocarbon deposit producing the anomaly remains undisturbed. The density of the sampling used depends on the detail desired for a project and on the type of deposit one expects to find. In regional geochemical reconnaissances, 1.6 to 4 samples per square kilometer have been used, with the sampling done on a square grid (where possible). For more detailed work, up to 50 samples per square kilometer have been taken. It is obvious that there is an optimum sampling density for each type of deposit (and production) and this can be determined by means of accumulated data from studies on areas of known production. It is noteworthy that several other factors that have been investigated as potential guides in the search for hydrocarbons also present anomalies in halolike forms (radioactivity, electrical conductance, and/or resistivity, and others), except for microbial density, which presents a full target as an anomaly. According to Sokolov (1971), the halo-shaped, arcuate, and straight anomaly patterns are found at different levels at depth in the stratigraphic section, thus indicating the existence of hydrocarbon deposits at greater depths. A hypothesis has not yet been proposed that explains with certainty the reason or the factors that, due to their activity, result in annular and semiannular or elliptical and semielliptical anomaly patterns often associated with subsurface petroleum and gas deposits. Attempts have been made to explain the origin of these halo anomalies by, for example, better aeration of the central upper part of a structure with resulting easier decomposition of organics, by the lesser permeability of the rocks over the gases from the center toward the sides of structures, and by a concentration of fractures in rocks at the peripheries of the structures. Diametrically opposed to the observation of continuity of the halolike and straight pattern dispersions with depth in the stratigraphic section was the observation of Stroganov (1969). Stroganov wrote that the dispersion and distribution of gases in rocks from depths of between tens and hundreds of meters rarely showed the development of halo anomalies and more often resulted in direct gas anomalies (full targets). In all cases the concentrations of hydrocarbon gases in rock cores taken in areas of oil and gas fields are much greater than those of rock cores taken from outside the limits of influence of the hydrocarbon fields. Thus, the Stroganov data presented a problem regarding the interpretations of the origin of the halo-shaped forms of the gasometric anomalies because it was necessary to explain how these annular or elliptical patterns had been developed from a full-target design that apparently existed at depth. As will be discussed later in this chapter, MacElvain (1963) presented an interesting theory to explain the observed differences.

Investigations are being made on the free and sorbed organic gases in materials taken from greater than near surface (10 meters) depths and the data deriving from these samples are being used in the exploration phase of geochemical programs in the Soviet Union. In the exploration stages of prospecting for hydrocarbons in regions highlighted first by geological and geophysical information, the Russian scientists use samples from depths of about 200 meters to try to localize hydrocarbon fields that may be present at less than 2000 meters from the surface; when the petroleum and/or natural gas fields are thought to be at depths greater than 2000 meters, the gasometric studies are made on materials taken from 200 to 600 meter depths. Geochemical prospecting by means of samples obtained from test wells is neither inexpensive nor rapid, but according to the Soviet information, results in critical reliable data when a program is already in the exploration stage that requires more detail than is provided by exploration using shallow materials. For example, in the South Caspian Basin, Sokolov and Dadashev (1972) showed that relatively deep gas geochemical methods (samples every 10 meters below 30 meters to a depth of about 100 meters) could provide an additional technique for the more complete evaluation of possible oil and gas fields and of different parts of structures; they demonstrated that oil and gas in the northeastern part of the Basin structure coincided with an increased percentage of gaseous components from rocks of the structure. Positive results were also reported by Geodekian and Stroganov (1973) who studied the hydrocarbon gas content of drill cores and drilling muds obtained from key stratigraphic horizons during the drilling of shallow structural wells. They found distinct methane anomalies over oil and natural gas pools trapped in both anticlinal and monoclinal structures tested in the Ukraine, South Mangyshlak, and West Uzbekistan, U.S.S.R. There are two types of deep samples that are being studied by the Soviet exploration geochemists: the heavy drilling mud and drilling cores; rock chips brought to the surface during drilling may also be used. Initially, when the researchers wanted a sample of gases from a specific part of the stratigraphic section, drilling and mud circulation were detained for a determined time (constant for each different program) in order that the gases that escape from the specific part of the section selected for study can enter into and bond with the drilling mud; since the velocity of mud circulation was known, the mud adjacent to the study section was brought to the surface for degassing and chromatographic analysis. Thus, the distributions and dispersions of different gases could be studied in three dimensions. In the Soviet Union detailed exploration studies are made on grids that vary from 100 to 200 meters, or a density to 25 test wells per square kilometer. However, in the beginning of the use of such techniques, there were problems of contamination due to filtrations of gas into the mud from other parts of the stratigraphic sequence through which the mud was being brought to the surface; also, there were losses of gas from the mud during its rise towards the surface. One of these problems was overcome by the invention of degassers that extracted gases more completely and by the measurement of gases remaining in the muds reinjected into a well. Complementing this technique is another in which an analyzer is lowered into a well through static drilling mud and continuously identifies and measures the gases in the

Table 9-2 Hydrocarbon Gas Concentrations in Cores Sampled by an Open Method and by the in situ Hermetically Sealed Method Using the KC Core Sampler (from Sokolov et al., 1971)

Type of Rock	Depth of Sampling (meters)	Mode of Sampling	Concentration (c) of Hydrocarbon Gases (10^{-4} cm^3/kg)	Relative Losses $\dfrac{c\ KC}{c\ open}$
Sandstone	385	KC	106,243	893
		Open	119	
Shale	575	KC	2,431	47
		Open	52	
Shale	620	KC	1,610	46
		Open	35	
Sandstone	640	KC	36,473	529
		Open	69	

Source. Reprinted from *Geochemical Exploration* (Boyle, tech. ed.), 1971, Special Volume No. 11, The Canadian Institute of Mining and Metallurgy.

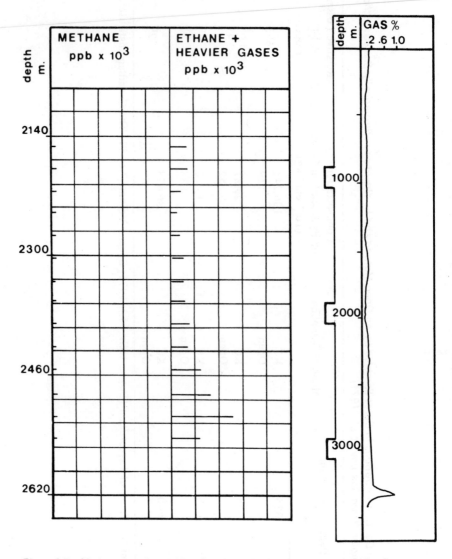

Figure 9-7. Mud gas log after drilling (reprinted from Sokolov et al., 1971, *Geochemical Exploration*, (Boyle, tech. ed.), Special Volume No. 11, The Canadian Institute of Mining and Metallurgy) and graphic representation of methane and ethane plus heavier hydrocarbons in drilling (rock) chips from a well in the Martha oil field, Liberty County, Texas. (from Horvitz 1959).

at all stratigraphic levels, automatically sending the analytical data to the surface; this technique has been called "diffusion gas logging" or "mud gas logging." Gasometry has also been used in the study of drilling chips and cores in the exploration phase of prospecting programs by drilling. The quantities of gases extracted from rocks were useful but generally low, due apparently to the loss of gas during the ascent of samples to the surface. In an attempt to improve the results, an apparatus was invented that sealed a sample hermetically *in situ*, later bringing it to the surface; the data generated from such samples are remarkable because of the tremendous quantity of gas present (compared with gas quantities obtained from samples taken in other ways) as is clearly illustrated in Table 9-2. This apparatus is called the "KC core lifter" (Sokolov et al., 1971; Sokolov, 1971). The analysis of gases in deep samples is not new; Horvitz (1959) showed the result of an investigation in the Martha field, Liberty County, Texas, of gases in rock chips from depths down to 2520 meters. It should also be noted that petroleum companies maintain a continuous control of the gas pressure in drilling muds in order to avoid blowouts, although these gas data are not used in prospecting. Figure 9-7 presents a graph from Horvitz (1959) together with one from Sokolov et al. (1971) that shows the results of gas logging in mud after the halting of drilling. Although the techniques described in this paragraph are very costly in relation to prospecting by surface or near-surface samples, the results of the dispersions of distinct gas components and/or groups of gases at different stratigraphic levels merit consideration for investments during the detail phases of an exploration program. In addition, it has been noted that productive horizons that are indicated by means of deep gasometric studies can be missed completely by electric drilling logs. As a somewhat unique approach to geochemical prospecting for hydrocarbons, Mekhtijev et al. (1972) worked on mud (gas–oil) volcanism, widespread over the South Caspian Basin continental and marine areas, as a surface sign of subsurface oil and gas accumulations. These authors emphasize that given positive structural form and/or rock dislocation in the subsurface, geochemical studies on solid, liquid, and gaseous products of the volcanism may clarify the geochemical aspects of the subsurface and allow different stratigraphic intervals and structures to be evaluated with respect to oil and gas. They also ascertained that most oil and gas fields in the South Caspian Basin area studied are characterized by being confined to zones of warm formational waters and high formational pressures, observations presently being evaluated.

Although gasometric analyses of soils from 2 to 10 meters below the surface (more realistic than deep samples with regard to time and investment) and of drilling muds and rocks taken from up to 600 meters in depth are often used in programs of geochemical prospecting for hydrocarbons, several other measureable factors also play a role. The measurement of patterns of relative surface radiation density resulted in halo-shaped patterns similar to those obtained in gasometric studies. Merritt (1952) conceived of a situation that would explain this coincidence by giving fundamental importance to the escape

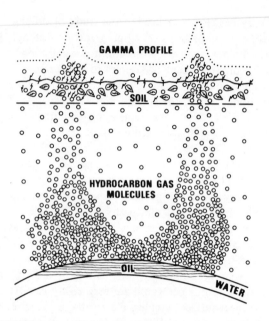

Figure 9-8. Hypothetical section across an oil-bearing structure. The pattern of molecular movement of hydrocarbon gases from the oil mass towards the surface and the movement of subsurface water with its natural radioactive component are portrayed. There are concentrations of radioactivity and hydrocarbon gases at the surface position vertical projection peripheral to the subsurface petroleum accumulation. (From Merritt 1959.)

of gases from a trap by peripheral gas filtration (see Figure 9-8) instead of escape as a single coherent mass. He believed that the gases passed through subsurface water containing its natural component of radioactive elements and carried higher concentrations of these elements towards the surface where the greatest escape of gas was taking place. Thus, the radiation can be measured at the surface on a predetermined grid to cover all the possibilities of shape, size, and condition of marginal anomalies and others in the search for hydrocarbon deposits in a study area. Merritt (1959) wrote that a station spacing of 100 meters in traverses 200 meters apart was sufficient in radiation prospecting for small fields and that the traverses could be separated by 400 meters in the radiation study of larger hydrocarbon deposits. In the search for stratigraphic traps and very small fields, Merritt recommended a grid of 100 by 100 meters. From his experience in this field of prospecting, Merritt was able to preview various causes of contamination, such as radiation from luminous-dial wristwatches, fallout of radioactive material, rain that can carry radioactive particles from the atmosphere in important quantities with respect to the anomalies being sought, meteorological cold fronts, and the influence of radioactivity in drilling mud, drilling chips, and saline drilling waters. He also gave a list of the natural causes of false scintillometer readings: daily variation of cosmic readings; thin soils over mother rocks result in lower readings than those obtained from thick soils; rock fragments in soils influence the determinations of relative radiation; weathered shales do not respond in the same way as do soils; granitic material mixed with soils give readings with greater radiation values; swampy areas in which the subsurface water can be cleansed of its critical radiation cannot be confidently studied; dry basins (of lakes) give very high radiation values and must be avoided in radiation prospecting; work in irrigated areas during or shortly after irrigation is not advised due to leaching, dilution, and transport of radioactive materials soluble in water; and, finally, the interference patterns caused by soils tend to present problems. Nevertheless, using a good control and the corrections necessary to eliminate the effects just mentioned, this method can be applied as a complement to the gasometric method if the director of a prospecting program deems it necessary. The technique is rapid on land and can also be used from aircraft. Other leaders in surface radiation survey studies have reviewed and generated data that indicate that the principal information interpretable from the radiometric surveys is the differentiation of the outcropping rock classes or soils derived therefrom, so that the surface radioactivity is indicative of surface geology and not related to hydrocarbons at depth. This inability to relate radiation data to petroleum is caused by the many interferences such as topography, vegetation, rivers, and rainfall that enhance or dilute radioactivity measured so as to render any trends caused by subsurface hydrocarbons virtually unobservable (Alekseev, 1965). Nonetheless, Foote (1969) reviewed radiometric techniques and equipment and feels that the separation of radiation interferences (for example, of highly variable atmospheric fractions and of the pedological alteration) is now technically possible so that if interference factors are identified and accurately accounted for, there should be a positive correlation between surface gamma-ray patterns and petroleum at depth.

Indirect and not truly geochemical methods have been used to obtain guides or clues to the presence of hydrocarbons at depth, but it is thought that the measurements given of these nongeochemical methods are a result of interplay with gases emanating from the petroliferous or gaseous deposits. For example, electrical conductivity in geologically and geochemically known areas has been evaluated by sending electrical charges into the Earth and reading their transmissibility at some determined distance away. The observed conductivity differences have been attributed to the effect of concentrations of gas in buffering the electrical conductivity of natural material, establishing zones of lesser conductivity where there are higher concentrations of organic gases in the ground. Although a relation betweeen conductivity and the location (at depth) of hydrocarbon deposits seems to exist, the method did not work after a rain and was relegated to the category of a method being researched because the electrical effects are apparently a surficial phenomenon instead of really representing concentrations of hydrocarbons in the subsurface (Merritt, 1959). Another premise was based on the hypothesis that in a petroliferous region concentrations of soluble salts should exist close to the surface where more gas has filtered to the surface; this could be because gas that traverses subsurface water will facilitate the process of evaporation of the water with the consequent deposition of salts close to the surface. A greater concentration of salts should be found where more gas filters through towards the surface, and where there are more salts, there should be greater electrical conductance or less electrical resistivity. Hence, the zones of lesser resistivity or greater conductivity should coincide with zones of halos of greater concentration of gases coming from deposits of hydrocarbons. This technique was checked out and it was proven that the effect of the salt presence is not surficial and should exist at depth (in wells) (Merritt, 1959). One might suppose that the quantity of soluble salts (in waters and acids) in soil samples could be determined and maps very similar to those made on the basis of resistivity measurements could be prepared. Although resistivity logs are used to indicate productive zones during drilling, the method is not applied much in the surface exploration stage except in the Soviet Union where various classes of electroexploration have been successfully employed. Along similar lines, Pirson (1969) found that the redox potential of soils delimits the extent of hydrocarbon accumulation and production successfully and theorized that the presence of hydrocarbons in the soil is the cause of the low redox potential. Caldwell (1969) gives a thorough description of the Soviet electroexploration methods used in potentially petroliferous provinces for mapping the basement rock and for tracing given sedimentary rock horizons. Such methods have been used by Soviet geoscientists in areas where seismic work is impractical (swamps, for example) or impossible but which can be studied electrically using helicopter support for logistics and as experimentation platforms.

In Chapter 6 it was noted that there are some local indicator plants that point to the presence of bitumen in the soil (see Table 6-3). Bitumen is a good pathfinder for petroliferous deposits at depth. It is also known that other plants that grow in bitumen-bearing soil could show morphological and/or physiological changes as rare forms of gigantism or deformation and sometimes a repeated, abnormal flowering. However, these changes are not very often noted in the field and are noted still less when the field man is not academically prepared in botany or has not consulted a botanist before going to the field.

Microbial activity in a soil or its associated waters develops strongly or weakly depending on available nutrients. Where there are more nutrients, there will be denser microbial colonies. In some cases, a class of microbes can live only if a specific type of nutrient is present. Such is the case with bacteria that derive their vital energy from the oxidation of propane (C_3H_8). Propane is formed naturally at the Earth's surface in very trace amounts and originates mainly from hydrocarbon accumulations so that its presence at the surface rather specifically indicates hydrocarbons at depth. A similar situation exists for ethane. It should be mentioned that Smith and Ellis (1963) reported that saturated and unsaturated hydrocarbons resulted from vegetation in the form of grass and roots (no quantitative analytical data supplied) so that so-called anomalies did not really originate from subsurface hydrocarbon accumulations. However, Horvitz (1972) challenged their data with research results showing that 1 gram of grass or roots contributed less than 1 ppb by weight of saturated hydrocarbons, in the range from ethane through pentane, to 200 grams of soil, analytical data opposing the cited 1963 claim of Smith and Ellis. Smith and Ellis also reported an inability to find anomalous hydrocarbon values over oil fields, but Horvitz (1972) explained this easily by showing that their 1963 soil-gas extraction technique was incapable of removing the major part of the saturated hydrocarbons adsorbed on the soil, making it difficult to discern an anomaly. If we accept the Horvitz data, then one can determine if propane-oxidizing microbes exist in a sample by means of the investigation of a microbial culture developed in an atmosphere of propane and sterile air, because only they can survive and multiply in such an environment. Because gases coming from hydrocarbon deposits at depth show different concentrations at the surface over deposits, it might be expected that the microorganisms that live from the oxidation of these gases would have colony densities (number of microbes per unit volume of soil or soil water) in direct relation to the concentrations of the gases. With the aid of microbiologist, specific microbes can be identified and their density in a sample determined. Davis (1969) has summarized the role of microbiology in petroleum exploration and recalled that Bokova et al. (1947) isolated ethane- and propane-oxidizing bacteria and identified these as the mycobacteria *Mycobacterium perrogosum var. ethanicum* and *Mycobacterium rubrum var. propanicum.* These are considered to be key microflora for detecting hydrocarbon gases in soils and waters. Field investigations showed that anomalies in the abundance of ethane oxidizers were invariably found to coincide with hydrocarbon production. Davis (1969) further stated that based on laboratory comparisons, the microbial oxidation of propane does not appear to be superior to the microbial oxidation of ethane as a petroleum (gas seepage) prospecting guide. It is Davis' opinion that microbiological methods undoubtedly can establish whether or not hydrocarbon gas seepages occur in a given area, but instead of being used for reconnaissance, the technique should be used with others (geological, geophysical, and geochemical) to upgrade a prospect by detecting gas seepages along structures or edges of stratigraphic traps. MacElvain (1964) shows an excellent coincidence between the surface dispositions of microbial densities and the positions of petroleum fields (see Figure 9-9 for an example). The patterns of microbe density are inverse to the anomaly patterns derived from gasometric studies and are found as full targets with greatest density over the hydrocarbon deposit and not in halo forms that are associated with gas distributions and other factors measured as part of programs in the geochemical search for hydrocarbons. Russian scientists have investigated the presence of bacteria that oxidize hydrocarbons, as indicators of gases migrating from depth; bacterial anomalies were found in soils of oil provinces, but according to Sokolov (1971), the possibility of finding them is limited and depends on many factors such as temperature, humidity, and composition of organic matter present. However, MacElvain (1963) used his observations to develop a hypothesis on the origin of the halo-shaped patterns found in the distributions of gases and radioactivity over accumulations of petroleum and natural gas; we shall shortly examine this hypothesis. Before doing so, however, it should be noted that Levin (1969) reported that there was petroleum in 65% of the wells drilled on the basis of geological and microbiological evidence, in some cases in important quantities that later resulted in the completion of oil fields.

As has already been described, the anomalies associated with the factors evaluated as geochemical guides in the search for hydrocarbons give complete and partial halo-type patterns (hydrocarbon gases, radioactivity, electrical conductance and/or resistivity, modifications of soils, unidentified components of soils, and others), except for the anomaly caused by microbial activity that manifested itself as a full target. Using a concept proposed by MacElvain (1963), the origin of the anomaly patterns observed in microbial exploration may be explained in what some consider a reasonable way. If the premise is made that the gases (and perhaps the vapors of the light liquids) that escape from a subsurface accumulation of hydrocarbons rise towards the Earth's surface as a more or less coherent mass (see Figure 9-10) instead of as an annular or linear mass related to the edges or borders of a deposit (see Figure 9-8), the following proposal can be made: If the single coherent mass arrives at the surface without being altered, it would be expected that the results of gasometric geochemical prospecting would show a full target or anomaly pattern; but, as has been found in many studies, this is not the common case and most often the halo-type patterns are developed. Therefore, a mechanism must exist by which the coherent mass of gas can be modified to give such halo-shaped anomalies. As was cited above, a microbial life system exists in soil and its associated waters. This life generates its vital energy by means of the oxidation of organic material such as methane, ethane, propane, and butane, and in accordance with this process, the colonies of bacteria are going to develop depending on the availability of a food supply. During the continuous oxidation of methane, for example, $CH_4 + 2O_2 = CO_2 + 2H_2O$, the oxidation of one volume of

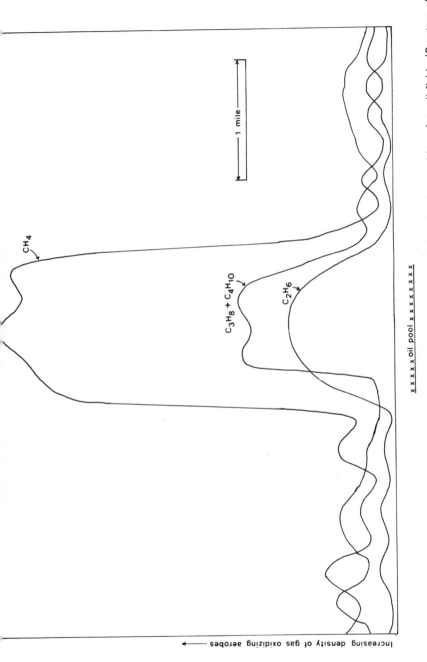

Figure 9-9. Biomicrograph showing the relation between specific microbial density and the subsurface position of an oil field. (Courtesy of Robert MacElvain, Eufaula, Alabama.)

249

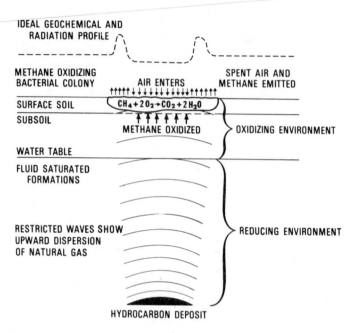

IDEAL GEOCHEMICAL AND
RADIATION PROFILE

METHANE OXIDIZING
BACTERIAL COLONY

AIR ENTERS

SPENT AIR AND
METHANE EMITTED

SURFACE SOIL

$$CH_4 + 2O_2 \rightarrow CO_2 + 2H_2O$$

SUBSOIL

METHANE OXIDIZED

OXIDIZING ENVIRONMENT

WATER TABLE

FLUID SATURATED
FORMATIONS

RESTRICTED WAVES SHOW
UPWARD DISPERSION
OF NATURAL GAS

REDUCING ENVIRONMENT

HYDROCARBON DEPOSIT

Figure 9-10. Methane may pass upward through a bacterial colony wherever bacteria are inactive. Bacteria at the colony's edge are constantly swept with oxygen-devoid air in a diffusion convection current that prevents oxidizing activity. (Reproduced from MacElvain, 1963, *Oil Gas J.,* February 18 and 25.

methane requires two volumes of oxygen. The oxygen comes mainly from the atmsophere with the result that a colony of microbes acts as a pump, drawing in the atmosphere to use its content of oxygen in the oxidation process. Because oxygen comprises 2 parts in 10 of air, each 10 volumes of air with its 2 volume units of oxygen can serve to oxidize 1 volume unit of methane. Other hydrocarbon gases require more oxygen and thus, more air, for their oxidation. Nitrogen and other components of the atmosphere, plus products of the oxidation reaction, return to the atmosphere driven by the pressure gradient produced by them between the soil air and the atmosphere, creating a rhythm of entrance and exit. The major rate of atmosphere inflow will be directed towards the area of soil where the maximum concentration of microbes live and multiply, that is, the central part of the colony. At the peripheral zones of the colony where there is spent air (without or with very little oxygen), the microbe density is less (due to the lack of oxygen for the life process of oxidation), and thus, the hydrocarbon gases at the edges of the colony maintain their concentrations because they were not oxidized. With an action mechanism such as this, the anomaly patterns observed can be satisfactorily explained as: the halos and semihalos of gases and of parameters related to the flow of or presence of gases around the positions of petroleum and gas fields, and also the full pattern of relative density of bacteria over the hydrocarbon deposits. What are not explained directly are the radioactive halos often associated with hydrocarbon deposits. According to MacElvain (1963), these halos of greater radioactivity can develop because the hydrocarbon gases are in continuous contact with radioactive material during their rise to the surface; much of this material has a very short half-life and does not represent radioactivity derived from the hydrocarbons. Therefore, the gases are adsorbing decaying radioactive matter until a long half-life isotope is formed, an isotope that could be responsible for the source of radioactivity measured at the surface. If the major concentration of radioactivity is related to the major concentration of organic gases, as Merritt (1959) and MacElvain (1963) believe, only the coincident halolike patterns found would be explained. Also, if the process of concentration of radioactive materials has proceeded for a sufficient time, it might be expected that the final product of radioactive decay of uranium and thorium, that is, lead, would be present in anomalous quantities in the annular zone defined by the greater concentrations of gases and radioactivity as suggested by MacElvain (1963).

In completing this chapter, it should first be mentioned that the content of hydrocarbon gases and their partial pressures in waters from formations associated with accumulations of petroleum and/or natural gas have been studied (Zorkin, 1969). It was observed that the concentrations and partial pressures of methane and heavier gases increased towards the accumulations of hydrocarbons and that nitrogen showed an inverse effect with a decrease in its concentration in formation waters closer to the hydrocarbon accumulations. This type of study can be made only during the detail stage of exploration when there are also other methods available.

Zarella (1969) emphasized that certain of the lower-molecular-weight hydrocarbons in petroleum are measurably soluble in water and that there is a partition of hydrocarbons between oil and water phases in the subsurface. Therefore, as the hydrocarbon content of waters adjacent to the oil pool attains equilibrium conditions, there is diffusion of part of the hydrocarbon into the waters of the formation away from the petroleum accumulation. The hydrocarbon content in subsurface brines associated with petroleum accumulation shows a regular concentration decrease with increasing distance from the oil occurrence (Zarella et al., 1967). In a direct application of such data from drill stem test samples (on benzene, the most soluble water-soluble hydrocarbon found in petroleum) that were used to complement geological and geophysical information, a new oil pool was discovered (Zarella, 1969).

Finally, the content of hydrocarbon gases in marine waters and the sediments immediately below them are being used in the search for petroleum by means of special gas chromatography units that can be lowered to a desired depth in the ocean to make the necessary analyses (Gerard and Feugère, 1969; Siegel, 1971). As a corollary to marine prospecting investigations, Gursky et al. (1972) reported on a comparison of oil-field gases formed in sedimentary rocks as a result of thermocatalytic transformations of organic matter and gases of biochemical origin found in marine sediments. The hydrocarbon gases of oil and gas pools contain a considerable amount of heavy, saturated hydrocarbons but do not contain significant quantities of unsaturated hydrocarbons. Gas analyses of recent sediments show that methane is present with an admixture of a very small amount of heavier hydrocarbons plus the gases CO_2, H_2S, N_2, H_2, NH_3, and others. In agreement with current ideas, the authors write that because of the vertical migration of hydrocarbons from oil and gas pools, the gas anomalies are formed throughout the section and move into recent sediments. They believe that from gas surveys of marine shelf sediments (and soils and rocks), the presence of oil and gas in the subsurface can be determined from one or more of the following: methane anomalies, total concentrations of the heavier gaseous saturated hydrocarbons, concentrations of individual saturated hydrocarbons, and correlation of methane with other hydrocarbons and saturated and unsaturated hydrocarbons.

Prospecting for petroleum and natural gas is at best a high-risk venture. On the basis of the work described in this chapter, it should be clear that geochemistry, if used correctly together with geological and geophysical data, can serve to reduce the high risk of exploration for hydrocarbons and definitely has an important contemporary and future role in the detection of these fossil fuel accumulations.

REFERENCES

Alekseev, F. A. (ed.), 1965, *Soviet Advances in Nuclear Geophysics,* Consultants Bureau, New York, 189 pp.

Bokova, E. N. and others, 1947, Oxidation of gaseous hydrocarbons by bacteria as a basis of microbiological prospecting for petroleum (in Russian), *Akad. Nauk SSSR, Dokaldy,* **56,** 755–757.

Bonoli, L., and Witherspoon, P. A., 1969, Diffusion of paraffin, cycloparaffin and aromatic hydrocarbons in water, *Advances in Organic Geochemistry,* Pergamon Press, New York, pp. 373–384.

Caldwell, R. A., 1969, Electrical methods of exploration in the Soviet Union: in *Unconventional Methods in Exploration for Petroleum and Natural Gas* (Heroy, ed.), Southern Methodist University Press, Dallas, pp. 87–104.

Chester, R., 1965, Geochemical criteria for differentiating reef from non-reef facies in carbonate rocks, *Bull. Am. Assoc. Petrol. Geologists,* **49,** 258–276.

Davidson, M. J., 1963, Geochemistry can help find oil if properly used, *World Oil,* July, 94–100.

Davidson, M. J., 1967, Petroleum geochemical exploration from space – a radical new technology, NASA, Manned Space Craft Center, Houston, 78 pp.

Davis, J. B., 1969, Microbiology in petroleum exploration, in *Unconventional Methods in Exploration for Petroleum and Natural Gas* (Heroy, ed.), Southern Methodist University Press, Dallas, Texas, pp. 139-157.

Debnam, A. H., 1969, Geochemical prospecting for petroleum and natural gas in Canada, *Geol. Surv. Can., Bull.* 177, 26 pp.

Drew, L. J., 1967, Grid Drilling exploration and its application to the search for petroleum, *Econ. Geol.,* **62,** 698-710.

Foote, R. S., 1969, Review of radiometric techniques in petroleum exploration, in *Unconventional Methods in Exploration for Petroleum and Natural Gas* (Heroy, ed.), Southern Methodist University Press, Dallas, pp. 43-55.

Geodekian, A. A., and Stroganov, V. A., 1973, Geochemical prospecting for oil and gas using reference horizons, *J. Geochem. Exploration,* **2,** 1-9.

Gerard, R., and Feugère, G., 1969, Results of experimental offshore geochemical prospection study, *Advances in Organic Geochemistry,* Pergamon Press, New York, pp. 355-372.

Gerard, R. E., 1971, Application of data processing methods in geochemical prospecting for petroleum, *Geochemical Exploration* (Boyle, tech. ed.), Special Vol. No. 11, Canadian Institute of Mining and Metallurgy, pp. 521-522.

Gursky, Y. N., Tichomolova, T. V., Chertkova, L. V., and Verchovskaja, Z. I., 1972, Geochemical peculiarities of hydrocarbon gases in recent and ancient sediments, Handbook 4th International Geochemical Exploration Symposium, The Institute of Mining and Metallurgy, London, p. 53.

Horvitz, L., 1939, On geochemical prospecting, *Geophysics,* **4,** 210-225.

Horvitz, L., 1945, Recent developments in geochemical prospecting for petroleum, *Geophysics,* **10,** 487-493.

Horvitz, L., 1959, Geochemical prospecting for petroleum, *Intern. Geol. Congr. 22nd Symp. Exploracion Geoquim. Ciudad de Mexico,* Tomo II, pp. 303-320.

Horvitz, L., 1969, Hydrocarbon geochemical prospecting after thirty years, in *Unconventional Methods in Exploration for Petroleum and Natural Gas* (Heroy, ed.), Southern Methodist University Press, Dallas, pp. 205-218.

Horvitz, L., 1972, Vegetation and geochemical prospecting for petroleum, *Am. Assoc. Petrol. Geologists,* **56,** 925-940.

Hunt, J. M., 1963, Geochemical data on organic matter in sediments, *Intern. Geochem. Conf. 3rd, Budapest.*

Laubmeyer, G., 1933, A new geophysical prospecting method, especially for deposits of petroleum, *Petroleum,* **29,** 1-4.

Levin, F., 1969, Oil exploration technology, *Sci. Technol.,* February, 15-25.

McCrossan, R. G., and Ball, N. L., 1971, An evaluation of surface geochemical prospecting for petroleum, *Geochem. Exploration,* Special Volume No. 11, Canadian Institute of Mining and Metallurgy, pp. 529-536.

McDermott, E., 1939, Concentrations of hydrocarbons in the earth, *Geophysics,* **4,** 195-209.

MacElvain, R. C., 1963, What do near surface signs really mean in oil finding? *Oil Gas J.,* February 18 and 25, 9 pp.

MacElvain, R. C., 1964, The birth and death of oil pools, Unpublished report, University, Alabama, 22 pp.

Mekhtijev, Sh. F., Dadashev, F. G., Buniat-Zade, Z. A., and Husejnov, R. A., 1972, Geochemical criteria for oil and gas in the South Caspian Basin, Handbook 4th International Geochemical Exploration Symposium, The Institute of Mining and Metallurgy, London, p. 55.

Merritt, J. W., 1952, Radioactive oil survey techniques, *World Oil*, July, 78–82.

Merritt, J. W., 1959, Geochemical and radiation surveying for oil and gas, *Intern. Geol. Congr., 22nd Symp. Exploracion Geoquim., Ciudad de Mexico*, Tomo II, pp. 283–302.

Pirson, S. J., 1969, Geological, geophysical and chemical modifications of sediments in the environments of oil fields, in *Unconventional Methods in Exploration for Petroleum and Natural Gas* (Heroy, ed.), Southern Methodist University Press, Dallas, pp. 159–186.

Ronov, A. B., 1958, Organic carbon in sedimentary rocks (in relation to the presence of petroleum), *Geochemistry*, 510–534.

Rosaire, E. E., 1938, Shallow stratigraphic variations over Gulf Coast structures, *Geophysics*, 12, 384–392.

Rosaire, E. E., 1939, The Handbook of Geochemical Prospecting, Author's publication, Houston, 119 pp.

Rosaire, E. E., 1940, Geochemical prospecting for petroleum, *Bull. Am. Assoc. Petrol. Geologists*, 24, 1401–1433.

Rosaire, E. E., McDermott, E. and Fash, R. M., 1940, Discussion of geochemical exploration (soil analysis), *Bull. Am. Assoc. Petrol. Geologists*, 24, 1434–4163.

Smith, G. H., and Ellis, M. M., 1963, Chromatographic analysis of gases from soils and vegetation related to geochemical prospecting for petroleum, *Am. Assoc. Petrol. Geologists*, 47, 1897–1903.

Sokolov, V. A., 1934, Summary of the experimental work of the gas survey, *Neft. Khoz.*, 27, 28–34.

Sokolov, V. A., 1971, The theoretical foundations of geochemical prospecting for petroleum and natural gas and the tendencies of its development, *Geochemical Exploration* (Boyle, tech. ed.), Special Volume No. 11, Canadian Institute of Mining and Metallurgy, pp. 544–549.

Sokolov, V. A. and Dadashev, F. G., 1972, Deep gas survey of the South Caspian Basin, Abstracts 4th International Geochemical Exploration Symposium, The Institute of Mining and Metallurgy, London, p. 60.

Sokolov, V. A., Geodekyan, A. A., Grigoryev, G. G., Krems, A. Ya., Stroganov, V. A., Zorkin, L. M., Zeidelson, M. I. and Vainbaum, S. Ja., 1971, The new methods of gas surveys, gas investigations of wells and some practical results, *Geochemical Exploration* (Boyle, tech. ed.), Special Volume No. 11, Canadian Institute of Mining and Metallurgy, pp. 538–543.

Sokolov, V. A., and Grigoriev, G. G., 1962, Methods and results of gas geochemical prospecting for petroleum and natural gas, *Gostoptechisdat*, Moscow.

Sokolov, V. A., Geodekjan, A. A. and Buniat-Zade, A. A., 1969, The general scheme of petroleum formation, alteration and migration, *Advances in Organic Geochemistry*, Pergamon Press, New York, pp. 279–288.

Stroganov, V. A., 1969, Some recommendations of the employment of geochemical prospecting, *Petrol. Gas Geol. Geophys.*, No. 7.

Weaver, C. A., 1968, Geochemical study of a reef complex: *Bull. Am. Assoc. Petrol. Geologists,* 52, 2153–2169.

Zarella, W. M., 1969, Applications of geochemistry to petroleum exploration, in *Unconventional Methods in Exploration for Petroleum and Natural Gas* (Heroy, ed.), Southern Methodist University Press, Dallas, pp. 29–41.

Zarella, W. M., Mousseau, R. J., Coggeshall, N. D., Norris, M. S., and Schrayer, G. J., 1967, Analysis and significance of hydrocarbons in subsurface brines, *Geochim. Cosmochim. Acta,* 31, 1155–1166.

Zorkin, L. M., 1969, Regional regularities of underground water gas contents in petroleum-gas basins, *Geol. USSR,* No. 2.

10

GEOCHEMICAL PROSPECTING
IN THE MARINE ENVIRONMENT

In perusing the past International Geochemical Exploration Symposia and the International Geological Congress sections dealing with geochemistry (since 1956), I have found a notable lack of papers concerning geochemical prospecting for minerals in the marine environment. In fact, although a call went out for such papers for the Fourth International Geochemical Exploration Symposium (1972), only one was presented (Holmes and Tooms, 1972). The economic potential and the social and to a lesser degree political importance of marine commodities are now at a stage that enjoins that the existing limitations of geochemical prospecting (generally restricted to continental prospecting) be eliminated and that applied geochemistry be extended to include marine geochemical prospecting by using and improving present exploration techniques and developing new ones.

The economic value of selected commodities extracted from sea water, beaches, sea floor and subfloor sites is given in Table 10-1 in two groupings; as is evident, there are discrepancies between the figures given for any single commodity common to both lists. The actual figures themselves are not important, but the order of magnitude is. For 14 important products shown in Group A (Table 10-1), 7.9% of their total worth was derived from offshore (continental shelf) mining operations (excluding beach sands) and have a value of more than 4170 million dollars; figures given in Group B indicate that closer to 8000 million dollars of mineral wealth were taken from the marine environment during 1969. It is obvious that these totals are dominated by petroleum and/or oil and gas production, 16 to 20% of which is derived from offshore wells.

The importance of initiating meaningful marine geochemical prospecting is further emphasized if one considers the work of Brooks and Lloyd (1968) on mineral economics and the oceans and that of McKelvey (1968) on the mineral potential of the submerged parts of the continents (Figure 10-1). Graphs are presented that show, for example, that if the world reserves of manganese are plotted against the cumulative world demand, the two projections will cross at

Table 10-1 Values of Selected Commodities Extracted from Seawater, Beaches, Sea Floor and Subfloor (from Siegel, 1971)

Group A	Annual Value from Sea (millions of dollars)	Percent of Total World Production	Group B	Annual Value from Sea (millions of dollars)	Percent of Total World Production
Petroleum	3900	15.0	Oil and gas	6100	16
Sand and gravel	150	10.8	Shell	30	80
Sulfur	37	16.6	Sulfur	26	8
Tin	24	5.2	Tin	24	4
Diamonds	4	1.4	Diamonds	9	Small
Gold	0.5	Small	Heavy minerals	13	13
Platinum	0.5	Small			
Ilmenite	0.5	Small			
Rutile	0.5	Small			
Zircon	0.5	Small			
Monazite	0.5	Small			
Precious coral	2	100	Salt	173	29
Iron ore	17	0.4	Magnesium	75	61
Coal	35	0.2	Water	51	59
			Bromine	45	70
			Magnesium compounds	41	6
			Heavy water	27	20
			Others	1	

Source. Group A from *Chem. Eng. News*, May 26, 1969, p. 15 and Group B from *Undersea Technol.*, January, 1970, p. 45.

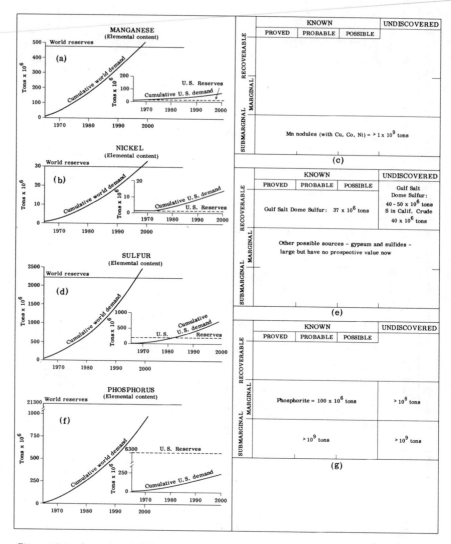

Figure 10-1. (a, b, d, and f) Projections of cumulative world demand for manganese, nickel, sulfur, and phosphorus versus world reserves. (From Brooks and Lloyd, 1968.) (c, e, and g) Estimated potential of the submerged parts of the continents for manganese nodules, sulfur, and phosphorus. (From McKelvey, 1968.)

about 1997 (Figure 10-1a). That is, by 1997 we will be using as much manganese as would be available in the (present) world reserves. The United States is far worse off than much of the rest of the world, as the cumulative United States demand now approximates United States reserves. For nickel, approxi-

mately the same figures hold.
will equal present world reserve. The cumulative world demand for nickel
United States demand approximat 1997, and in 1971 the cumulative
10-1b). McKelvey (1968) has publiched United States reserves (Figure
amounts of recoverable reserves and nta that he believes represent the
eral elements in the submerged parts of and submarginal resources of sev-
ganese nodules (containing nickel, cobalt, itinents. He feels that the man-
abound over significant areas of the deep oct and other metals) that
10^9 tons of ore (Figure 10-1c), and this ore he cor contain more than 1 X
terials recoverable at prices greater than 1.5 times ts as submarginal (i.e. ma-
have some foreseeable use and prospective value). Tatrevailing now but that
detailed estimates of the reserves of metals in manganese 2 (Mero, 1972) gives
Ocean; reserves of selected elements such as titanium, vanad les of the Pacific
nickel, cobalt, and copper are so great that they are large enou manganese,
ly unlimited at present rates of consumption. The availability of be essential-
is obviously dependent on the economics of the ore mining and met se reserves
processes and world prices of not one, but an assemblage of elements extraction
extracted from the manganese nodules. However, the extension of the d tribu-
tion of the manganese nodules is such (Figure 10-2) as to warrant continue research and development on the density of their distribution, mining, and metal-
lurgical extraction technique.

Figure 10-1 also shows comparative data from Brooks and Lloyd (1968) and
McKelvey (1968) for sulfur (Figure 10-1d and 10-1e) and phosphorus (Figure
10-1f and 10.1g). Thus the stage is definitely set for major geological and geochemical mapping and prospecting in the oceans.

Table 10-3 (Mero, 1967) indicates the type of deposit and the location in the
marine environment where a particular type of deposit may be found. I have modified this table by adding minor items, for example, precious coral, extractable potable water, heavy water essential for the continued development of nuclear technology for peaceful uses, and metal-bearing hot brines and associated sediments.

Considerable amounts of money are being put into the marine science programs of many of the nations of the world. As an example, data can be cited
for the United States marine science program. During the 1969 fiscal year, approximately 463 million dollars were expended by the federal government.
For the fiscal year 1971, the budget called for an expenditure of $537.2 mil-
lion, but of this sum only about $9.2 million (less than 1.7% of the total sum)
was spent in fields directly related to mineral prospecting.

It is estimated (Booda, 1972) that for the fiscal year 1972, 605 million dollars was allotted to the United States marine science program of which $16
million ($\sim$ 2.6% of the total sum) was used for studies in nonliving resources.
In Fiscal 1973, about 3% of the requested marine science program budget of
$667 million is to be directed to the nonliving resources studies. Although

Table 10-2 Estimated Reserves of Metals in Manganese Nodules of the Pacific Ocean (all tonnages in metric units) (from Mero, 1972)

Element	Amount in Nodules (billions of tons)	Reserves in Nodules at 1964 Rate of Consumption (years)	Approximate World Land Reserves (years)	reserves in nodules / reserves on land	Rate of U.S. Consumption in 1964 (millions of tons/year)	Rate of accumulation (millions tons/year)	rate of accumulation / rate of U.S. consumption	world consumption / U.S. consumption
Mg	25	600,000	_a	–	0.04	0.18		
Al	43	20,000	100	200	2.0	0.30		
Ti	9.9	2,000,000	_a	–	0.30	0.069		2.5
V	0.8	400,000	_a	–	0.002	0.0056	2.8	2.0
Mn	358	400,000	100	4,000	0.8	2.5	3.0	4.0
Fe	207	2,000	500	4	100	1.4	0.01	4.0
Co	5.2	200,000	40	5,000	0.008	0.036	4.5	2.0
Ni	14.7	150,000	100	1,500	0.11	0.102	1.0	
Cu	7.9	6,000	40	150	1.2	0.055	0.05	4.0
Zn	0.7	1,000	100	10	0.9	0.0048	0.005	3.5
Ga	0.015	150,000	–	–	0.001	0.001	1.0	–
Zr	0.93	+100,000	+100	1,000	0.0013	0.0065	5.0	–
Mo	0.77	30,000	500	60	0.025	0.0054	0.2	2.0
Ag	0.001	100	100	1	0.006	0.00003	0.005	–
Pb	1.3	1,000	40	50	1.0	0.009	0.009	2.5

^{a}Present reserves are so large as to be essentially unlimited at present rate of consumption.

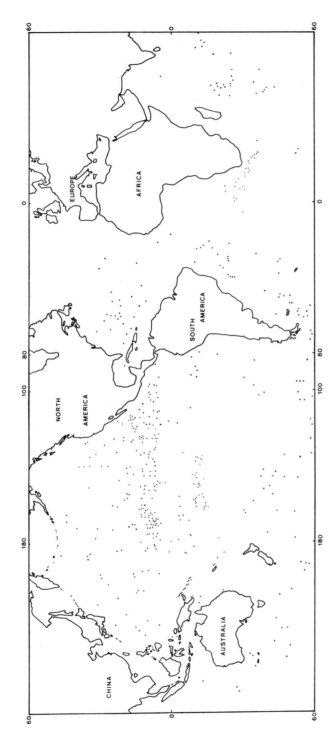

Figure 10-2. Distribution of manganese nodules in the world ocean. (Compiled from Figures 1–4 of Horn, Horn, and Delach, 1972.)

Table 10-3 *Mineral Deposits and Their Location(s) in the Marine Environment*
(slightly modified from Mero, 1967)

Zone	Commodities of Interest
Beaches	Placer deposits of Au, Pt, diamonds, magnetite, ilmenite, zircon, rutile, columbite, chromite, cassiterite, scheelite, wolframite, monazite, quartz, calcium carbonate, sand, and gravel
Continental shelves	Calcareous shell deposits, precious coral, phosphorite, glauconite, barium sulfate nodules, sand, and gravel; placer deposits in drowned river valleys or cassiterite, Pt, Au, and other minerals
Subsea-floor rocks	Oil, gas, sulfur, salt, coal, iron ore, and possibly other mineral deposits in veins and other forms as in the rocks on land
Seawater	Common salt, magnesium metal, magnesium compounds, bromine, potash, soda, gypsum, desalinized water, heavy water, and, potentially, sulfur, strontium and borax; most other elements are found in seawater, and given recently developed extraction techniques, seawater is a potential source of uranium, molybdenum, and other commodities
Deep-sea floor	Manganese nodules — as a source of manganese, iron, nickel, cobalt, copper, molybdenum, vanadium, and possibly several other metals Hot brines and associated sediments — as a source of gold, silver, copper, zinc, lead, and possibly several other metals Animal remains — as a possible source of phosphates and metals such as tin, lead, silver, and nickel Clays — for structural uses and possibly also for alumina, copper, cobalt, nickel, and other metals Calcareous oozes — as cement rock and other calcium carbonate applications Siliceous oozes — as silica and in diatomaceous earth applications Zeolites — as a source of potash

percentual increases are evident in the above-cited numbers, amounts that will be spent in fields directly related to mineral prospecting and the actual work that can be done is probably significantly less because of world-wide inflationary trends. One can only hope that other nations or international organizations will budget percentually larger amounts of their marine science programs in oceanographic mineral prospecting than has the United States and put this phase of marine programs at the immediate importance level it merits.

The Organization of American States (OAS), for example, awarded a grant for establishing an oceanography center (in Bahia Blanca, Argentina) for Argentina, Brazil, and Venezuela, but the initial award was only $200,000 and the program did not include any research on marine mineral prospecting. This field of investigation should be in the research program if one considers that Argentina and Brazil now claim sovereignty for a distance of 200 miles seaward from their coasts, a zone of much potential resource richness. Some private institutions, such as the Imperial School of Mines, London, have growing research programs in the area of marine geochemical prospecting.

Ingerson (1970) reported that geochemical prospecting for petroleum was being used by at least four major companies offshore in the Gulf of Mexico and that successes using the geochemical techniques had been reported in the North Sea and off Gabon. Deep-sea exploration has been given impetus by the indication of oil associated with deep-sea sediments and salt domes in the Gulf of Mexico through data obtained in the work of the Glomar Challenger. Perhaps the greatest effort in evaluating the potential of the sea for many important commodities has been and will be made by the United States Geological Survey. For example, in 1970 there was a second printing (first printing in 1969) of preliminary maps with the title *World Sub-Sea Mineral Resources,* compiled and presented by McKelvey and Wang. In this particular group of maps, the authors were concerned with petroleum, saline minerals, subsea minerals and other offshore deposits. In 1969 Holmes published an article on the potential of the sea and gave examples of geochemical maps delineating percent of free titanium found in bottom samples that averaged 10 centimeters penetration in the Gulf of Mexico; a similar map showing the percent of zirconium in bottom samples was also presented, and other geochemical maps are being prepared (Holmes, personal communication). Holmes' (1969) data represent very large scale reconnaissance; about 1400 samples were taken from an area of 107,500 square miles in the northern part of the Gulf of Mexico, or approximately one sample per 77 square miles. This may be considered by some as satisfactory for preliminary large-scale surveying, but in the future the reconnaissance will have to be developed on a closer scale, as was done by Stone (1967) for an area of 250 square miles off the South Carolina coast where samples were taken on the average of one per 5 to 10 mi^2. The definition of potential ores with depth can be proven with cores or perhaps with *in situ* analysis penetration techniques that are now being researched.

Two things must be immediately obvious in the evaluation of mineral deposits on the continental shelf or deeper parts of the ocean. First, a surface sample tells us nothing or at best little about what is present with depth and thus it is absolutely necessary (at present) to take cores, an extremely costly process. Second, one's samples, whether they are grab samples of 10-cm skims of the surface sediment or rock, or cores penetrating several meters into the ocean bottom, represent a very limited area or volume on which analyses must be made. Since many potential marine metalliferous deposits represent a disseminated ore, a larger volume of sample from as large an area and as deep a penetration as possible would be most desired. When the samples are transported to an onshore laboratory and an economically interesting find is made, the ship must return to sea for detailed sampling. From a standpoint of cost-effectiveness, this is wasteful. The process has been made more efficient by setting up chemical or geochemical laboratories on the ship using wet chemistry, colorimetry, and atomic absorption spectrometry; direct-reading emission spectrometry should soon become part of the shipboard analytical equipment. However, each of these methods, although they can give accurate and precise data on the elemental distribution in a sample, requires that the sample undergo preparation. Ship time is expensive and militates against such laboratory work. Thus, scientists must develop a potential for rapid, accurate *in situ* geochemical analysis requiring no sample preparation, and possibly representative of a large volume–area sample, or presenting continuous analyses of a systematically changing large volume–area suite.

Deffeyes (1969) reported on the development of a scanning electron macroprobe for the rapid chemical analysis of deep-sea cores on shipboard and presented examples of the results of analyses for calcium, iron, and silicon. The unit developed by Deffeyes does not require a vacuum chamber and gives the analysis of an area of 1 cm^2. The technique is based on a measurement of the K-α x-ray lines resulting from the electron bombardment of the sample using the β-emission from a ^{63}Ni source. Although Deffeyes believed that the core to be analyzed did not need special preparation, subsequent investigation showed that the surface to be analyzed had to be very smooth and care had to be taken that the core did not rotate during the analysis. The method gave qualitative results. Therefore, there still exists the need for *in situ* geochemical analyses and observations.

Figure 10-3 is taken from a report (and subsequent publication) by Sheets (1967) at a symposium on marine sciences and industrial potential and shows the evolution of submersibles for undersea exploration and exploitation. As is indicated in the diagram, submersibles are being developed that have a depth limitation of approximately 2000 meters. This is now considered the practical limit in the near future for both oil and gas exploration and exploitation and mining operations and gives access to about 15% of the ocean floor. With the

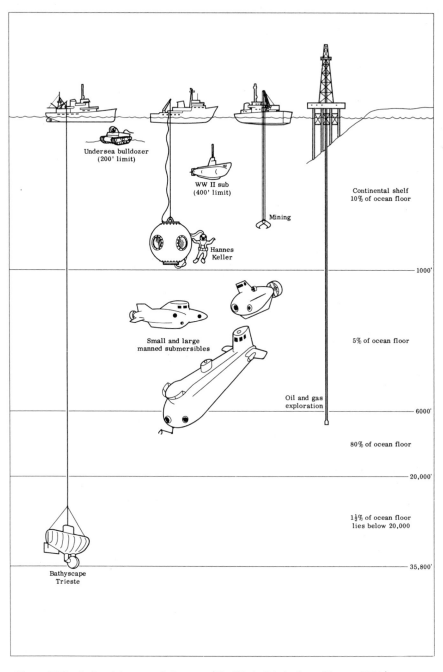

Figure 10-3. Industrial usage of the sea. (Modified slightly from Sheets, 1967.)

availability of such submersibles, one can view and directly photograph or transmit pictures of the environment via television camera to a mother ship. The use of undersea cameras and TV equipment is not new. The Deepsea Ventures, Inc. R/V PROSPECTOR has logged more than 30,000 stations in the world oceans taking samples and making TV reconnaissance of sampling sites. However, this indirect viewing is no substitute for on-site observations of a scientist who can maneuver a submersible to interesting specific areas as they may occur. Many of the deep-sea vessels are being manufactured with special manipulator arms that will allow a scientist to select the samples he considers most important for collecting and transport them to the surface for subsequent analyses. Again we have the problem that the sample must go to the surface for analyses.

This problem can be resolved by the continued improvement of a technique using thermal neutron activation with ^{252}Cf and interpretation of the resulting capture gamma spectrum. Senftle (1970) reported on the success of the activation type of approach applied as a mineral exploration tool on the Earth's surface (Senftle and Hoyte, 1966; Hoyte, Martinez, and Senftle, 1967), in bore holes (Keys and Boulogne, 1969), and in marine environments (Senftle, Philbin, and Sarigianis, 1969; Senftle, Duffey and Wiggins, 1969). The analysis of the capture gamma spectrum after sample activation with ^{252}Cf has been notably successful in tests with the transition elements, chlorine, gold, manganese, and mercury. Senftle (1970, personal communication) noted that nickel was readily detectable under 55 in. of overburden and proposes the use of *in situ* systems such as those presented in Figure 10-4a, 10-4b, and 10-4c for activating large volume–area samples, reading the capture gamma emissions with a Ge–Li detector, and using the computer for data reduction and subsequent mapping. Senftle (1970) proposed the use of a satellite to transmit the raw spectrum from actual field stations to the computer, and in field tests the entire spectrum was transmitted over telephone lines to the computer in about 30 seconds; in less than 5 minutes, the elemental concentrations were teletyped back to the field exploration party. Here we have a clear evolution from field accelerator activation analyses for silver in the terrestrial environment to the ^{252}Cf activation technique for surface, but more importantly, marine (geochemical) prospecting.

Most recently Battele-Northwest Laboratories has reported successful testing of the first self-contained instrument for on-site analysis of seabed minerals by a direct probing technique using neutron activation analysis with ^{252}Cf (Figure 10-5). It is projected that the instrument will detect and quantitatively analyze 30 different elements on the sea floor with concentrations being delivered as a computer printout. The instrument package as shown in Figure 10-5 is lowered from a surface ship for operation with depth capability limited only by the winch capability and cable length.

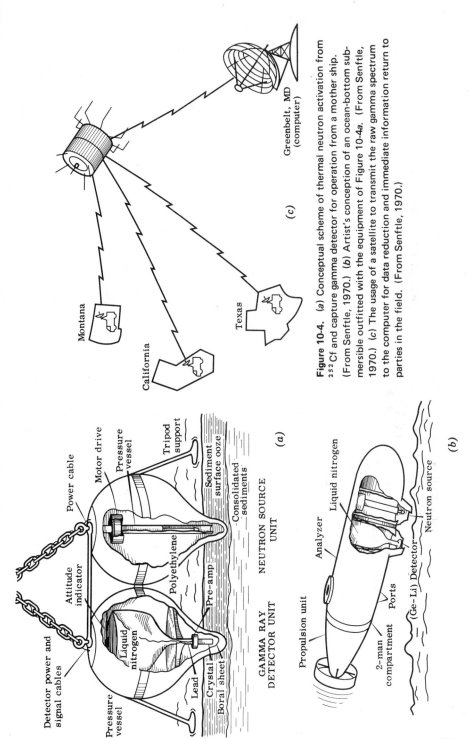

Figure 10-4. (*a*) Conceptual scheme of thermal neutron activation from ^{252}Cf and capture gamma detector for operation from a mother ship. (From Senftle, 1970.) (*b*) Artist's conception of an ocean-bottom submersible outfitted with the equipment of Figure 10-4*a*. (From Senftle, 1970.) (*c*) The usage of a satellite to transmit the raw gamma spectrum to the computer for data reduction and immediate information return to parties in the field. (From Senftle, 1970.)

267

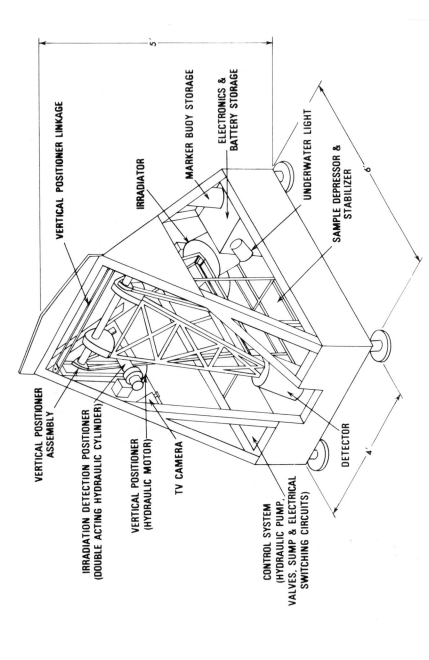

Figure 10-5. Nuclear instrument from Batelle that will detect and quantitatively analyze 30 different elements on the sea floor. (Reprinted from *UnderSea Technol.*, January 1973, p. 10, by permission of copyright owners, Compass Publications, Inc.)

VERTICAL POSITIONER LINKAGE

IRRADIATOR

MARKER BUOY STORAGE

ELECTRONICS & BATTERY STORAGE

UNDERWATER LIGHT

SAMPLE DEPRESSOR & STABILIZER

VERTICAL POSITIONER ASSEMBLY

IRRADIATION DETECTION POSITIONER (DOUBLE ACTING HYDRAULIC CYLINDER)

VERTICAL POSITIONER (HYDRAULIC MOTOR)

TV CAMERA

CONTROL SYSTEM (HYDRAULIC PUMP, VALVES, SUMP & ELECTRICAL SWITCHING CIRCUITS)

DETECTOR

5'

6'

4'

Summerhayes et al. (1970) demonstrated that measurements of radioactivity of samples can be used as guides to their phosphate contents. These authors made preliminary tests with a submergible scintillometer in waters up to 1000 meters deep, while the boat from which they were working was stopped. Their data were sufficiently positive so as to give impetus to the development of equipment with continuous recording capability that could be towed by a boat. As these researchers noted, such remote sensing techniques could be used rapidly and effectively to determine the existence and distribution of phosphate off the coasts of developing countries that have agricultural economies but little fertilizer resources.

Positioning can be one of the major problems in successful marine geochemical prospecting, and once a prospect is defined as exploitable, the on-site positioning becomes critical. The most precise locations can be attained using satellite navigation, which is good to 1:500,000 or 1 meter in 500 km; however, equipping a vessel for satellite navigation is expensive — on the order of $225,000 to $250,000. In a report at the Brighton, England marine technology conference, Summerhayes (1969) stated that a solution to the problem of effective on-site positioning of floating prospecting vessels was achieved (Saunders and Beck) by integrating a Decca Sea-Fix navigational system with a propulsion system consisting essentially of one Voith-Schneider propeller emplaced at each end of a ship; the position accuracy of this "Dyna-Fix" system was estimated at 1 meter in the 60-km range. Visual fixes can be achieved by using permanent relief maps of the ocean floor in depths to 1200 feet that can be made using side-scan sonar such as has been marketed by E. G. & G. International Inc.

More than 70% of the Earth's surface is covered by the seas, and of this a total covered area of 7.6% has water depths of less than 200 meters (approximately 656 feet) and is both explorable and exploitable. Table 10-1 shows the tremendous return from exploration in the 200-meter depth zone. Another 8.5% of the oceanic area is found in water with depths of between 200 and 2000 meters and is now explorable and in part exploitable. The Deepsea Ventures, Inc. R/V PROSPECTOR has conducted a successful prototype manganese nodule hydralic suction dredge mining pilot program on the Blake Plateau in 300 to 900 meters of water. A prototype continuous bucket line dredging system for mining manganese nodules was tested at over 3600 meters by a Japanese consortium and according to Padan (1971), at least two other large firms were continuing to develop techniques for evaluating and mining deep sea manganese nodules. Although Padan feels that deep-sea mining will not take place in the near future, the head of Deepsea Ventures, Inc. put together a working group during 1972-1973 with the objective of mining a million tons of nodules a year by 1974-1975 and having processed metals by 1975-1976. Because the prototype hydraulic suction dredge was operated successfully at 900 meters, the design of a similar system that will operate at about 6000 meters is proceeding. That the project can be put in operation by 1975 or 1976 is conceded by some since a process for metal winning (economically) from the nodules has been used successfully in a demonstration plant (UnderSea Technology, 1971) for the production of high-purity nickel, cobalt, copper, and manganese in an efficient manner (Flipse, 1972).

Very few companies are demonstratively planning for increased future involvement in ocean mining operations; however, if they are not to be left in arrears, major and minor operators must start now to organize consortia for such high-risk ventures and start taking risks.

Also in the deep sea, Degens and Ross (1970) speculate that more than 2 billion dollars worth of gold, silver, copper, and zinc are extractable from a 30-ft thickness of Red Sea hot brines and related sediments of the Atlantis II Deep, at depths greater than 2000 meters. This prospect is located in international waters between the Sudan Republic and Saudi Arabia, and according to Degens and Ross, at least one exploitation consortium has applied to the Sudanese government for mining concession in the zone. A few national governments are forward looking enough to assist their industries in such high-risk ventures. Germany, for example, is supporting Red Sea metalliferous sediment studies and Mn nodule investigations to 70% of all capital and operating costs (Holmes and Tooms, 1972). Also, according to Flipse (1972) the continuous-line bucket dredge system developed by Japanese scientists for deep-sea mining was sponsored by the Japanese government (conducted by Sumitomo) and because of the success of the pilot study will have a larger testing.

Another point that has been reported in the various ocean science trade journals requires discussion. This concerns the laws governing the exploration and exploitation of the seas. At the Geneva Conference on the Convention on the Continental Shelf in 1958, it was more or less decided that the smaller nations had to be protected and that each nation would be in control of operations in its territorial waters to depths of 200 meters. Also, a vote was held at the United Nations on a moratorium resolution calling for an immediate halt to exploitation of sea-bed resources beyond the limits of national jurisdiction. This moratorium was adopted by the United Nations General Assembly, during its session of December 15, 1969, by a vote of 52 to 27, with 35 countries abstaining. National jurisdiction, for all practical purposes, depends on the nation — a distance of 200 miles from shore has been adopted by many nations in lieu of the 200-meter depth or the "lawful" 3- or 12-mile limit. Clarifications concerning the resolution passed by the General Assembly and other resolutions are especially important from the point of view of the petroleum industry, which projects that by 1980, 33% of the world production of petroleum and gas will come from offshore operations and that by the year 2000 more petroleum will be extracted from the subsea floor area (to 2000

Table 10-4. National Jurisdictional Limits for Fishing Rights in the Marine Environment that Have Been or May be Extended by Each Nation to Include Mineral Rights

More than 12 Miles	12 Miles	6 Miles	3 Miles
Argentina	Albania	Ceylon	Australia
Brazil	Algeria	Dominican	China
Cameroon	Belgium	Republic	Costa Rica
Chile	Bulgaria	Finland	Cuba
Ecuador	Burma	Greece	Gabon
El Salvador	Cambodia	Haiti	Gambia
Guinea	Canada	Israel	Ivory Coast
Iceland	Colombia	Lebanon	Japan
India	Cyprus	Maldives	Kenya
Indonesia	Dahomey	Islands	Malasia
Korea	Denmark	Senegal	Malta
Mexico	Greenland	Somalia	Poland
Nicaragua	Faro Islands		Trinidad and
Panama	Ethiopia		Tobago
Peru	France		
Phillipines	Germany		
	Ghana		
	Guatemala		
	Honduras		
	Iran		
	Iraq		
	Ireland		
	Italy		
	Jamaica		
	Kuwait		
	Liberia		
	Libya		
	Malagasay		
	Republic		
	Mauritania		
	Morocco		
	Netherlands		
	New Zealand		
	Nigeria		
	Norway		
	Pakistan		
	Portugal		
	Rumania		
	Saudi Arabia		
	Sierra Leone		

Table 10-4. (Continued)

More than 12 Miles	12 Miles	6 Miles	3 Miles
	South Africa		
	Spain		
	Sudan		
	Sweden		
	Syria		
	Tanzania		
	Thailand		
	Togo		
	Tunesia		
	Turkey		
	Ukrania S.S.R.		
	United States		

meters) than will be extracted from land operations. We still lack a clear and precise definition of what will be done with the areas lying beyond national jurisdictions (for example, the manganese nodules of the Pacific Ocean or the Red Sea hot brines and associated sediments between the Sudan Republic and Saudi Arabia). When exploration of hydrocarbons and metalliferous or non-metalliferous mineral resources beyond national jurisdiction limits is initiated, who will prevent "claim-jumping"? — no laws cover these areas. It has been suggested that the United Nations have control of awarding concessions and collecting royalties on deep-sea deposits. This seems to be a good idea since the United Nations now represents a great majority ($\sim$ 95%) of the world population; it is hoped that the royalties would be utilized to benefit developing nations, members and nonmembers of the United Nations, or provide funds to aid areas where great disasters strike (due to natural causes or disagreements between nations). The wealth of the deep oceans belong to all humanity. So that the "have-not" access-to-the-sea nations and the lesser-developed nations all have an opportunity to benefit from the potential riches of the oceans belonging to the world. regardless of a nation's political philosophy or ability to put protective or threatening gunboats in the water, I suggest, idealistically, that the individual companies or consortia established to explore and possibly exploit the oceans offer 49% of their ownership to the federal governments of all the nations of the world. Thus, all could risk in the ventures and all could share in the profits. National jurisdiction varies by country and many countries have adopted a distance of 200 miles from their coasts instead of 3, 6, or 12 miles or a water depth of 200 meters (Table 10-4).

International conferences on national jurisdictions and sea rights outside na-

Table 10-5   Reclaimed Nonferrous Metals Used in the United States

Metal	Quantity (1967 consumption)	Units	1969 Value (millions of dollars)
Copper	1,243,000	Short tons	1,096
Aluminum	885,000	Short tons	403
Lead	554,000	Short tons	161
Zinc	263,000	Short tons	76
Antimony	25,568	Short tons	26
Tin	22,760	Short tons	71
Mercury	22,150	76-lb flasks	12
Silver	59,000,000	Ounces	101
Gold	2,000,000	Ounces	70

Source. *Chem. Eng. News,* April 6, 1970

tional limits with respect to mining and other topics have and will continue to take place and may serve to resolve many of the existing marine legal problems.

This problem of jurisdiction was treated at the second ministerial level meeting of the Third World countries at Lima, Peru (1971). These countries asked for recognition from developed nations on the principle of the right of developing nations to control the marine resources within their jurisdictions and insure their participation in the appreciable benefits that may be obtained from an international administration of economic marine deeps and subsurface, outside of national jurisdiction boundaries, giving special attention to the needs of states without access to the oceans.

As a final point, I should like to emphasize that projections as to mineral needs for the future (Figure 10-1a, 10-1b, 10-1d, and 10-1f) may well be understated. The population of the world by the year 2000 may be about double that of today; that is, between 6.4 and 8 billion people. More people mean more goods; more goods require more raw materials. Certain metals may be reclaimed from waste materials (Table 10-5), but not enough to satisfy the growing metal requirements. Indeed, 50% of the world's copper consumption and more than 50% of the world's lead consumption is reclaimed metal. Not only must we consider the population growth, but we must also understand and accept the fact that the so-called lesser-developed area of the world are, or will soon be, undergoing a significant industrial expansion. As the kinetics or rate of industrial expansion keeps increasing the need for raw material, minerals, and fuels becomes even more prepossessing. Space exploration broadens our knowledge and expands pride in man's ability to "reach for the stars" but does not improve our ability to provide the world with raw materials. More time, expanded efforts, and tremendously increased funds, both private and governmental, must now be invested in marine mineral-wealth surveys and techniques must be developed to make our geochemical evaluations of the oceanic area more efficient and more precise. The future is near, less than 30 years to the end of the century, and in the sea; our forward-looking mineral-oriented companies must become part of that future.

REFERENCES

Booda, L. L., 1972, Marine sciences and ASW receive big budget increases, *Undersea Technol.,* March, 30–32.

Brooks, D. B., and Lloyd, B. S., 1968, Mineral economics and the oceans, *Proc. Symp. Mineral Resources of the World Oceans,* Occasional Publication No. 4, School of Oceanography, University of Rhode Island, pp. 23–30.

Deffeyes, K. S., 1969, The electron macroprobe for logging deep sea cores, Abstracts, Annual Meeting of the Geological Society of America, Atlantic City, p. 264.

Degens, E. T., and Ross, D. A., 1970, The Red Sea hot brines, *Scientific American,* **222,** 32–42.

Duffey, D., El-Kady, A., and Senftle, F. E., 1970, Analytical sensitivities and energies of thermal-neutron-capture gamma rays, *Nucl. Inst. Methods,* **80,** 149–171.

Flipse, J. E., 1969, An engineering approach to ocean mining, *1st Annual Offshore Technol. Conf.,* Houston, Paper No. OTC 1035, 16 pp.

Flipse, J. E., 1972, Ocean mining – the sleeping giant, *Undersea Technol.,* January, 37.

Holmes, C. W., 1969, Potential of the sea – in the Gulf of Mexico – geochemical exploration produces exciting results, *Ocean Ind.,* **4,** 49–52.

Holmes, R., and Tooms, J. S., 1972, Dispersion from a submarine exhalative orebody, (abs.), Handbook 4th Intern. Geochemical Exploration Symposium, The Institute of Mining and Metallurgy, London, pp. 29–30.

Horn, D. R., Horn, B. M., and Delach, M. N., 1972, Distribution of ferromanganese deposits in the world ocean, in *Ferromanganese Deposits on the Ocean Floor* (Horn, ed.), National Science Foundation, Washington, D.C., pp. 9–17.

Hoyte, A. F., Martinez, P., and Senftle, F. E., 1967, Neutron activation method for silver exploration, *Trans. Society of Mining Engineers,* **238,** 1–8.

Ingerson, E., 1970, Geochemistry, *Geotimes,* **14,** 13–14.

Keys, W. S. and Boulogne, A. R., 1969, Well-logging with californium -252, *Proc., Soc. Prof. Well Log Analysts Symp.,* Houston, Texas, DP-MF-68-101, 25 pp.

McKelvey, V. E., 1968, Mineral potential of the submerged parts of the continents, In Proceedings of the Symposium on Mineral Resources of the World Ocean, Occasional Publ. No. 4, School of Oceanography, University of Rhode Island, pp. 31–38.

McKelvey, V. E., and Wang, F. F. H., 1970, World sub-sea mineral resources (preliminary maps), U.S. Geological Survey Miscell. Geologic Investigations, Map I-632 (second printing) plus discussion to accompany Map I-632.

Mero, J., 1965, *Mineral Resources of the Sea,* Elsevier, New York, 321 pp.

Mero, J., 1967, Marine Sciences and Industrial Potential Symposium, Houston, Texas, Transference of Technology Series No. 2, College of Business Administration, The University of Texas at Austin, 263 pp.

Mero, J. L., 1972, Potential economic value of ocean floor manganese nodule deposits, in *Ferromanganese Deposits of the Ocean Floor* (Horn, ed.), National Science Foundation, Washington, D. C., pp. 191–203.

Padan, J. W., 1971, Ocean mining begins to compete with sources, *Undersea Technol.*, January, **31**.

Senftle, F. E., 1970, Mineral exploration by nuclear techniques, *Mining Congr. J.*, 6 pp.

Senftle, F. E., Duffey, D., and Wiggins, P. F., 1969, Mineral exploration of the ocean floor by *in situ* neutron absorption using a californium-252 (^{252}Cf) source, *Marine Technol. Soc. J.*, 3, 9–16.

Senftle, F. E. and Hoyte, A. F., 1966, Mineral exploration and soil analysis using *in situ* neutron activation, *Nucl. Inst. Methods*, **42**, 93–103.

Senftle, F. E., Philbin, P. W., and Sarigianis, P., 1969, Use of ^{252}Cf for mineral exploration – comparison with accelerators for *in situ* activation of silver, American Nuclear Society Conference on Californium-252, TID Report 4500 (Conf.-681032), pp. 321–346.

Sheets, H. E., 1967, Marine Sciences and Industrial Potential Symposium, Houston, Texas, 27-66, Transference of Technology Series No. 2, College of Business Administration, The University of Texas at Austin, 263 pp.

Siegel, F. R., 1971, Marine geochemical prospecting – present and future, *Geochemical Exploration* (Boyle, tech. ed.), Special Volume No. 11, Canadain Institute of Mining and Metallurgy, pp. 251–258.

Stone, I. C., Jr., 1967, Geochemistry and mineralogy of continental shelf sediments off the South Carolina Coast, Ph.D. thesis, The George Washington University, Washington, D.C., 104 pp.

Summerhayes, C. P., 1969, Offshore minerals technology, *Underwater Sci. Technol. J.*, 25-27.

Summerhayes, C. P., Hazelhoff-Roelfzema, B. H., Tooms, J. S. and Smith, D. B., 1970, Phosphorite prospecting using a submersible scintillation counter, *Econ. Geol.*, **65**, 718–723.

Wakefield, B. D., 1969, Mining hard minerals three miles under water, *Iron Age*, 61–68.

Wooster, W. S., 1969, The ocean and man, in The Oceans, a Scientific American Book, Freeman, San Francisco, pp. 123–130.

11

GEOCHEMISTRY, HEALTH, AND
THE POLLUTION OF THE ENVIRONMENT

The discoloration of the atmosphere caused by smelting and foundry operations was recognized very early in man's history, and in modern times the famous (now nonexistent) London smog and smogs over Los Angeles, Tokyo, and other large and small cities of the world aroused public opinion and evoked legislative and health authority response because of the direct effect of pollutants on the respiratory system and eyes. Nevertheless, man "learned" to tolerate and live with these "inconveniences" before attaining ecological maturity a few decades ago. Contamination or pollution will not diminish in the future if stringent controls are not implemented to prevent or alleviate the basic causes (Table 11-1) of the problem. Pettersen (1966) considered air pollution and other atmospheric variables such as air stability, wind velocity, frequency and type of storms, and rain and prepared a map predicting which would be the regions of the world where dangerous pollution conditions could develop by the year 2000 (or sooner) because of industrial expansion (Figure 11-1). Many of these regions already suffer significant air pollution (Table 11-2), but governments are now alert to the dangers presented by pollution and have initiated action programs for its immediate alleviation, as well as demanding good planning for the future preservation of the environment before population and industry enter a region. Man is a living organism and the biological concept that an organism cannot survive and multiply in its own wastes is as applicable to man as to any other organism.

The realization that environmental pollution is increasing daily and is a real danger for man's continued existence received one of its initial impulses with the beginning of the era of nuclear testing. The fallout of radioactive particulate material was not restricted to the immediate surroundings of the explosion of an artifact, but extended for thousands of kilometers over great areas with very important human populations.

Pollution has many origins from the most obvious to the most subtle, with the degree of obvious to subtle varying according to the definition given for

*Table 11-1 Basic Causes of Pollution and Pollutants
(from Bryson and Kutzbach, 1968)*

Source of Pollution	Pollutants
Heavy industry — steel	Fe, Mn, Ti
Gasoline motors — automobiles	Pb, NO_3, and benzene-soluble particulates
Burning of fuel for energy — coal, petroleum	SO_4, Pb, suspended particulates
Refining of crude oil	Pb, Ni
Plating industry	Zn–Sn, Cr, Cu
Agriculture	Pesticides, fungicides (DDT, chlordane, dieldrin), NO_3, PO_4, plant and animal nutrients, other organic compounds
Populations	Organic wastes (which require oxygen for their decomposition), inorganic waste
Nuclear processes	Radioactive isotopes, heat
Construction industry	Sediment
Natural minerals	Pb, Cu, Zn, Cd, and others yet to be discovered

Source. Reprinted from Resource Paper No. 2, *Air Pollution,* with permission of the Association of American Geographers, Commission on College Geography.

pollution. Tully (1966) put forth a rather global definition: "Pollution can be considered as an alteration of natural environments, air, water, soil, which renders them offensive or deleterious to the aesthetic senses or for man's uses, or for the animals, fish, or crops which man wishes to preserve. We recognize that some degree of alteration of the environments is a necessary consequence of human activities. But such alterations are not considered as pollution until they reach a limit of tolerance"; we are still trying to establish tolerance limits for the many classes of pollution. The geochemist works with the natural environment and for geochemical prospecting works especially with the soil. Rennie (1966) gave a good definition of what would be considered as soil pollution: "Any substance that is common or foreign to soil systems, which by its presence causes adverse effects, directly or indirectly, on the productivity of the soil (the productivity includes the yield and quality parameters of the food products produced), is called a soil pollutant."

Pollution is due to two principal factors: urbanization and technology. Subfactors such as the population explosion, the concentration of large populations in relatively small areas, and industrialization and mechanization that

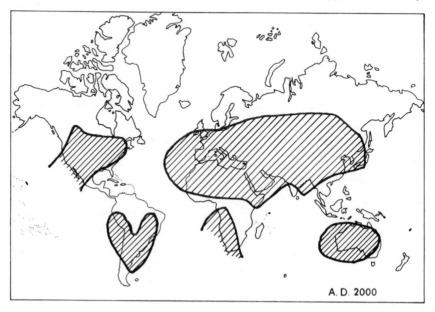

Figure 11-1. World regions in which dangerous pollution problems have developed, are developing, or could develop in the future, A.D. 2000. (Modified slightly from Pettersen, 1966.

Table 11-2   Air Pollutants in the Atmospheres of Los Angeles, California, and New Jersey Have Been Reported as Gases and Aerosols, or as Compounds (from Bryson and Kutzbach, 1968)

Los Angeles, California	Gases	Aerosols
	Carbon dioxide	Aluminum compounds
	Acrolein	Calcium compounds
	Lower aldehydes	Coal
	Formaldehyde	Iron compounds
	Hydrocarbons	Lead compounds
	Ozone	Soluble ethers
	Nitric acid	Silica
	Sulfur dioxide	Sulfuric acid
New Jersey	Compounds	
	Nitrogen oxides	
	Nitric oxide	
	Nitrogen dioxide	
	Sulfur dioxide	
	Oxidizing agents	
	Aldehydes	
	Carbon monoxide	
	Hydrocarbons	
	Carbon dioxide	

Source.   Reprinted from Resource Paper No. 2, *Air Pollution* with permission of The Association of American Geographers, Commission on College Geography.

serve to provide the necessities of the great populations are derived from urbanization and technology (Figure 11-2). There are four large, general categories of pollutants: inorganic compounds coming from industrial activities, organic materials, radioactivity, and minerals; there are a number of other categories such as noise and odors. The geochemist is strongly bound to study rocks and minerals, but on many occasions he has the opportunity to work in areas where his wide scientific knowledge and logical thought processes can contribute to the recognition of the existence of real or potential pollution; because of his training and experience, the geochemist, working in a team with other specialists, can consider pollution problems in a methodical, scientific manner, recognize the subtilties of the relations between the pollutants and other environmental factors, and in some cases suggest partial or complete solutions to eliminate or alleviate certain causes of pollution.

It is well known from classical geochemistry and prospecting studies that the chemical element distributions in rocks, soils, waters, and sometimes in the atmosphere can vary significantly in important areas of large zones that are considered relatively homogeneous with respect to the geology. Logically, vegetation may similarly have varying contents of the chemical elements because of its relation to soils and waters and because of its own ability to accumulate or select the elements that it does or does not need for the vital processes. These differences in the elemental content of rocks, sediments, soils, waters, plants, and air provide us with the basis for mineral exploration via geochemical prospecting. A leader in studies on the natural element distribution in plants and pollution and its relation to the earth sciences is Dr. Harry V. Warren, Professor of Geology and Geography, the University of British Columbia, Canada. Warren published a paper in 1954 on geology and health and has followed this first general report with many other more specific ones, some of which will be considered below.

In their study on the biogeochemistry of Pb in Canada, Warren and Delavault (1960) found that the ash of vegetation with which they were working contained between 10 and 100 ppm of Pb, but that near highways there were very often concentrations greater than 1000 ppm in some specimens of the vegetation (which by itself can vary markedly in its tendency to absorb Pb); in general, the soils in which the vegetation grew contained between 0.5 and 5 ppm of Pb, while associated rocks contained an average of 16 ppm. Previous geochemical investigations had demonstrated that only near some strong shows of Pb mineralization were similar concentrations of Pb (1000 ppm) found in the vegetation. Because there were no signs of mineralization near the sampling sites of the specimens studied, and because Pb concentrations in the samples dropped considerably at a short distance (less than 100 meters) from the highways, Warren and Delavault theorized that exhaust emissions from automobiles (using fuel containing tetraethyl lead) were responsible for the higher

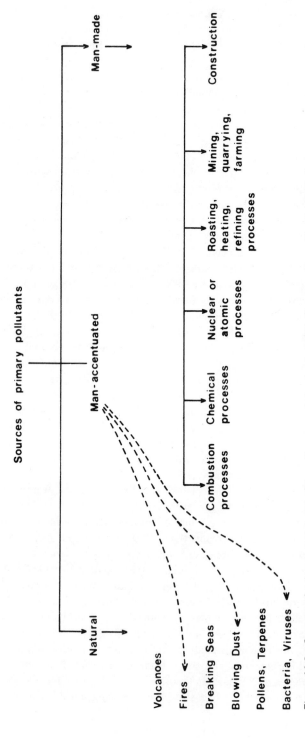

Figure 11-2. Sources of primary pollutants. (From Bryson and Kutzbach, 1968, Resource Paper No. 2, *Air Pollution*, by permission of Association of American Geographers, Commission on College Geography.)

amounts of Pb available for absorption (through the soils) by some plant species. Obviously, other sources of Pb could be found in highly industrialized areas (smoke emissions from smelters and other industries). Two publications in 1971 very clearly demonstrated the direct effect of human activity on pollution and geochemical prospecting. Hosking (1971) described the problems of pollution related to prospecting in Cornwall, England, an area from which economically important quantities of Sn, Cu, W, Pb, Zn, U, As, Fe, Sb, and Mn have been taken from more than 600 mines. The pollution caused by mining, mineral beneficiation, and smelting operations, complemented by that caused by domestic and industrial wastes and the use of metal-bearing fungicides, influenced geochemical sampling, especially in soils, and must be taken into consideration in the planning of geochemical programs in the area. In another paper, Hornbrook (1971) evaluated geochemical and biogeochemical methods for the detection of Ag vein deposits (Cobalt, Ontario, Canada) and found that surface contamination can frequently produce anomalies equal to or greater than natural anomalies in till associated with the veins. The pollution in this area comes from old mining gangue, trenches, open pits, abandoned mines, and gangue in lakes, all characteristic in the area because of mining activity during many decades. Also, the access roads to the mines in the area were constructed with waste rock from the mines and these rocks were sufficiently mineralized so as to contaminate the soils and vegetation to distances of tens of feet from the roads. In another area studied by Warren and Delavault, orchard soils that were treated with lead arsenate 10 years previous to their sample collection contained between 40 and 100 ppm of Pb, whereas other nearby orchards that were not treated with the fumigant had soils with less than 1 ppm of Pb. Thus, man's influence on geochemical samples, and by extension, their relation with certain diseases, should be quite obvious.

In 1961 Warren studied the incidence (where data were available) of multiple sclerosis as a function of the content of some metals in soils and noted that where the soil contained more Pb than might be considered "normal," there was a greater incidence of the disease; he inferred that this relation could be linked to the fact that foodstuffs that grow in the soils with "abnormal" Pb absorbed the metal, which later was ingested with the food. The normal ingestion of Pb is 0.25 to 0.50 mg daily, and 2 lb (about 900 grams) of food containing 1 ppm of Pb would result in the ingestion of 0.9 mg of Pb (Warren, 1966). Warren, Delavault, and Cross later (1967) cited the prevalence of multiple sclerosis in counties in Norway other than those observed by Swank (1950) and Swank et al. (1952); Swank and his colleagues related the prevalence of the sickness to the intake of fat in the diet, a factor also cited by others. However, there were a limited number of places where comparative epidemiological studies were available for use in evaluating the relations mentioned above. Warren, Delavault, and Cross presented data from

Norway, Sweden, North Scotland, and North Ireland to establish their hypothesis on the relation between geochemistry (presence of metals in soils) and multiple sclerosis. Initially, epidemiologists and geochemists could not come up with an explanation for the prevalence of multiple sclerosis in Norway and North Scotland on the basis of common geographic factors such as climate, topography, diet, occupation, and race. However, when epidemiological maps were considered together with geologic maps of Norway, North Scotland, and Sweden it was immediately recognized that areas or zones of low prevalence of the disease generally corresponded with areas where there were Precambrian gneisses and schists, all characterized by a lack of base metal mineralization. In North Ireland the areas of high prevalence of multiple sclerosis corresponded to Paleozoic rocks similar to those in other areas that are host rocks to Pb and Zn mineralization. After reevaluating the data from Norway, Sweden, and North Scotland and studying some small areas in detail, Warren, Delavault, and Cross (1967) noted that in areas underlain by schists and gneisses there were fairly small zones where the soil contained anomalous quantities of the base metals, and these zones were where the multiple sclerosis had an elevated incidence. They wrote that a relation seems to exist between the areas whose soils (and by inference, produce) have anomalously positive concentrations of Cu, Zn, and Pb and the districts with an elevated prevalence of multiple sclerosis, especially where there was much more Pb than Cu in the soils. This then might represent the natural pollution that could be detected by applied geochemistry.

Other authors showed the real danger represented by the ingestion, one way or another, of an abnormal quantity of Pb; Warren cited the work of Dubrovskaia (1960) who found that the air in some streets of the cities Chelysbinsk and Tula (U.S.S.R.) is so contaminated with Pb that medical examinations of the traffic controllers clearly showed some changes in their nervous and cardiovascular systems and in their gastrointestinal tracts. Blood analyses of these controllers revealed the presence of reticulocites and eritrocites containing basophilic granules.

With this in mind, we will examine the interesting results published by Warren and Delavault (1968) on the content of Pb in vegetables (Table 11-3). If lettuce is taken as an example of a common vegetable with a high content of Pb, one may calculate that it is necessary to eat little more than the equivalent of 50 grams of dried lettuce to absorb 0.9 mg of Pb, a quantity far in excess of what has been established as the normal ingestion. One may eat the equivalent of 50 grams of dried lettuce, but probably not on a regular basis. However, it is necessary to eat the equivalent in dried vegetable of 335 grams of carrots, 600 grams of potatoes, or 908 grams of cabbage to ingest 0.9 mg of Pb. The figures published by Warren and Delavault (1968) represent a limited sampling from England, Scotland, Wales, and British Columbia, but these figures

Table 11-3 Summary of the Pb Content of Selected Vegetables after Oven
Drying (Values are in ppm); Contents of Pb in Ashed Vegetables are Given in
Parentheses. The Samples Come from Several Places in England, Scotland,
Wales, and British Columbia (Canada) (from Warren and Delavault, 1968)

Vegetable	No. of Samples	Average	Mean	Range
Lettuce	22	18 (119)	11 (59)	0.3-50 (3-570)
Potatoes	29	1.5 (40)	0.4 (10)	0.2-17 (4-510)
Cabbage	27	1.0 (11)	1.0 (10)	0.2-3 (2-28)
Carrots	11	2.7 (37)	2.0 (20)	0.2-11 (5-140)

definitely indicate that scientists worldwide should be interested in similar analyses or other classes of analyses that together could give us sufficient data to
statistically evaluate this type of natural pollution and possibly find a correlation between trace elements and epidemiological abnormalities. For example,
it has already been proposed that anomalous concentrations of Zn and Pb in
the soils of at least two localities are related to a high degree of mortality
from cancer of the stomach. Also, geochemical evidence provided the basis
for the hypothesis that in some areas, ^{210}Pb, inhaled or ingested, can have a
significant effect regarding sicknesses such as cancer or multiple sclerosis.

Geochemical information may also give indications of areas where there are
deficiencies of one or more chemical elements in the soil and thus provide
farmers with vital information since a deficiency can be as damaging as an excess. One factor that researchers must determine is the chemical form in which
an element is ingested because there are forms that are going to be assimilated
and will accumulate in tissues and specific organs and others that are not going
to accumulate and be retained by the body. Warren and his research team and
other investigators are studying several elements (Pb, Hg, As, Ag, Mo, Cr, Cu,
Zn, Cd, and others) with geochemical and epidemiological aims and collaborat-
ing with medical doctors, agriculturists, and other specialists, whenever possible,
placing special emphasis on the knowledge of the relations between the geochemistry of soils and/or waters and the areal distributions of diseases such as
cancer, colitis, anemia, rheumatism, and arthritis.

From medical studies, it has been established that small concentrations of
Cu and Zn are essential in man's diet and that they are ingested as food and
water. However, the optimum intake quantity so as to be neither toxic nor
deficient has not yet been determined. For example, the medical researchers
Halsted (1968) and Smith and Halsted (1970) commented on geophagia, or
geophagy, the habit of eating earth, including clay and other types of soil. Ele-
ven patients from Iranian villages had long-standing geophagia with associated
profound anemia, dwarfism, hypogonadism, and marked hepatosplenomegaly.
The anemia was demonstrated to be of a pure iron-deficiency type with

prompt reticulocytosis and return of hemoglobin and hematocrit to normal after a suitable period of iron therapy. There were suggestions from later studies that the endocrine abnormalities might be related to an associated zinc deficiency. It is possible that the ingested clay prevents the absorption of iron
by cation exchange; the same may be true for zinc. The awareness of the
geologist–geochemist consulting with medical researchers to the types of materials that may be ingested and some of their physicochemical characteristics,
such as ion exchange, could result in a more rapid hypothesis for explaining an
epidemiological phenomenon related to natural materials and assist in suggest-
ing the proper therapy for alleviating the problem.

There are many situations in which there is a possible relation between
health and the ingestion of metals through food and water. While working in
the Central Cordillera of Colombia near the town of La Ceja, Department of
Antioquia, this author encountered a possible example of natural poisoning by
base-metal ingestion. A vein called La Cristalina was being opened for further
exploration and subsequent exploitation and two miners commented to me
that since they began working at the vein they had been suffering stomach
pains. For drinking purposes and for preparing their meals, the miners used
water from a small fall that flows over the vein. Field analyses of these waters
gave strong positive tests for the total heavy-metal assemblage of Cu, Zn, and
Pb. I told the miners what I knew about the problems that could develop
from the ingestion of an excess amount of these heavy metals and then found
a small nearby creek with waters that gave a negative field test for the same
metals; I then suggested to the miners that they possibly would not have the
same problem potential if they used this latter source of water for drinking
and cooking. A month later, when I returned to the area, the two miners stated that they were no longer suffering the stomach pains. Obviously, the public health authorities of the area have to study the particular situation and de-
cide if my observations were correct. The mineralization of the vein involved
is chalcopyrite and pyrrhotite, but the direct source of the natural pollution of
the heavy metals was copper sulfate that had developed from the chalcopyrite
and was in contact with the water flowing over the vein.

Mineralization of the base-metal elements as sulfides can be the cause of
natural poisoning due to their supplying an excess of Cd to an environment.
Cadmium is an element that is geochemically related to Zn in minerals such as
sphalerite. After the second world war, which ended in 1945, the strange
osteomalacia itai-itai disease appeared among farmers and their families in a
very restricted district along the Jintsu River in Toyama Prefecture, Japan.
The disease causes a weakening of the bones to the point that they fracture
easily and require that the afflicted person remain bedridden; the pain associated with the disease is intense and to x-rays, the bones of fractured legs studied appear as transparent as soft tissues. The Japanese scientists Kobayashi

and Hagino (1965) found excess amounts of Cd, Pb, and Zn in the human tissues and bones analyzed and then found excesses of these metals in the drinking water and agricultural products of the epidemological district, notably the Japanese bran rice. Geological work showed that a mineralized zone existed near the sickness area and was the site of Pb and Zn mines. Sphalerite was the principal zinc ore from the mines and during the beneficiation process metal-bearing fine particles were concentrated by flotation, and due to poor treatment of the wastes muddy waste waters were emptied into the upper stream of the Jintsu River and flowed into the agricultural field downstream polluting the waters used for drinking, cooking, and agriculture. The itai-itai disease suffered by the families of the restricted district and the damage to the rice crop from the excess heavy-metal intake or absorption were gradually corrected and finally eliminated after 1955 when the mine operator constructed a dam and conducted all the muddy waste water from the flotation plant to the dam where the sediment containing the metals settled and was not introduced into the drainage system, thus freeing the Jintsu River from this source of pollution.

The excesses and deficiencies of nutrient and other elements in the diet regime affects all organisms. Such effects on various types of vegetation have already been presented (Table 6-2). For an example of the effect of a diet lacking certain trace elements, refer to a book by Millar, Turk, and Foth (1966); a photograph is presented of a calf in an advanced stage of salt sickness or nutritional anemia that shows the animal emaciated and in poor physical condition and there is evidence of diarrhea. Another photograph of the same animal after its diet was supplemented with Fe-Cu shows a healthy animal that had recovered beautifully. There is also a photograph of a cow in a condition similar to that of the sick calf just described, but the cow was suffering from a Co deficiency; once Co was added to the animal's food ration, it recovered its good state of health. Warren (1961) found that a deficiency of Cu in soil ($<$ 1 ppm) or in vegetation ash ($<$ 25 ppm) resulted in the birth of unhealthy animals and that an addition of zinc (increased from 16 to 46 ppm) to the food of pigs greatly aided their growth. We have already seen how plants are affected by excesses and deficiencies of nutrients, but sometimes it is not only the lack or excess of a nutrient that controls healthy growth systems. Millar, Turk, and Foth (1966) showed a photographic sequence of white bean plants that received between 0 and 720 lb of phosphate in fertilizers over a period of 3 years; these plants had a poor growth and a delayed maturity that was attributed to a zinc deficiency caused by the use of phosphorus-bearing fertilizer. When such nutrient deficiencies exist in soil for one reason or another, one of two things can be done to eliminate this natural pollution and make the soil usable: the nutrient or nutrients deficient in the soil can be added to the soil in forms that would assure their availability to the vegetation or another crop that does not require the deficient nutrient for its healthy growth can be planted.

Another relatively common metal that has surged forward as a dangerous pollutant (introduced into the environment by human activities) is mercury, and researchers have already determined which is the toxic chemical form of Hg that can cause poisoning. Mercury, like Pb, does not seem to have an essential role in the vital functions of plants and animals, and like Pb, is a ubiquitous element. A permissible limit of 0.5 ppm of Hg had been established for foodstuffs and when in 1970 Hg analyses of fish from Lake Ontario (Canada and U.S.A.) gave quantities very much in excess of this figure, 7 ppm, health authorities were shocked. Subsequent analyses of canned tuna fish also showed some cases of Hg contents in excess of the permissible limit, with the result that great quantities of canned tuna fish were taken off market shelves. It was also demonstrated that swordfish meat accumulated important concentrations of Hg and its sale has been practically eliminated in the North American market. CEK (1970) prepared a summary of mercury in the environment and reported that the surface soils of the northeastern United States had an upper limit of 0.04 ± 0.02 ppm of Hg and waters of this area had an average content of 0.06 ppb of the element; the atmosphere contains a few nanograms (10^{-9} grams) per cubic meter. According to CEK (1970), the illnesses caused by the intake of an excess of Hg can be cured or can be fatal, and because Hg poisoning is progressive and depends on the accumulation of the element in the organic system, it has a slow effect on the nervous system. Doctors are aware of a series of signs for the diagnosis of Hg poisoning such as tiredness, headache, loss of ability to concentrate, and failing memory, all of which are reversible if contact with Hg is eliminated. There are other symptoms that are not reversible, such as blurred vision, loss of feeling in the fingers, and emotional irritability. As is the case with other pollutants, the chemical form in which the mercury is absorbed is fundamental to its potential for poisoning an organic system; for Hg, the methyl compound and other alkyl forms are toxic. Phenyl mercury and inorganic mercury are less damaging, but in water they can be converted into methyl mercury; also, some organisms can oxidize the Hg^{2+} to methyl mercury under aerobic conditions. The data compiled by CEK (1970) indicate that the methyl mercury enters into organisms at the beginning of the food chain by the ingestion of fine sediments and that almost all the Hg found in animal tissues is in the form of methyl mercury (98%). The major part of the mercury enters the environment via industrial effluent discharge: the chlorine-alkali industry originally discharged 100 to 200 grams of Hg into the environment for each ton of caustic soda produced; Hg is used in the manufacture of urethane and vinyl chloride plastics; the electrical industry uses more than a million pounds of Hg annually; the agricultural industry uses Hg compounds in the treatment of seeds; sewerage treatment plants use Hg in filters and several types of seals. The effluents and the unforeseen losses from the sources mentioned above and others, artificial and natural, contaminate the environment. If the use of mercury were to be prohibited completely, many industries would fail, but if the use of mercury were to be regulated and the effluents treated, the discharge into the environment could be controlled. In the United States the industrial factories reduced the discharge of Hg by some 86% in a few months after the discovery of the concentration of Hg in the fish and sediments of Lake Ontario and other areas.

Arsenic is another poisonous metal that can be added to the environment by human activities. Spencer (1970) noted the use of As in pesticides, and Angino et al. (1970) discovered environmental pollution from the As in detergents carried into the hydrological system by discharge of spent, dirty water from washing machines. Angino and his colleagues discussed the problem with sanitary engineers and later performed some laboratory experiments on waters that carrieed an experimental concentration of 200 ppb of As using two common water treatment methods; the cold-lime softening treatment removed 85% of the As and the charcoal filtration treatment removed 70% of the As. Therefore, the correct treatment of the waste waters that carry As alleviates the danger posed by the possible accumulation of the metal in waters or sediments.

We can thus understand that many of the elements that the geochemist studies and for which he searches or which he uses as pathfinder elements for economic deposits are actually or potentially natural or artificial pollutants in the environment. The data and concepts put forward by the geochemist can be used by other specialists in pollution investigations to better understand and alleviate or resolve a given problem.

Chlorinated hydrocarbon pollution in the environment is well documented, especially that attributed to DDT, which in many countries can no longer be used or can be used only in limited quantities and/or on specific crops. Other organic compounds such as chlordane and dieldrin may also negatively affect the environment and are being carefully studied (along with a host of other chlorinated hydrocarbons). However, when considering this particular source of pollution (real or potential), governments must make value judgements: where there is hunger that can be eased by the use of pesticides, fungicides, herbicides, and rodenticides to boost agricultural production and protect the harvested product, the question must be asked if it would be better to provide possibly contaminated or damaged food over a period of time, or not provide food at all — the hungry person will want to have the food available. Also, it must not be forgotten that chlorinated hydrocarbon insecticides have saved millions of lives by being employed in the control and eradication of vector-borne diseases such as malaria, viral encephalitis, cholera, Rocky Mountain fever, and tularemia (O'Brien, 1967).

Chlorinated hydrocarbons are man-made chemicals that persist when introduced into an environment and which by their toxic nature pose a health hazard to man and other organisms. Residues of these chemicals are found in the atmosphere, hydrosphere, lithosphere, and biosphere (in our food web and our bodies) both in accessible and relatively inhabited areas and in deep oceans and other remote areas. They are toxic not only to insects, but to a broad representation of life including annelids, arthropods, mollusks, and vertebrates by acting primarily on the nervous system and triggering spontaneous firing of nerve axons resulting in tremors, acute convulsions, and eventually death. At sublethal levels, the chlorinated hydrocarbons cause a variety of physiological effects in test animals, such as nutritional abnormalities, weight loss, susceptibility to disease, and change in behavior. By acting through hepatic enzyme induction, the chlorinated hydrocarbon insecticides can cause serious disturbances in the metabolism of sex hormones (Fouts, 1970). This mechanism has been suggested as the reason for the thinning of eggshells of birds and interference in breeding (Peakall, 1967).

Many investigations have been made on the inclusion of dietary chlorinated hydrocarbons, and the observed effects on test specimens have included decreased hatchability, lack of feathering, parrot beak, shortening and deformities of legs and spine, edema, growth retardation, and syndactylism. Conflicting or noncorroborating results have derived from some investigations. Kenny, Dacke, and Wagstaff (1972) have shown that commercial DDT added to the diet of Japanese quail in quantities of 3, 10, 30, and 100 ppm causes a diminution in shell thickness (by weight/unit area) and egg production. Especially notable was the observation that a lower dose rate of 10 ppm DDT in the diet had a more detrimental effect than a high dose rate of 30 ppm. Since the quail diet had adequate Ca, it was suggested that the shell thinning could be due to decreased activity of carbonic anhydrase, probably in the shell gland. A 1970 report cited by Gadhok and Kenny (1973) showed a marked decrease in carbonic anhydrase levels after quail were on a 3-month diet containing either p, p'-DDT (22% decrease) or p, p'-DDE (44% decrease). However, Japanese quail fed diets with 0.1, 0.3, 1.0, 3.0, 10, 30, and 100 ppm of a technical preparation of DDT for 84 to 112 days, showed no significant change in the blood carbonic anhydrase levels. Gadhok and Kenny (1973) also found that with dietary chlorinated hydrocarbon pesticides, negative responses appear in foul after a certain length of time on the diet only to disappear with further continued feeding. That the effect of dietary chlorinated hydrocarbons varies with the life form being studied was emphasized in the work of Davidson and Sell (1972) who reported that chickens fed at levels of 20 ppm of dieldrin or 200 ppm of p, p'-DDT as part of a nutritionally balanced diet did not suffer from a thinning of eggshells.

Man may ingest or inhale pesticides. Small amounts of ingested insecticides are degraded and effectively excreted, but much of the total quantity ingested is stored and accumulated in the body fat. As regards the potential for ingestion of such chlorinated hydrocarbon pollutants, Adams (1972) has emphasized that soil is the link or barrier in the physical transfer of much of these pollutants from one point in the environment to another. Geochemists are very much involved in similar evaluations for selected elements for prospecting studies. The movement and activity of pesticides in soil is limited by a number of soil variables of which the organic matter content is the most important in temperate climates. By absorption, organic matter retards pesticide movement, reduces pesticide activity, and delays pesticide degradation. The toxicity potential of pesticide decreases as organic content in a soil increases due to absorption and a resulting decrease in biological availability. Although organic matter content of a soil may increase pesticide persistence in a soil, it may also contribute to the degradation of pesticide in the environment and prevent pesticide diffusion throughout the environment because the presence of increased organic matter also indicates a greater microflora population which itself works to decrease degradation time.

Thus far, however, all studies indicate that the present levels of pesticide residues in man's food and environment produce no adverse effects on his health. It has also been noted that because the fertilizers, insecticides, herbicides, fungicides, and rodenticides may reach an environment as mixtures rather than singly, research into the effect(s) of such mixtures on life forms was warranted. It has been demonstrated that mixtures of pesticides were more teratogenic (cause more and serious malformations or deviations from normal structure) than single pesticides even when the single pesticide was administered to laboratory subjects at 50 times the concentration found in human foods. It was also found that different mixtures may show either cumulative or countering effects.

Experimentation on humans has not been done. However, the effects of human exposure to mixed chlorinated hydrocarbons, organophosphate, carbamate, and other pesticides (plus fungicides and herbicides) because of occupation have been described by Sandifer and Keil (1972). These researchers worked with an exposed group and a control group of humans matched by age, race, sex, physical activity, weight, and height and found that there were significant increases in systolic and diastolic blood pressures as well as in serum cholesterol in the 30 subjects with intensive occupational exposure to pesticides and that these increases correlated well with DDT blood levels. There was also unusual evidence of hypertension and elevated blood glucose among the exposed population, with the Negro component of the group showing a relatively high prevalence of hypertension and total lipids as well as elevated plasma DDT levels. Sandifer and Keil (1972) emphasized that there were no differences in medical history between exposed or control subjects or between races.

In addition to the organic pollutants artificially introduced into the environment, naturally occurring organic compounds in clean and "unpolluted" waters are being identified and attempts are being made to correlate type and concentrantion of these with disease. Hanya et al. (1973) cited the Takizawa proposal that some organics in drinking water, such as p-coumaric acid, are possibly responsible for the Kaschin Beck disease prevalent in some areas of Asia, notably in Eastern Siberia and Northern China. The bioassay of rats proved that the drinking water found in areas of Japan with Kaschin Beck disease contains certain substances that cause pathological changes in tissues of rats similar to those of the Kaschin Beck disease; these substances are p-coumaric acid and ferulic acid. The identification of trace amounts of such organic compounds at natural levels in drinking water is difficult but possible, and Hanya et al. (1973) have identified C_{11}–C_{33} straight chain hydrocarbons and some aliphatic cyclic hydrocarbons in drinking water from the Tamagawa River as well as tentatively identifying 25 aromatic hydrocarbons.

The problem of environmental pollution by organic compounds is not restricted to those compounds used directly or indirectly in the agricultural industry nor to those occurring naturally.

There is a class of organic compounds that is potentially dangerous and the use of which should either be prohibited or limited to specific controlled usage for which no viable substitute is available; these compounds are the polychlorinated biphenyls (PCBs) (Gustafson, 1970). These were first found in the natural environment in Sweden in 1966 and subsequently in the United States in 1967 while scientists were investigating several environments for the presence of residues of chlorinated pesticides; later they were found in England, the other Scandanavian countries, Central America, Holland, and the Antarctic. Harvey, Steinhauer, and Teal (1973) found PCBs at the surface and at various depths in the waters of the North Atlantic Ocean between 23° and 63°N latitude. Concentrations averaged about 20 parts per trillion (ppt) in the upper 200 meters of water (passed through a 0.3 μm filter); unfiltered seawater samples had 10% more PCBs. The PCB concentration decreases with depth in the upper 200 meters and higher concentrations are prevented from accumulating in the biologically productive zone by vertical migration of the compounds plus their transport downward by sinking shells, feces, and dead organisms. Martell (1972, personal communication) found traces of PCBs in Atlantic waters between the Virgin Islands and Guadalupe.

Gustafson (1970) cited various experiments with shellfish and foul that showed negative effects of the toxicity with PCBs and observed that because of the presence of these compounds, the concentrations of other chemicals very important for the proper functioning of life forms could be depressed until malfunctions developed in the vital systems; for example, one such effect can be the degradation of steroids that can then lead to altered endocrine relations. However, research on toxicity of PCBs is as yet incomplete. At sublethal exposure, they are physiologically active, inducing enzyme activity, causing liver and kidney damage, reduced spleen, and affecting reproduction in laboratory test animals (Peakall, 1970; Friend and Trainer, 1970; Lincer and Peakall, 1970; Vos, 1972). In another investigation, the conditions on San Miguel Island, off Southern California, where premature births have been reported in the sea lion population, have been studied. The concentration of PCBs and of DDT and its derivatives was 2 to 8 times as high in premature pups and their mothers as it was in full-term controls (*Sea Technology*, 1973). Although 16 Japanese died and over 1000 others were affected (in 1968) when they had eaten rice oil contaminated with polychlorinated biphenyls that leaked from a heat exchanger (Kuratsune et al., 1972) PCBs (as with DDT and other related chlorinated hydrocarbons) are yet to be proved harmful to humans, except in massive doses (> 0.3 grams/day). The PCBs have been used in carbonless reproduction paper and are used in many plasticizing operations, but their most important application is related to their electrical properties as refrigerant fluids and transformer insulation. PCBs are very resistant to oxidation and other types of chemical degradation and therefore are persistent and tend to accumulate in the environment; in Lake Anne, Northern Virginia, PCBs were present at approximately equal concentrations in surface sediments and sediments about 4 in. below the surface, whereas DDT concentration decreased with depth while DDT metabolites increased in concentration with depth (Martell, 1973). The PCBs enter the natural environment mainly by incineration, a process in which they are vaporized, carried to the atmosphere, and accumulated on atmospheric particles that later fall to the natural surface environment (a fact that explains widespread PCB distribution especially in the oceans); they also derive from industrial wastes and effluents and from water runoff on PCB-containing materials, or they may escape from the factories of their manufacture.

Federal agencies in the United States have suggested that stringent curbs rather than an outright ban be put on the use of PCBs so that all uses of PCBs, except in electrical capacitors and transformers where an acceptable substitute is not available, should be discontinued. The Environmental Protection Agency states that it will limit industrial effluents so that levels of PCBs in lakes and rivers do not exceed 0.01 ppb. However, Martell (1973) studied a real estate lake in Northern Virginia and found PCBs in water 10 times in excess of the EPA maximum permissible concentration; lake sediments had 500 times as much and fish 3000 times as much PCBs as the associated waters, thus indicating the potential for PCB accumulation out of the ecosystem by sediments and into the ecosystem by fish and lower members of the food web.

Monsanto Chemical Company has responded correctly to the possible hazards posed by PCBs and will now not sell PCBs for operations in which wastes from the final products of an industrial process or application are not controlled, but chemical industries in Europe and Asia have not responded to the necessity of working for the environment and do not maintain such a control.

Pollution can come from organic sources even though the contaminants are inorganic compounds such as the nutrients PO_4 and NO_3. In Suffolk County, New York, there are great expanses of brackish waters (salinity of 22 to 23 parts per thousand) separated from the ocean by barrier islands that in some places are open to the ocean for the interchange of waters with the brackish bays. Not too many years ago, the areas called Great South Bay and Moriches Bay were the sites of a large shellfish industry, but more recently, more than 40% of the areas of these bays have been classified as contaminated and no shellfish could be taken for food sales. The contamination is ultimately caused by the availability of a large quantity of nutrients on which algae thrive; the algae may bloom tremendously and when their life cycle ends, they sediment onto the bay bottom and use available oxygen for their organic decomposition, depriving other life forms of the normal oxygen supply. Because of this, the shellfish do not develop well and have thin shells and unacceptable meats. Unfortunately, the interchange of waters between the ocean and the bays is not strong enough to provide the oxygen necessary for the proper functioning of the bay environment. The polluting nutrients come from two sources. The first source is waste waters from sewerage overflow and from cracks in cesspools and cisterns; the population of the zone increased fantastically from 250,000 inhabitants in 1963 to 750,000 in 1968, with projections of 2 million inhabitants by 1980, but only 5% of the population is tied in to sewerage collection and treatment facilities. Second, 60% of the ducks consumed in the United States annually are grown in the area and for a long time, the excrement of more than 3 million ducks (which lived an average of 6 weeks) entered directly into the bay waters. This last problem was alleviated only by legal means and threats of the application of large fines, but the sediment composed of the residues of the duck excrement (sludge) still covers a large part of the bays and the beds of streams and rivers flowing into the bays. Here then we have a good example of pollution tied directly to the population and the industry of an area, pollution that partially ruined the ecology of what had been a live, vibrant, and flourishing environment. What is possibly equally as interesting to the reader and the appraiser of the problem is the fact that the state and federal governments had offerred to pay about 85% of the costs for construction of sewerage collection and treatment plants for the areas with the remaining 15% to be provided by the communities via the selling of bonds and corresponding slight rise in taxes to pay the bonds off; however, in a referendum on the issue held in February, 1967, the population voted 6 to 1 against the project. Sometimes it is easier to understand nature than man. The problem has not been resolved as yet.

The presence of solid and liquid animal wastes in the environment provides a danger of serious contamination. In some cattle-producing countries it is very common to raise the herds in giant lots in which thousands of animals may inhabit relatively restricted areas. The excrements are sources of major quantities of nitrate. If the nitrate is able to enter the hydrogeological system and is absorbed into sources of potable water, a dangerous situation would exist for infants because the intake of excess amounts of nitrate can result in the blue baby sickness. There is much less of a problem with this type of pollution of the cattle feed on the prairees or large verdant areas.

Although the geochemist does not work that much on topics related to organic material or the organic vital system such as the last few mentioned, he must be aware of them, take them into consideration during his normal activities, and be alert to their influences on any problem in the natural environment being investigated.

Finally, let us consider radioactive pollution. Radionuclides come from several sources:

1. From the primeval thermonucleosynthesis (for example, ^{235}U, ^{238}U, ^{232}Th, ^{40}K, and their many decay products).

2. From the interactions of cosmic rays (high-energy protons) and/or solar protons with elements in the atmosphere (N_2, O_2, Ar) and extraterrestrial material (for example, ^3H, ^7Be, ^{10}Be, ^{14}C, ^{26}Al, and ^{32}Si); the radionuclides that originate in this way reach the Earth's surface adsorbed onto precipitation or as particulates (Mason, 1966, calculated that up to 10,000 metric tons of cosmic dust are being deposited daily on the Earth).

3. From human activity via fission and fusion processes (for example, ^{90}Sr, ^{137}Cs, and 40 other radioisotopes have been introduced into or added to the environment from nuclear explosions, nuclear reactors for electrical power, fuel production, ship and space vehicle propulsion, nuclear fuel processing, waste disposal, auxiliary power generators, and radioactive tracers).

Many of these radioactive chemical elements have already established or approach a secular equilibrium with the natural environment, that is, the quantity that is produced and added to the Earth environment is approximately equal to the amount that is deposited or diminished by natural process decays. Examples of this group of radioisotopes include ^{14}C, ^{10}Be, and ^3H.

A great part of the research on pollution from radionuclides has centered in the marine environment because the seas cover more than 70% of the planetary surface. In addition, in the marine environment and samples taken from it, there is less interference and dilution of materials due to natural radioactivity than there is on land and there are fewer problems in the separation of the elements for analysis. Complementary data from many years of studies in geological–biological–physical–chemical oceanography exist and serve to assist us

Table 11-4 Radionuclides Identified in the Marine Environment (compiled from Koszy and Rosholt, 1962, Fairbridge, 1966; Lal and Peters, 1967; Joseph et al., 1971)

Nuclide	Half-Life	Concentration in Ocean (grams/liter)	Concentration in Sediments (dry) (grams/grams)	Arms	Other Wastes	Natural[a]
^{3}H	12.26 years	1.7×10^{-18}		X		X[c]
^{7}Be	53 days	$<4.9 \times 10^{-17}$				X[c]
^{10}Be	2.5×10^6 years	2.2×10^{-17}	$(1-3) \times 10^{-13}$			X[c]
^{14}C	5570 years	$(2-3) \times 10^{-14}$	$(0.1-1) \times 10^{-13}$	X		X[c]
^{24}Na	2.6 years					X[c]
^{26}Al	7.4×10^5 years	2.9×10^{-19}	$(0.15-1.5) \times 10^{-15}$			X[c]
^{32}Si	710 years	5×10^{-19}	$(0-2) \times 10^{-16}$			X[c]
^{32}P	14.3 days	$<1.5 \times 10^{-18}$		X		X[c]
^{33}P	25 days	$<3.1 \times 10^{-18}$				X[c]
^{35}S	87 days	$<1.8 \times 10^{-18}$				X[c]
^{35}Cl	3.1×10^5 years	7.7×10^{-17}				X[c]
^{39}Cl	1 hour					X[c]
^{37}Ar	35 days					X[c]
^{39}Ar	270 years	3.8×10^{-20}				X[c]
^{40}K	1.25×10^9 years	4.7×10^{-5}	$(0.8-4.5) \times 10^{-6}$			X[t]
^{46}Sc	83.9 days				X	
^{51}Cr	27.8 days				X	
^{54}Mn	291 days				X	
^{55}Fe	2.6 years			X	X	
^{59}Fe	45.1 days			X	X	
^{57}Co	270 days			X	X	

Nuclide	Half-life					
^{58}Co	71.3 days			X	X	
^{60}Co	5.24 years			X	X	
^{65}Zn	245 days			X	X	
^{76}As	26.4 hours			X		
^{85}Kr	10.3 years				X	X^ت
^{87}Rb	4.7×10^{10} years	3.4×10^{-5}	$(2.3-5.7) \times 10^{-6}$			
^{89}Sr	50.5 days			X	X	
^{90}Sr	27.7 years	$(0.63-9.5) \times 10^{-19}$		X	X	
^{90}Y	64.2 hours			X	X	
^{91}Y	57.5 days			X		
^{95}Zr	65 days			X	X	
^{95}Nb	35 days			X	X	
^{99}Mo	66 hours			X	X	
^{103}Ru	39.8 days			X	X	
^{106}Ru	1 year			X	X	
^{106}Rh	130 minutes			X	X	
Cd-113m	5.1 years					
Cd-115m	43 days			X		
^{115}In	6.0×10^{14} years				X	X^t
^{125}Sb	2 years				X	X^t
^{129}I	1.7×10^{7} years	1.6×10^{-10}		X	X	
^{131}I	8.08 days			X	X	
^{132}Te	77.7 hours			X	X	
^{137}Cs	26.6 years	$(0.52-2.6) \times 10^{-1}$			X	X^t
^{138}La	2.0×10^{11} years					
^{140}La	40.22 hours			X		
^{140}Ba	12.8 days			X		

Table 11-4 (Continued)

Nuclide	Half-Life	Concentration in Ocean (grams/liter)	Concentration in Sediments (dry) (grams/grams)	Derivation		
				Arms	Other Wastes	Natural[a]
^{141}Ce	33.1 days			X	X	
^{144}Ce	285 days			X	X	
^{144}Nd	5.0×10^{15} years					X^ت
^{147}Sa	1.3×10^{11} years					X^t
^{176}Lu	2.4×10^{10} years					X^t
^{180}W	10^{14} years					X^t
^{185}W	75.8 days			X		
^{187}Re	5.0×10^{10} years					X^t
^{190}Pt	10^{12} years					X^t
^{207}Bi	8.0 years			X		
^{207}Tl	4.79 minutes	$<1.2 \times 10^{-23}$	2.1×10^{-21}			X^t
^{208}Tl	3.10 minutes	4.1×10^{-24}	6.7×10^{-22}			X^t
^{210}Pb	19.4 years	1.1×10^{-15}	4.5×10^{-14}		X	
^{211}Pb	36.1 minutes	$<9 \times 10^{-23}$	1.6×10^{-20}			X^t
^{212}Pb	10.6 hours	2.4×10^{-21}	3.9×10^{-19}			X^t
^{214}Pb	26.8 minutes	2.9×10^{-21}	1.2×10^{-19}			X^t
^{210}Bi	5.01 days	7.8×10^{-19}	3.1×10^{-17}			X^t
^{211}Bi	2.16 minutes	$<5.6 \times 10^{-24}$	1.0×10^{-21}			X^t
^{212}Bi	60.5 minutes	2.2×10^{-22}	3.7×10^{-24}			X^t
^{214}Bi	19.7 minutes	2.1×10^{-21}	8.8×10^{-20}			X^t
^{210}Po	138.4 days	2.2×10^{-17}	8.8×10^{-16}			X^t
^{211}Po	0.52 seconds	$<6.8 \times 10^{-29}$	1.2×10^{-26}			X^t

Isotope	Half-life				
212 Po	3.04×10^{-7} seconds	1.2×10^{-32}	2.4×10^{-29}		X^t
214 Po	1.64×10^{-4} seconds	3.0×10^{-28}	1.1×10^{-27}		X^t
215 Po	1.83×10^{-3} seconds	$<8.1 \times 10^{-29}$	1.4×10^{-26}		X^t
216 Po	0.158 seconds	1.0×10^{-26}	1.7×10^{-24}		X^t
218 Po	3.05 minutes	3.4×10^{-22}	1.4×10^{-20}		X^t
219 Rn	3.92 seconds	$<1.7 \times 10^{-25}$	3.1×10^{-23}		X^t
220 Rn	51.5 seconds	3.3×10^{-24}	5.4×10^{-22}		X^t
222 Rn	3.8 days	6.3×10^{-19}	2.5×10^{-17}		X^t
223 Fr	22 minutes	$<7 \times 10^{-24}$	1.4×10^{-21}		X^t
223 Ra	11.68 days	$<4.4 \times 10^{-20}$	8.5×10^{-18}		X^t
224 Ra	3.64 days	2.1×10^{-20}	3.4×10^{-18}		X^t
226 Ra	1622 years	1.0×10^{-13}	4.0×10^{-12}		X^t
228 Ra	6.7 years	1.4×10^{-16}	2.3×10^{-15}		X^t
227 Ac	21.6 years	$<1.0 \times 10^{-15}$	5.9×10^{-15}		X^t
228 Ac	6.13 hours	1.5×10^{-20}	2.4×10^{-19}		X^t
227 Th	18.17 days	$<7.0 \times 10^{-20}$	1.3×10^{-17}		X^t
228 Th	1.91 years	$<4.0 \times 10^{-17}$	7.0×10^{-16}		X^t
230 Th	7.52×10^4 years	$<3.0 \times 10^{-13}$	2.0×10^{-10}		X^t
231 Th	25.6 hours	8.6×10^{-20}	2.9×10^{-20}		X^t
232 Th	1.42×10^{10} years	1.0×10^{-10}	5.0×10^{-6}		X^t
234 Th	24.1 days	4.3×10^{-17}	1.4×10^{-11}		X^t
231 Pa	3.43×10^4 years	$<2 \times 10^{-12}$	1.0×10^{-11}		X^t
234 Pa	1.14 minutes	1.4×10^{-19}	4.7×10^{-20}		X^t
234 U	2.48×10^5 years	1.9×10^{-10}	8.1×10^{-11}	X	X^t
235 U	7.13×10^8 years	2.1×10^{-8}	7.1×10^{-9}	X	X^t
238 U	4.51×10^9 years	3.0×10^{-6}	1.0×10^{-6}	X	X^t
239 Np	2.35 days			X	
239 Pu	24.36 years			X	

[a] X^c indicates natural cosmic origin and X^t indicates natural terrigenous origin.

in recognizing and explaining some of the processes related to radioisotopes (Table 11-4). For example, a factor called residence time in the oceans has been estimated for many chemical elements; this factor gives the researcher an idea of how long a chemical element will remain in marine waters once it is introduced into the ocean environment. It is defined by the expression:

$$\tau = \frac{A}{dA/dt}$$

in which A = the quantity of an element that is introduced into the marine environment

dA/dt = the quantity of the element that is deposited or removed from the system in a unit of time

The τ depends on many variables such as the chemical reactivity of an element, the quantity of the element that enters in the aqueous system, the manner and location of introduction, the element's chemical or physical form (particulate, liquid, gaseous), the water depth, the rate of sedimentation, and the biological effect.

We know that for more than 25 years, man's role in the radioactive pollution of the oceans has been increasing even though the disposal of radioactive materials and wastes is controlled by provincial (or state), national, and to some degree international laws. For example, there is a value that has been designated as the maximum permissible concentration (MPC) of a radionuclide that can be present in the effluents from factories or industries that use or process radionuclides. However, the figures used for these concentrations are very empirical since we do not have statistical data representing a long enough time period to be able to evaluate the effects of these MPCs on the human body with complete confidence. It has been established that in an ecosystem, different forms of life have different radiosensitivities and that there is a gene selection against the radioresistant genes (Bowen et al., 1971). Templeton et al. (1971) wrote that the effects of radiation from within and outside a life form are cumulative, that is, the risk of inducing disease or damage is greater with accumulated dose and may be linear with dose. They found that radiation effects are most dramatically expressed and include direct lethality, reduced vigor, shortened life span, and diminished reproductive rate; genetically, they are expressed by the transmission of radiation-altered genes to offspring. The lethal dose varies with many factors, among which are species, age (older forms are more resistant), size of body, physiological state, temperature, dissolved oxygen, salinity, and general chemical composition of the environment. Any radiation level exposure increase (over normal background) is genetically detrimental and although individual abnormalities resulting from increased radiation exposure are not repairable, natural selection repairs genetic damage in a

Table 11-5 Acceptable Dose Limits for Members of the Public as Recommended by the International Commission on Radiological Protection (1959). Similar Limitations are Recommended by the United States Federal Radiation Council (1960,1961) (from Foster et al., 1971)

Part of Body	ICRP Dose Limit
For individuals	
Gonads, red bone marrow, whole body (uniformly irradiated)	0.5 rem/year
Skin, bone	3.0 rem/year
Thyroid	3.0 rem/year (1.5 rem/year for children under 16)
Hands, forearms, feet ankles	7.5 rem/year
Other single organs (including gastrointestinal tract)	1.5 rem/year
For populations	
Whole body	5.0 rem/person for 30 year

population. Templeton et al. (1971) reported that at Oak Ridge, Tennessee, where natural populations were exposed to radiation levels higher than background for many years, increased chromosomal aberrations of lake sediment larvae of *Chironomus tentans* were eliminated by natural selection, and that on Bikini and Eniwetok islands, populations of plants, animals, and insects were reestablished after suffering the ravages of nuclear arms testings; also, in the very radioactive-contaminated coastal zone of the Irish Sea adjacent to the Windscale atomic facility, there were effectively no changes in bottom organisms in the region of the Windscale discharge.

In the United States there are some states not in agreement with the maximum permissible concentrations established by the National Atomic Energy Commission; these states have set their own permissible concentration levels with the result that there exist judicial proceedings between state and federal legal departments. The International Radiological Protection Commission of UNESCO initiated the establishing of the so-called maximum permissible concentrations (Table 11-5), but until there are more clinical data on several generations of humans exposed to the maximum and greater concentrations (in addition to the normal background radiation and medical exposure), the figures suggested for the MPCs must be considered only as guides. The MPC is defined as the concentration of a radioisotope that can be present in inhaled air and in drinking water for workers at their sites of employment—the general public has a maximum permissible concentration level that is one tenth of the MPC. Unfortunately, there is much flexibility in the establishment of such figures and this may be damaging for the generations of human beings that follow us. In some countries many uranium miners recently developed cancer and it is supposed that the cancerous condition developed from continual exposure of the miners to radiation or to the inhalation of radioactive gases; this suggests that the levels of exposure protection and perhaps the means of control were not determined carefully enough to afford the maximum protection to the now afflicted miners. After detection and diagnosis of the disease, the controls were made more stringent. The sources of radiation are critical, as are the frequency and length of time that people are exposed to the source (as well as the number of people that may be exposed) and careful measures and records must be kept for future comparative studies.

There are two classes of radionuclides that have been added to the global inventory by human activities. These are the radionuclides created by neutron reactions and include the isotopes ^{32}P, ^{46}Sc, ^{51}Cr, ^{54}Mn, ^{55}Fe, ^{59}Fe, ^{60}Co, ^{65}Zn, ^{76}As, ^{125}Sb, and ^{131}I, and the radionuclides produced by fission reactions which include ^{89}Sr, ^{90}Sr, ^{90}Y, ^{95}Zr, ^{95}Nb, ^{103}Ru, ^{106}Ru, ^{106}Rh, ^{137}Cs, ^{144}Ce, ^{210}Po, ^{239}Pu, ^{234}U, ^{235}U, and ^{238}U. These isotopes originate for the most part from the sources cited below, of which the first three are important now, and the remaining are becoming increasingly important:

1. Nuclear arms testing
2. Extraction plants for sources of spontaneous fission
3. Reactors for thermoelectric nuclear power
4. Nuclear powered ships
5. Tracers
6. Auxiliary power generators

Ninety-eight (98) radioisotopes representing 54 chemical elements have been identified in seawater and in marine organisms, and of the 98, 62 (many of which are daughter products from the decay of uranium and thorium) occur naturally (Table 11-4). The elements from the sources cited above enter into the human environment and may be dispersed or accumulated depending on their chemical and/or physical forms and the physical–chemical–biological conditions of the environment. The following are found in particulate form: ^{54}Mn, ^{55}Fe, ^{59}Fe, ^{57}Co, ^{58}Co, ^{60}Co, ^{90}Y, ^{95}Zr, ^{95}Nb, ^{99}Mo, ^{103}Ru, ^{106}Ru, ^{106}Rh, ^{141}Ce, ^{144}Ce, ^{144}Pr, ^{207}Bi, and ^{239}Pu. The isotopes ^{32}P, ^{51}Cr, ^{89}Sr, ^{90}Sr, ^{125}Sb, ^{131}I, and ^{137}Cs, are found in soluble form. These isotopes may be concentrated locally by physical–chemical processes such as adsorption, ion-exchange, and coprecipitation (Figure 11-3), processes especially important in relatively quiet water. When they enter a very energetic physical marine environment, they may be dispersed and effectively diluted depending on the physical and chemical states of the substances involved, the manner of introduction (point source or blanket fallout), and the location of introduction into an environment. The isotopes may be advected away from the point of introduc-

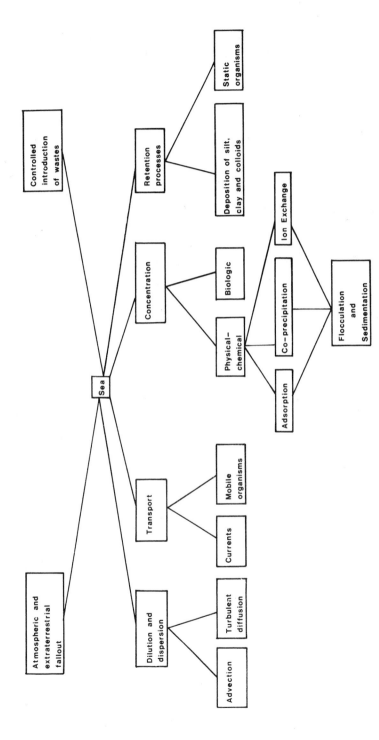

Figure 11-3. A schematic showing the processes that influence the dispersion of an affluent injected into the marine environment. (From Waldichuk, 1961.)

Table 11-6 Average Concentration Factors of Chemical Elements with Respect to Seawater for Benthic Algae, Plankton, and Mollusk, Crustacean, and Fish Muscle (from Lowman, Rice, and Richards, 1971)

Element	Benthic Algae	Phytoplankton	Zooplankton	Mollusk Muscle or Soft Parts	Crustacean Muscle	Fish Muscle
Li				0.28	1.2	0.47
Na				0.2	0.3	0.13
K				8	13	13
Rb				16	13	17
Cs				8	23	15
Fr						
Cu	100	30,000	6,000	5,000		1,000
Ag		23,000	9,000	7,100		
Au	470			400	7	60
Be	110	1,000	15		400	
Mg	2	2	4	1		0.2
Ca	2	2.5	5	0.4	120	1.5
Sr	96			1	3	0.1
Ba	1,400	17,000	900	1,300		8
Ra	410	12,000	190		140	130
Zn	200	15,000	8,000	11,000	2,000	500
Cd						1,000
Hg						
B	<1		1.6			
Al	15,000	100,000	100,000	9,000	12,000	2
Ga	1,300	8,000	7,000	2,000	2,000	10,000
In						
Tl						
Sc	2,000	2,000	1,000		300	750

Element						
Y	480	1,000	105	12		250
La						
Ce	670	90,000	1,000	360	2	0.3
Pu	1,300	2,600	2,600	260	3	3
C	~4,000	3,600	2,800	4,700	3,600	5,400
Si	100	2,000	300	50		20
Ge	50					
Sn						
Pb	700	6,000	450	40		
Ti	4,100	40,000	3,000			
Zr	2,200	25,000	17,000			
Hf		60,000	25,000	2	2	<1
N	10,000	36,000	24,000	47,500	44,000	65,000
P	2,000	34,000	13,000	6,000	24,000	33,000
As				650	400	700
Sb						
Bi						
V	600	600	700	1,700	330	110
Nb	1,000	1,000		7	3	100
Ta	1					
S	1			0.3		1
Se						
Te						
Po						
Cr	1,000	2,400	1,900	440	100	70
Mo	1,600		26	60	10	10
W	8			20	2	3
F	5					
Cl	1	1	1	1	1	1
Br	1					

Table 11-6 *(Continued)*

Element	Benthic Algae	Phytoplankton	Zooplankton	Mollusk Muscle or Soft Parts	Crustacean Muscle	Fish Muscle
I	5,000		3,000	50	30	12
A						
Mn	2,300	4,000	1,500	12,000	1,900	80
Tc						
Re						
Fe	4,800	45,000	25,000	9,600	2,400	1,600
Co	800	1,500	700	600	500	10
Ni	1,000	5,000	3,000			
Ru	390	200,000	34,000	3	100	0.05
Rh						
Pd						
Os						
Ir						
Pt						

tion as a contaminated cloud by local currents and subsequently dispersed by nonadvective mechanisms (diffusive action) and by turbulent mixing processes. For example, eddies larger than a contaminated cloud move the entire cloud, whereas eddies smaller than the cloud tend to shear and stir the cloud, effectively breaking it up. In addition to a rapid mixing of injected substances that may take place in the coastal ocean (includes estuaries, lagoons, continental shelfs, and marginal seas), circulation patterns (especially of soluble materials that are mixed rapidly) may favor the near-coast retention of substances introduced into the coastal ocean, and in such a situation, isotope distribution may replicate oceanic flow patterns.

Very important to a discussion of concentration of radionuclides in the environment is the fact that some organisms we may use for food can concentrate chemical elements without distinguishing between the nonradioactive and radioactive forms, thus creating a potential hazard for the population that consumes them. The estuarine, coastal, and near shore zones (to depths of 1000 meters) are areas of intense biological activity (result of good light, strong nutrient concentrations, bottom rich in biota feeding from photic zone above) from which we obtain our major harvests of sea food and are areas in which the phyto- and zoo-plankton are most likely to influence radionuclide distribution in the ocean environment. The plankton represent the largest biomass in the ocean, have the largest exposed biological surface area, have the highest chemical element concentration factors of marine life forms (Table 11-6), and are quick to reach chemical equilibrium with radionuclides. Marine organisms may accumulate radionuclides that are fission products (e.g., ^{90}Mo, ^{95}Zr, ^{103}Ru, ^{106}Ru, ^{131}I, ^{140}Ba, ^{141}Ce and ^{144}Ce) or neutron-induced radionuclides (e.g., ^{32}P, ^{51}Cr, ^{54}Mn, ^{55}Fe, ^{59}Fe, ^{57}Co, ^{65}Zn, ^{110}Ag, ^{181}W, ^{185}W, ^{187}W, and ^{40}K) that may be biologically important elements (e.g., ^{32}P, ^{54}Mn, ^{55}Fe, ^{59}Fe, ^{57}Co, ^{58}Co, ^{60}Co, ^{58}Fe, ^{60}Fe, and ^{65}Zn) of shells, enzymes, vitamins, and molecules that participate in oxygen and electron transfer, or may not be biologically important (Lowman, Rice, and Richards, 1971). In addition, Maucheline and Templeton (1964) have shown that organisms may concentrate the radioisotopes in specific parts of their anatomy (Table 11-7). Both ionic feeders and filter feeders accumulate the radionuclides with the lower trophic levels (smaller size, greater volume, greater surface area), concentrating them significantly depending on their biological availability (whether in particulate form, what chemical form, and what type of matrix to which the isotope is attached or in which it is imbedded). Filter feeders achieve radionuclide equilibrium concentration levels more rapidly than nonfilter feeders for those elements adsorbed onto sediments by epiphyton (bacteria) that may accumulate several trace elements (and isotopes) efficiently (Lowman, Rice, and Richards 1971). The radionuclides may pass along the food web from the photosynthesizing plants to the grazers or herbivores to the predators on herbivores to the

Table 11-7 Concentration Factors for Radioisotopes in Tissues of Marine Organisms (modified from Mauchline and Templeton, 1964; after Krumholz, Goldberg, and Boroughs, 1957)

Isotope	Algae	Invertebrates		Vertebrates		Plankton
		Soft Parts	Skeleton	Soft Parts	Skeleton	
[32]P	10,000	10,000	10,000	40,000	2×10^6	
[46]Sc	1,500–2,600					
[51]Cr	300–100,000					
[54]Mn	6500	70–1,000		2,000	2,000	750
[55]Fe, [59]Fe	20,000–35,000	10,000		1,000	5,000	2,000–140,000
[57]Co, [58]Co, [60]Co	450	10,000	100,000	30–1,000	1,500	
[65]Zn				160	300–3,000	1,000
[76]As	200–6,000	1–25	180–1,800	1,000		
[89]Sr, [90]Sr	1–40			0.1–2.0	100–200	9
[90]Y, [91]Y	100–1,000	20–100	300	5		
[95]Zr	350–1,000					
[95]Nb	450–1,000					1,500–3,000
[99]Mo	10–100	100		20		
[103]Ru, [106]Ru, [106]Rh	15–2,000	2,000		3		
[125]Sb		100(?)		100(?)		600–3,000
[131]I	3,000–10,000	50–100	50	7–10		50
[137]Cs	1–100	10–100	0	10–100		
[141]Ce, [144]Ce	300–900	300–2,000	45	12		0–5
[209]Bi		500		50		7,500
[234]U, [235]U, [238]U			400	<20		
[239]Pu	10			13		

predators on predators until "man" the predator is reached at the apex of the web. Whether or not high (relative) concentrations of radionuclides reach man depends on various factors of which the biological concentration factor (BCF) may be the most important. The biological concentration factor is defined as the concentration of the element in the organism divided by the element concentration in seawater or in the preceding food link. In some publications, a similar quantity called the "biological accumulation factor" (BAF) of the elements in an organism or its tissues is used and is the number and concentration of radioisotopes by weight in excess of the quantity of radioisotopes in an equal weight of water from the research environment, assuming that equilibrium conditions have been established. Both of these factors are influenced by the biological half-life of an element in an organism, the accumulation of an element in different body pools, and whether an entire organism or selected organs are ingested along the food web. The vital processes are as important, therefore, as the inorganic chemical processes. Depending on feeding rates, food utilization, and relative utilization, there may be a large accumulation of radioisotopes (Lowman et al., 1971; Bowen et al., 1971); however, if these are of a short biological half-life, as is the case for ^{32}P, ^{51}Cr, ^{59}Fe, ^{76}As, ^{95}Nb, and ^{131}I, the problems of concentration may be less grave (potentially) because elimination of the radioactive isotopes plus their natural decay work to diminish the danger of human pollution, especially when the radionuclides move down in fecal pellets and are adsorbed onto them and enter into the deposition phase of the sedimentological cycle where they may suffer mixing at the sediment–water interface, but where they may be expected to be rapidly taken out of the ecosystem. Lowman et al. (1971) have determined that the radionuclides expected to remain in fecal pellets and not be released again into the marine environment include those that form relatively insoluble salts of hydroxides in seawater and are accompanied by carrier amounts of the corresponding (or a similar) stable element, those that form strong complexes with biological surfaces, those that coprecipitate with inorganic scavengers, and those that are incorporated into biological structural compounds (e.g., C, H, N, S, P, Ca and Si into skeletal material, lipids, proteins, and polymerized carbohydrates). The potentially most dangerous radioisotopes present in marine waters are ^{90}Sr and ^{137}Cs, but ^{137}Cs has a short biological half-life and is rapidly evacuated, whereas ^{90}Sr is retained by and keeps accumulating in certain forms of life. On the basis of data such as that presented in Tables 11-4, 11-6, and 11-7 and evaluations of other factors discussed above, it may be possible to determine critical pathways of introduction of radioactive pollution into our ecosystem and both attempt to short circuit these pathways and be able to correctly estimate or predict and control doses of exposure from within and from outside an immediate life form. With these capabilities and assembled facts, efforts must be maintained to control low-level radioactive waste disposal in the marine environment so that the ocean activity will not reach unacceptable levels. With this in mind, we must insist that over-harvesting of lower trophic levels not be allowed because depleting such levels may reduce the ocean capacity to safely accept radioactivity. Bowen et al. (1971) believe that safe harvesting of these lower trophic levels can supply enough animal protein to feed six billion people.

In addition to these contaminants, there are others that must be disposed of and they include the tools and equipment used in the manipulation of radioactive materials and in other processes taking place in nuclear plants. It is essential to correctly determine where these highly radioactive materials may be stored until the radioactivity diminishes, or if the radioactivity is of a long half-life, we must be careful as to where they are stored or dumped so that their radioactivity does not affect the ecosystem. Each country has its own norms for the resolution of the radioactive waste disposal problem, and little by little the controls on the manufacture and dumping of radioactive materials and products are increasing. For some time the United States and the Soviet Union have confined their nuclear arms testing to underground environments and have been in close contact on nuclear arms limitations (SALT talks). However, within the community of nations, France and the People's Republic of China continue introducing nuclear radioactive particles and gases into the atmosphere via arms testing, although the nations and peoples of the world condemn this activity and at the least are insisting on an ending of nuclear arms tests in the atmosphere; man cannot continue to poison man in the name of deterrent and national defense.

Most recently, Kubo and Rose (1973) have discussed the options that can be employed in the problem of disposal of nuclear wastes from nuclear fission reactors. They believe that at an increased but still modest cost (compared to present disposal techniques), the outlook for nuclear waste disposal can be improved. The major options to be considered (given in Figure 11-4 and summarized in Table 11-8) include a storage in mausolea, disposal in mines of various sorts (salt, for example) and perhaps in ice (possibly in Antarctica), fusion in *in situ* melts, and via further chemical separations other than those developed to the present; all these options are interdependent.

In accord with national laws, thermonuclear plants constructed for the generation of electrical energy must treat their effluents. The radioactive wastes are introduced into retention tanks where they may be neutralized, filtered through preformed sludge layers, and continuously controlled until the radioactivity is less than the MPC, after which the wastes are carried to the ocean. The question still persists as to whether the MPCs are really permissible with respect to the global environment. The directors of the Windscale, England, thermonuclear electrical generating plant have an additional control on the level of emission of radionuclides in the plant's effluents; they maintain a strict observation

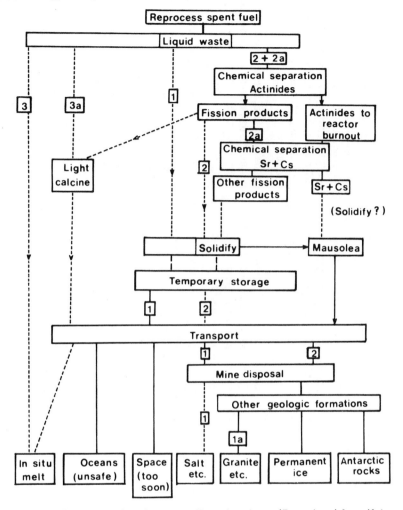

Figure 11-4. Taxonomy of nuclear waste disposal options. (Reproduced from Kubo and Rose, 1973, *Science.* Copyright © 1973 by The American Association for the Advancement of Science.)

on the concentration of the isotope ^{106}Ru that accumulates in marine algae used in the preparation of a special bread called laver bread. While a permissible level of radioactivity is present in the alga, fish from the zone can be eaten and the alga can be used for the preparation of laver bread. The chance of an accident or other sudden release of radioactive nuclides to the environment always exists. If this were to occur in an estuarine or coastal area, for example, it could prove hazardous to large population concentrations. Capa-

Table 11-8 Summary of Radioactive Waste Disposal Options. The Routes Refer to the Diagram in Figure 11-4 (from Kubo and Rose, 1973)

Route	Option	Cost (mill/kwhe)	Advantages	Disadvantages
1	Salt mine	0.045–0.050	Most technical work to date; plastic media with good thermal properties occur in seismically stable regions	Corrosive media; highly susceptible to water; normally associated with other valuable minerals; difficult to monitor and retrieve wastes
1A	Granite	0.050–0.055	Crystalline rock; low porosity if sound; comparable to salt in thermal properties; retrievable wastes	Nonplastic media; presence of groundwater; difficult to monitor
2	Further chemical separation; recycle actinides	0.065–0.320	Reduced long-term toxicity; technology feasible; increases future options	Additional handling and processing; more toxic materials in fuel inventory; waste dilution due to processing; fission products remain
2A	Further chemical separation; remove Sr and Cs; recycle actinides	0.140–1.100	Reduced long-term toxicity; reduced short-term thermal power; some reduction of fission product toxicity; increases future options	Additional handling and processing; more toxic materials in fuel inventory; waste dilution due to processing; storage and disposal of Sr and Cs extract; fission products remain

3	Melt *in situ*	0.011–0.016	*In situ* creation of insoluble rock-waste matrix; no transportation; reduced handling	Highly mobile wastes during 25-year boiling phase; presence of groundwater; irretrievable wastes; proliferation of disposal sites; difficult to monitor
3A	Melt *in situ*, central repository	0.031–0.036	*In situ* creation of insoluble rock-waste matrix; short boiling period; no proliferation of sites	Presence of groundwater; irretrievable wastes; difficult to monitor
2	Antarctic rocks		Immobile water	Very narrow temperature limits; not a permanent geologic feature; difficult environment
2	Continental ice sheets		Immobile water	Cannot dispose of actinides; limited amount of ice; not a permanent geologic feature; difficult environment

Source. Reprinted from *Science*, Copyright 1973 by the American Association for the Advancement of Science.

bilities for capturing and cleaning up an environment after a possible accident have been studied and it has been proposed that sediments or other solids with sorption capacities could rapidly scavenge the radionuclide pollutants from the waters and cause them to deposit on the bottom. It has been demonstrated that radionuclide loss from water depends on the distribution coefficients of the elements involved (e.g., in oxygenated versus anoxic conditions), the particles present, the rates of sorption, and settling velocities. The sorption seems to be affected by the chemical properties of the radionuclides and by the physical–chemical and biological conditions of the sediments (pH, Eh, ionic potential, and bacterial activity) more than by their bulk composition. Although complete sorption equilibrium may not be readily reached, an 80% equilibrium level is attained in a few hours. Duursma and Gross (1971) have treated the sorption topic in excellent detail.

The number of vessels that sail under nuclear power is limited, but can be expected to increase as global fuel limitations become more severe. Both fission products (^{89}Sr, ^{90}Sr, ^{131}I, ^{140}Ba, ^{137}Cs and ^{144}Ce) and corrosion products (^{18}F, ^{24}Na, ^{51}Cr, ^{56}Mn, ^{59}Fe, ^{60}Co, ^{64}Cu, ^{65}Ni, ^{182}Ta, and ^{187}W) have been identified in the reactor coolant water of the USS NAUTILUS (Joseph et al., 1971). Onboard the waste products from nuclear reactions may be passed through an ion-exchange resin or other suitable medium where the radioisotopes are trapped; the resins are sealed in special packets and then dumped into the ocean where they settle on the floor and are presumably not going to harm the immediate environment. Otherwise, the reactor wastes are contained in the unit and treated only when the vessel makes port. The plants at which sources of spontaneous fission are extracted have controls and continuous monitoring on the radioactive wastes in the liquid effluents, chimney emissions, and local working area atmospheres. Although radioactive tracers are used in medical and nonmedical research programs, the quantity used is small and not expected to increase significantly and the isotopes employed are generally those with short half-lives. Medical therapy programs may use radioactive materials, but monitoring of radiation in areas of such use is rigidly controlled both in the patient and in the medical personnel administering the therapy. Considering all this, the aspects of pollution of the environment by radioactivity are not all negative. In marine geochemistry, for example, the radionuclides aid in the study of many fundamental processes that help us to understand the environment and permit us to make more precise evaluations of potential utilization of the environment and subenvironments and their contained materials; among these processes are diffusion in the sea, circulation and mixing of marine waters, overturn, ages and boundaries of water masses, may also be studied in this way.

Pollution is with us and in some cases is a necessary evil of our times. Previously it was thought that the solution to pollution was dilution; but it is clear that such a solution will not resolve man's problems in the (near) future.

The ocean system is large but has been seriously polluted locally; the great scientist Cousteau, and others, all of whom have passed a great part of their lives studying oceanographic topics, can remember the past, relate the problems of the past with those of the present, and project what will be the future problems in the marine environment. They fear that even the sea, despite its immense volume, can suffer serious pollution because humanity as a whole believes that the sea is a virtual limitless dumping ground, whereas in truth, the sea has its limits. In the same way, the atmosphere has its limits in its capacity to receive and dilute the gases and aerosols coming from human activities. Therefore, the solution for pollution problems can be found only in proper planning for the present and future and in technological capability and advances. The geochemist has strong ties with some classes of pollution, but interest in all types of pollution. He is a naturalist and should be one of the principal specialists involved in the environmental planning because of his scientific training, his intimacy with several natural processes (geologic, hydrologic, oceanographic, and atmospheric), his knowledge of the relations between these processes and man's needs, and his capability to communicate intelligently, precisely, and clearly with other specialists in pollution problems.

REFERENCES

Adams, R. S. Jr., 1972, Pesticides in the environment, in *Trace Substances in Environmental Health-V,* (Hemphill, ed.), University of Missouri, Columbia, pp. 81–93.

Angino, E. E., Magnuson, L. M., Waugh, T. C., Galle, O. K., and Bredfeldt, J., 1970, Arsenic in detergents – possible danger and pollution hazard, *Science,* **168**, 389–390.

Bowen, V. T., Olsen, J. S., Osterberg, C. L., and Ravera, J., 1971, Ecological interactions of marine radioactivity, in *Radioactivity in the Marine Environment,* National Academy of Sciences, Washington, D.C., pp. 200–222.

Bryson, R. A., and Kutzbach, J. G., 1968, *Air Pollution,* Assoc. American Geographers, Resource Paper No. 2, Washington, D. C., 42 pp.

Cannon, H. L. and Hopps, H. C., 1971, Environmental Geochemistry in Health and Disease, *Geol. Soc. Am., Mem.,* **123**, 230 pp.

Cannon, H. L., and Hopps, H. C., (eds.), 1972, *Geochemical Environment in Relation to Health and Disease,* Geological Society of America, Special Paper 140, 77 pp.

Chemical and Engineering News, 1968, Pollution – causes, costs, control, Special Report, 33–64.

Davidson, K. L., and Sell, J. L., 1972, Dieldrin and p,p'-DDT effects on eggshell characteristics in chickens, in *Trace Substances in Environmental Health-V,* (Hemphill, ed.), University of Missouri, Columbia, pp. 257–268.

Dubrovskaia, F. I., 1960, Pollution of the air in urban streets by vehicles operating on ethylated gasoline, *Hygiene Health,* **25**, 15–18 (in Russian with English summary).

Duursma, E. K., and Gross, M. G., 1971, Marine sediments and radioactivity, in *Radioactivity in the Marine Environment,* National Academy of Sciences, pp. 137–146.

Fairbridge, R. W. (ed.), 1966, *The Encyclopedia of Oceanography,* Reinhold, New York, 1021 pp.

Foster, R. F., Ophel, I. L., and Preston, A., 1971, Evaluation of human radiation exposure, in *Radioactivity in the Marine Environment,* National Academy of Sciences, pp. 240–260.

Fouts, J. R., 1970, Stimulation and inhibition of hepatic microsomal drug metabolizing enzymes with special reference to effects of environmental contaminants, *Toxicol. Appl. Pharmacol.,* **17,** 804–809.

Friend, M., and Trainer, D. O., 1970, Polychlorinated biphenyl, Interaction with duch hepatitis virus, *Science,* **170,** 1314–1316.

Gadhok, R. S., and Kenny, A. D., 1973, DDT and blood carbonic anhydrase in Japanese quail, in *Trace Substances in Environmental Health-VI,* (Hemphill, ed.), University of Missouri, Columbia, pp. 169–172.

Gustafson, C. G., 1970, PCB's – prevalent and persistent, *Environ. Sci. Techol.,* **4,** 814–819.

Halsted, J. A., 1968, Geophagia in man: its nature and nutritional effects, *Am. J. Clin. Nutrition,* **21,** 1384–1393.

Hanya, T., Ishiwatari, R., Katase, T., and Takada, T., 1973, Identification of a trace amount of organics in natural and polluted waters, in *Trace Substances in Environmental Health-VI,* (Hemphill, ed.), University of Missouri, Columbia, pp. 355–360.

Harvey, G. R., Steinhauer, W. G., and Teal, J. M., 1973, Polichlorobiphenyls in North Atlantic Ocean water, *Science,* **180,** 643–644.

Hemphill, D. D., (ed.), 1973 (1972, 1971, 1970, 1969, 1968), *Trace Substances in Environmental Health,* University of Missouri, Columbia, 399 pp. (559 pp., 456 pp., 391 pp., 318 pp., 236 pp.)

Hopps, H. C., and Cannon, H. L. (eds.), 1972, Geochemical environment in relation to health and disease, *Ann. N. Y. Acad. Sci.,* **199,** 352 pp.

Hornbrook, E. H. W., 1971, Effectiveness of geochemical and biogeochemical exploration methods in the Cobalt Area, Ontario, *Geochemical Exploration,* (Boyle, tech. ed.), Special Volume No. 11, Canadian Institute of Mining and Metallurgy, pp. 435–443.

Hosking, K. F. G., 1971, Problems associated with the application of geochemical methods of exploration in Cornwall, England, *Geochemical Exploration* (Boyle, tech. ed.), Special Volume No. 11, Canadian Institute Mining and Metallurgy, pp. 176–189.

Joseph, A. B., Gustafson, P. F., Russell, I. R., Schuert, E. A., Volchok, H. L., and Tamplin, A., 1971, Sources of radioactivity and their characteristics, in *Radioactivity in the Marine Environment,* National Academy of Sciences, pp. 6–41.

Kenny, A. D., Dacke, C. D., and Wagstaff, D. J., 1972, Effects of DDT on calcium metabolism in the Japanese quail, in *Trace Substances in Environmental Health-V,* (Hemphill, ed.), University of Missouri, Columbia, pp. 247–256.

Kobayashi, J., and Hagino, N., 1965, Strange osteomalacia by pollution from cadmium mining, Progress Report WP 00359, Okayama University, pp. 10–24.

Koczy, F. F. and Rosholt, J. N., 1962, Radioactivity in oceanography, In Nuclear Radiation in Geophysics, Springer-Verlag, pp. 18–46.

Koss, J., Carvajal, J. F., and Solano, J. A., 1969, Relacion suelo-planta por cobre y zinc en once suelos del area cafetalera de Costa Rica, *X Congr. Latinoamericano Quim., Res.,* pp. 124–125.

Krumholz, L. A., Goldberg, E. D., and Boroughs, H., 1957, in *The Effects of Radiation on Oceanography and Fisheries, U.S. Nat. Acad. of Sci. Nat. Res. Council Pub.,* **551,** 69–79.

K, C. E., 1970, Mercury in the environment, *Environ. Sci. Techol.*, 4, 890–892.

Kubo, A. S. and Rose, D. J., 1973, Disposal of nuclear wastes, *Science*, 182, 1205–1211.

Kuratsune, M., Yoshimura, T., Matsuzaka, J., and Yamagochi, A., 1972, Epidemiological study on yosho, a poisoning caused by ingestion of rice oil contaminated with a commercial brand of PCB's, *Environ. Health Perspectives*, 1, 119–128.

Lal, D. and Peters, B., 1967, Cosmic ray produced radioactivity on the earth, In Encyclopedia of Physics, Springer-Verlag, pp. 600–601.

Lincer, J. L., and Peakall, D. B., 1970, Metabolic effects of polychlorinated biphenyls in the American kestral, *Nature*, 228, 783–784.

Lowman, F. G., Rice, T. R., and Richards, F. A., 1971, Accumulation and redistribution of radionuclides by marine organisms, in *Radioactivity in the Marine Environment*, National Academy of Sciences, pp. 161–199.

Martell, J. M., 1973, Distribution of chlorinated hydrocarbons in Lake Anne, Virginia, M.S. Thesis, The George Washington University, Washington, D.C., 85 pp.

Martell, J. M., Rickert, D., and Siegel, F. R., 1973, Distribution of chlorinated hydrocarbons in Lake Anne, Virginia, Abstracts, Mid-Atlantic Section Meeting, American Chemical Society, Washington, D. C., B-45, p. 34.

Mauchline, J., and Templeton, W. L., 1964, Artificial and natural radioisotopes in the marine environment, Oceanography and Marine Biology; an Annual Review, Vol. 2, (Barnes, ed.), Allen and Unwin, London, pp. 229–279.

Millar, C. E., Turk, L. M. and Foth, H. D., 1966, *Fundamental of Soil Science*, Wiley, New York, 491 pp.

O'Brien, R. D., 1967, *Insecticides, Action and Metabolism*, Academic, New York, 322 pp.

Peakall, D. B., 1967, Pesticide-induced enzyme breakdown of steroids in birds, *Nature*, 216, 505–506.

Peakall, D. B., 1970, Pesticides and the reproduction of birds, *Sci. Am.*, 222, 72–78.

Peakall, D. B., and Lincer, J. L., 1970, Polychlorinated biphenyls, another long-life, widespread chemical in the environment, *Bio Sci.*, 80, 958–964.

Pettersen, S., 1966, Recent demographic trends and future meteorological services, *Bull. Am. Meteorol. Soc.*, 47, 950–963.

Rennie, D. A., 1966, Pollution and Our Environment, The Canadian Council of Resource Ministers, a National Conference, Background Paper C. 22-3, p. 1.

Sandifer, S. H., and Keil, J. E., 1972, Pesticide effects on cardiovascular risk factors, in *Trace Substances in Environmental Health-V*, (Hemphill, ed.), pp. 329–334.

Smith, J. C., and Halsted, J. A., 1970, Clay ingestion (Geophagia) as a source of zinc for rats, *J. Nutrition*, 100, 973–980.

Spencer, D. A., 1970, Trends in pesticide use, *Environ. Sci. Techol.*, 4, 480.

Swank, R. L., 1950, Multiple sclerosis – a correlation of its incidence with dietary fat, *Am. J. Med. Sci.*, 220, 421–430.

Swank, R. L., Lerstad, O., Strøm, A., and Backer, J., 1952, Multiple sclerosis in rural Norway, *New Engl. J. Med.*, 246, 721–728.

Templeton, W. L., Nakatani, R. E., and Held, E. E., 1971, Radiation effects, in *Radioactivity in the Marine Environment*, National Academy of Sciences, pp. 223–239.

Tully, J. P., 1966, Pollution and Our Environment, The Canadian Council of Resource Ministers, a National Conference, Background Paper D. 6, pp. 1–2.

Van Brackle, R. D., 1967, The farmer's stake in air pollution, *Crops Soils*, October.

Vos, J. G., 1972, Toxicology of PCBs for mammals and for birds, *Environ. Health Perspectives*, 1, 105–117.

Waldichuk, M., 1961, *Fisheries Res. Board Can., Biol. Stat. Circ.,* **59,** 24 pp.

Warren, H. V., 1954, Geology and health, *Sci. Monthly,* **78,** 339–345.

Warren, H. V., 1961, Some aspects of the relationship between health and geology, *Can. J. Publ. Health,* **52,** 157–164.

Warren, H. V., 1966, Some aspects of air pollution in perspective, *J. Coll. Gen. Practit.,* **11,** 135–142.

Warren, H. V., and Delavault, R. E., 1960, Observations on the biogeochemistry of lead in Canada, *Trans. Royal Soc. Canada, Sect. 3,* 54, 11–20.

Warren, H. V. and Delavault, R. E., 1967, A geologist looks at pollution – mineral variety, *Western Miner,* **40,** pp. 22–32.

Warren, H. V., and Delavault, R. E., 1968, Lead in vegetables, *Lancet,* June 8, p. 1252.

Warren, H. V., Delavault, R. E., and Cross, C. H., 1965, Some geological factors in medical geography, Soc. Mexicana de Geografia y Estadistica, Reunion Especial de la Comision de Geografia Medica, VI, 47–53.

Warren, H. V., Delavault, R. E., and Cross, C. H., 1966, Mineral contamination in soil and vegetation and its possible relation to public health, Pollution and Our Environment, The Canadian Council of Resource Ministers, a National Conference, Background Paper A3-3, 11 pp.

Warren, H. V., Delavault, R. E., and Cross, C. H., 1967, Possible correlations between geology and some disease patterns: *Ann. N. Y. Acad. Sci.,* **136,** 657–710.

12

GEOCHRONOLOGY:
RECENT ADVANCES,
INNOVATIONS, AND IDEAS

Of all the paleodata that can be extracted from geologic materials, absolute age (or time) may be the most fundamental. A knowledge of the absolute age of specimens from a study suite is as necessary to the geoscientist as is the microscope to the biologist, the telescope to the astronomer, and the accelerator and cloud chamber to the nuclear scientist. The measurement of absolute geologic time is based on the premise that a statistically random process will take place during a given time span and that the results of such a process are quantitatively determinable.

For example, in isotopic age determinations, the number of radionuclei in a sample is related to the number of those nuclei the sample contained when the decay process began by the equation:

$$N = N_0 e^{-\lambda t}$$

in which N = the total number of nuclei (being measured) present in a sample

N_0 = the total number of nuclei present when the process began at zero time

λ = the decay constant that defines the number of radioactive nuclei that will decay per unit time

If λ is known, the half-life of a radioactive element can be given by:

$$\text{Half-life} = \frac{0.693}{\lambda}$$

An example of the classical isotope determination method is the ^{14}C system. In 1946 Libby proposed that by measuring the ^{14}C content of a material and relating it to an equilibrium $^{14}C/^{12}C$ system, age determinations of

Table 12-1 Summary and Comparison of Some Current Methods Used for the Determination of Geologic Ages
(from Zeller, 1965) (1 My = 10^6 years; 1 Gy = 10^9 years)

Method	Kind of Sample	Parameter Measured	Range and Accuracy	Remarks
		Isotopic or Radioactive Yield Methods		
U–Pb	Zircon, xenotime, and other igneous and metamorphic minerals with original U, but with no original Pb	U and Pb by analyses and α-counts per unit of mass per unit of time	3 to 4×10^9 years; not as accurate as with isotopic analysis	There should be no leaching of U during life of mineral and no original Pb in mineral at time of crystallization
$^{238}U - ^{206}Pb$	Any igneous or metamorphic mineral or whole rock with original ^{238}U (especially uraninite, pitchblende, samarskite, cyrtolite, thorianite, zircon, and thorite)	Quantities of ^{238}U, ^{206}Pb, and ^{204}Pb by mass spectrometer analysis; ^{204}Pb gives indication of amount of original Pb, of all isotopes, that is present	Approximately 10×10^9 years; more accurate than the chemical U–Pb method	^{238}U is more abundant than ^{235}U so less chance of analytical error; ^{222}Rn half-life is long enough (3.82 days) to suggest that there may be error through Rn escape
$^{235}U - ^{207}Pb$	Any igneous or metamorphic mineral or whole rock with original ^{235}U (same minerals as above)	Quantities of ^{235}U, ^{207}Pb, and ^{204}Pb by mass spectrometer analysis; ^{204}Pb gives indication of amount of original Pb, of all isotopes, that is present	3 to 4×10^9 years; more accurate than the chemical U–Pb method	^{219}Rn in series has such short half-life (3.92 seconds) that escape is much less likely than in $^{238}U/^{206}Pb$ series
Pb–Th	Accessory minerals in igneous rocks	^{232}Th, ^{208}Pb by mass spectrometer analysis	Range accuracy 10%	^{208}Pb may be removed selectively from thorium minerals resulting in ages that appear too low
Pb–Pb	Galena or uranium minerals	^{206}Pb, ^{207}Pb, ^{208}Pb, ^{204}Pb by mass spectrometer analysis	3.5×10^9 years; 1% over 10^8 years (less than 10^8 may be calculated with accuracy of 10^6 years)	Has the advantage that it involves an isotopic ratio instead of absolute measurements

Method	Materials	Measured by	Range and error	Remarks
^{210}Pb	Fresh and marine water and glacial ice; recently deposited sediments	Count rate of ^{210}Bi derived, by radioactive decay of separated ^{210}Pb	100 years ± Sequence of analyses give fairly accurate estimate of rates of accumulation	Part of chain of $^{238}U-^{206}Pb$; ^{222}Rn, to atmosphere from rocks, yields ^{210}Pb; ^{210}Pb washed from atmosphere by rain; this disintegrates to $^{210}Bi-^{210}Po-^{206}Pb$; ^{210}Pb is rapidly removed from water by precipitation or adsorption on sedimentary particles; rate or removal from water and original amount of ^{210}Pb in sediments are necessary data for age determination
K–Ar	Micas and feldspars	^{40}K, ^{40}Ar; K measured by flame photometer, Ar by mass spectrometer	$1 \times 10^6 - 2.7 \times 10^9$ years; error in micas less than 10%	Feldspars lose Ar (up to 60%); metamorphism affects the results if rocks are heated
Rb–Sr	Mica lepidolite (first used) muscovite biotite; feldspar; whole rock; glauconite (in sediments)	Amount of ^{86}Sr; amount of ^{87}Sr; amount of ^{87}Rb; ratios determined by mass spectrometry	Not suited for ages below 50 $\times 10^6$ years due to long ^{87}Rb half-life; accuracy of method dependent on ^{87}Rb half-life which is not accurately determined	Decay constant of ^{87}Rb uncertain due to low energy of β-particles emitted; metamorphism may cause rehomogenation of Rb and Sr isotopes so method may be used to date metamorphic events
He	Magnetite crystal cores; nonmetamict zircon	Abundance of He; ^{238}U, ^{235}U, and ^{232}Th abundance by their alpha activity; U/Th ratio	Meteorites to approximately Miocene or Pliocene (modern minerals have a deficiency in radioactive materials may have excess He and give too old an age); accuracy in dispute; if U/Th ratio is unknown, error increases from ± 1% at 300 My to ± 3.5% at 1 Gy to ± 11% at 2.6 Gy	Method is not extensively used at present due to discovery of the ease of diffusion by He

Table 12-1 (Continued)

Method	Kind of Sample	Parameter Measured	Range and Accuracy	Remarks
Re–Os	Iron meteorites; Molybdenite and possibly gadolinite, zircon, thortveitite, chromite, tantalate and columbate minerals	^{187}Re/^{187}Os measurements made by β-count and mass spectrometry	Only very ancient minerals are datable; accuracy undetermined due to uncertain half-life of ^{187}Re	Rhenium is a very rare element thus few minerals are usable; the minimum amount of ^{187}Os that can be measured with 10% accuracy is 2×10^{-7} grams
^{14}C	Charcoal, wood, shell	β-Activity of carbon in sample by β-counting techniques	Range 35,000 years; accuracy good for last 20,000 years	Assumes a steady state for the production of ^{14}C in the atmosphere by cosmic radiation; assumes uniform distribution of ^{14}C throughout the Earth's atmosphere
^{3}H	Water samples, 2–6 liters	β-Activity of tritium	Last 30 years	Recent bomb testing has greatly affected the usefulness of this method
Io–Th	Deep sea sediment	Io/Th ratio and depth of sediment; Io and Th measured by alpha particle spectroscopy	Sediment must be younger than 350,000 years; method may also be used to measure rate of sedimentation	Assumptions of Io–Th method: ^{230}Th/^{232}Th ratio in water mass next to sediment constant over time considered. Th and Io precipitate from sea water in same proportion; analyzed material has no detrital substances of continental or volcanic origin with significant contributions of Io or Th; no migration of Th in sediment

Radiation Damage Methods

Metamict	Zircon (best); euxenite; sphene; uraninite; davidite	Amount of radiation damage (i.e. metamict condition) in mineral usually by x-ray diffraction; amount of radioactive material by α-count	Now considered a comparatively inaccurate method	Heating will cause crystal to become crystalline again, therefore method may be used to date metamorphic or igneous events
Pleochroic halo	Biotite with an inclusion of zircon, apatite, or monazite	Intensity of the coloration of the biotite; width of the coloration; α-activity of the inclusion	Late Precambrian to approximately Eocene; useful for relative age determination	Coloration is temperature sensitive and may reverse at a high level of radiation; assumes identical irradiation sensitivity of all biotites and that the coloration, once formed, does not fade through time
Thermoluminescence	Carbonate rocks, fluorite, quartz, feldspars	Luminescence of sample when heated to temperatures up to 400°C; glow-curve is measured by a recording microphotometer	1 Gy maximum; accuracy $\pm 15\%$	May be used to determine periods of metamorphism; has specific uses for paleoclimatological measurements
Radiation pitting	Mica; other minerals may also be usable	Particle tracks counted by optical microscopy	1 Gy; accuracy has not yet been tested fully	Very promising new method; may be sensitive to heating in metamorphic processes

Table 12-1 (Continued)

Method	Kind of Sample	Parameter Measured	Range and Accuracy	Remarks
		Chemical Methods		
Devitrification of glass	Glass, either natural obsidian or man made	Crystallinity of the glass especially on the surface	Useful for relative age determinations in past 5 My for obsidian	Rate of crystallization dependent upon chemical composition, temperature, and water present in proximity of the glass
Degradation of amino acids	Shell or bone fragments	Amino acids and peptide bonding	Accuracy best in last 1 My but amino acids are detectable in rocks over 500 My old	Involves a chemical process and rate is temperature dependent
Protein nitrogen	Shells of mature forms of mollusks	Organic nitrogen content of water-soluble residues (by micro Kjeldahl method).	Probably limited to Pleistocene and Late Tertiary	Only relative ages may be determined

Source. Reprinted by courtesy of the California Mineral Information Service.

carbon-bearing materials could be accurately made. The ^{14}C is produced in the upper atmosphere by the interaction of neutrons with ^{14}N; in nuclear notation, ^{14}N$(n, p^+)^{14}$C. If the assumption is made that the rate of production of ^{14}C is constant and that an equilibrium is established between ^{14}C and the other isotopes of C in the atmosphere, hydrosphere, and biosphere, it is possible to measure the ^{14}C content of a material that has developed in the equilibrium system and subsequently been removed from it and relate the departure from equilibrium to the half-life of ^{14}C (approximately 5730 years, but constantly being modified by researchers), with the final result being the determination of the absolute age of the sample. Radionuclide pollution from nuclear weapons testing has altered the recent equilibrium system significantly and is the cause of careful planning by present day experts for the benefit of future researchers.

The most successful of the absolute age dating techniques developed to date are those related to the study of radionuclides and/or their daughter products found in geological materials. Other methods continue to be evaluated and refined in an effort to better and extend geochronological capabilities. The most prominent and promising of these are those in which the damage of a crystal lattice can be measured; when combined with a knowledge of the radioactive element content of a crystal, the measured radiation damage can be used to determine absolute age (e.g., fission track, recoil nucleus, thermoluminescence, and electron spin resonance). A theoretical consideration of magnetization of sediments and sedimentary rocks has led to laboratory experimentation giving results indicating that the viscous magnetization of samples increases in proportion to the logarithm of time and thus might be a measureable quantity for dating sediments younger than the last inversion of the Earth's geomagnetic field (700,000 years BP). Also, a timetable for reversals of the Earth's magnetic field has been accurately documented by independent techniques for the last 3.5 million years and extrapolated to 11 million years, possibly offering another method for age determinations.

To review the well-established techniques would be to review reviews that have been excellently done or cite well-prepared texts (Hamilton, 1965; Faul 1966; Wetherill and Tilton, 1967; Pasteels, 1968; U.S. Geological Survey, Data of Geochemistry, Section HH, Geochronology, in preparation; Zeller, 1965) that are readily available to most readers. These techniques have been summarized in general form by Zeller (1965) and are presented in Table 12-1. The presentation here will be of the less-well-known and documented methods that may not yet have been developed to the point where accuracy and/or precision are comparable to other well-established techniques or that have limited time-span capability. These may not now be considered reliable (except for fission track methods) by many evaluators, but do show promise for the future as basic theory catches up with empirical applications or as instrumentation advances are made.

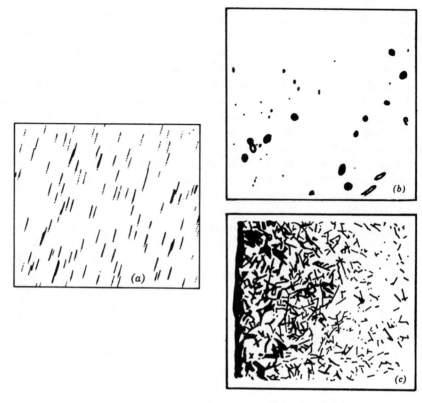

Figure 12-1. (a) Track registration in muscovite; etched tracks of 40-MeV Ar ions that entered the mica at a 15° angle. (b) Natural tracks in volcanic glass (macusanite) from Peru (etched for 10 seconds in 48% HF). (c) Natural tracks in zircon from Australia (etched for 1 minute in boiling phosphoric acid at 500°C); the increase in track density as the free surface of the zircon is approached indicates that a nonuniform uranium distribution existed in the past. Sketches are representations of photomicrographs. (Reproduced from Fleischer, Price, and Walker, 1965, *Science.* Copyright © 1965 by the American Association for the Advancement of Science.)

One of the rather exciting developments in the quest for new age-determination methods during the past decade is related to the spontaneous fission of ^{238}U. The fission products from this radioactive element are energetic, heavy, charged particles that create narrow, well-defined damage tracks (up to 10 μm in length) in crystalline solids as they tear through a crystal lattice. By correct preferential chemical etching techniques that have now been developed for several minerals, these fission tracks can be made visible (to the petrographic microscope) in many naturally occurring (and manmade) materials and provide the geologist with a tool for investigating the age of study materials. The

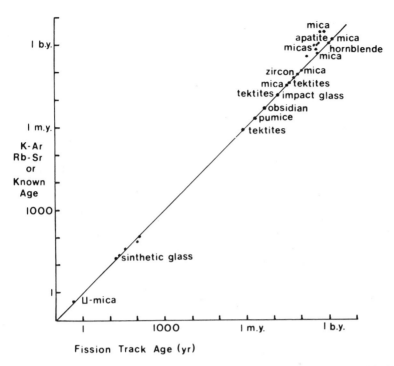

Figure 12-2. Comparison of ages found by fission-track dating with those established by other means. The samples less than 200 years old are man-made; the older samples are geological. (Reproduced from Fleischer, Price, and Walker 1965, *Science.* Copyright © 1965 by the American Association for the Advancement of Science.)

appearance of fission tracks in selected minerals is given graphically in Figure 12-1. Fleisher, Price, and Walker (1965) who present an excellent review of "Tracks of Charged Particles in Solids," cite the basic assumption that fission tracks may be stored over extended time spans (and that one fissioning of ^{238}U occurs about each 2,300,000 decays), show that a concentration of as little as 1 ppm of ^{238}U (a common natural concentration) is sufficient to make natural materials greater than 500,000 years old dateable, and give a graph with a comparison of ages found by fission-track dating with those ages established by other means (K–Ar, Rb–Sr, ^{14}C, or known age) (see Figure 12-2). Samples less than 200 years old to more than 1000 million years old have been successfully dated (see Figure 12-2) and meteorite ages confirmed as accurate with fission-track dating. Applied fission track dating of volcanic glass from Bed I of the Olduvai Gorge sequence, Tanganyika, site of the hominid remains Zinjanthropus and Homo habilis, resulted in an absolute age value of 2.03 ± 0.28 million years; this age is in agreement with the 1.8 million years obtained on an anorthoclase from Bed I using the K–Ar method (Fleisher et al., 1965).

The mathematical relation by which an age A can be calculated from fission-track experimental data is given by

$$\frac{\rho_s}{\rho_i} = [\exp(\lambda_D A) - 1] (\lambda_F / \lambda_D f)$$

in which ρ_s = the density of the natural spontaneous fission tracks (which are counted visually using the petrographic microscope)

ρ_i = the new density of fission tracks induced by radiation (after annealing of the naturally occurring tracks)

f = the fraction of the total ^{238}U fissioned in the reactor irradiation that was used to measure the U content of the sample

λ_F = the spontaneous fission constant of ^{238}U

λ_D = the total decay constant of ^{238}U

While investigating the potential of the fission-track technique on various types of carbonate minerals, Schroeder (personal communication) found that at the base of scleractinian septae, there were anomalously high concentrations of uranium, and he is now investigating the possibility of using these zones of uranium concentration in a manner similar to growth rings in trees to determine absolute ages of corals and their growth rates. Attempts will be made to correlate ages thus determined with ^{90}Sr activity of corals that are known to concentrate up to about 1% Sr in their carbonate skeletons. In a similar approach, visual traces attributable to older events have been studied in reef-forming corals in an attempt to date growth bands and determine growth rates; Knutson et al. (1972) analyzed autoradiographs and x-radiographs made from vertical sections through the center of reef corals from Eniwetok Island (site of much nuclear weapons testing). Radioactive bands caused by the inclusion of ^{90}Sr showed up in the coral structure and are related to specific series of nuclear testing, thus permitting calculations of long-term (since 1955) growth rates. The data presented in the report by Knutson et al. (1972) indicate that the cyclic variations in radial density revealed by x-radiography are annual. Bender (1973) studied He/U dating of aragonitic fossils for Cenozoic geochronology in testing the reliability of such ages on fossil corals. He found that for He/U ages to be valid, corals must remain closed systems for U and its radioactive daughters and must retain He. Thus He/U ages of corals are reliable provided that there is no anomalous disequilibrium of ^{238}U daughters, a correction is made for initial He, a correction is made for alpha stopping loss when necessary, and the samples are free of U- and He-bearing contaminants.

An outgrowth of the fission-track-dating method was the recognition by Huang and Walker (1967) of another type of fossil nuclear track in mica that resulted from the recoil nucleus accompanying alpha-particle decay of contained trace quantities of U and Th. Measurement of the fossil alpha-recoil nucleus track density of a mica sample with the phase contrast microscope after chemical etching, plus U and Th analyses, can supply the necessary data for absolute dating, analogous to the fission-track technique. The alpha-recoil nucleus method may be advantageous, however, because there is a several thousandfold increase in sensitivity over the fission-track approach. No dates have been published yet based on the alpha-recoil nucleus age-determination method. The mathematical relation for an absolute-age calculation is given by:

$$\rho_\alpha = N_0 \, C_U \, \lambda_\alpha \, (U) \, TR_\alpha + N_0 \, C_{Th} \, \lambda_\alpha \, (Th) \, TR_\alpha$$

in which ρ_α = the density of alpha-recoil nucleus tracks as counted using the phase contrast microscope after proper chemical etching of a sample

N_0 = the number of atoms per cubic centimeter

C_U = the U concentration

C_{Th} = the Th concentration

$\lambda_\alpha(U)$ = the decay constant related to recoils of alpha from U

$\lambda_\alpha(Th)$ = the decay constant related to recoils of alpha from Th

R_α = the etchable range of alpha-recoil nuclei (about 125Å in mica)

T = the time during which the recoil process has acted—the AGE of a sample.

In an evaluation of track registration efficiency from alpha-recoil in mica, Katcoff (1969) found that recoil reactions coming from the decay of radioactive nuclei occurring in nature have energies of between 70 and 169 KeV. He showed that the tracks from recoil reactions, seen as points after acid attack, have a registration efficiency approaching 80% in mica. When the recoil reaction energy falls to about 40 KeV, the efficiency falls to about 50%. Nevertheless, because the decay of each impurity atom of U or Th in mica is followed by a cascade of six or eight alpha particles, Katcoff believes that the total track registration efficiency should be almost 100%. Thus, the dating method based on alpha-recoil nuclei in mica should be as good as fission-track dating but more sensitive by three orders of magnitude.

The fission-track and, potentially, the alpha-recoil nuclei techniques are especially attractive to laboratories or areas where an atomic energy commission has facilities available for reactor irradiation, but where funds are not available for the purchase of costly equipment used in the standard isotope techniques. To date, more than 600 scientific papers have been published on fission-track studies.

Along these same lines, Zimmerman (1971) studied small mineral grains (10 to 50 μm in diameter) of high U concentrations (100 to 3000 ppm) that were found in polished sections of eight potsherds from Cyprus, England, Greece, and Mexico, by induced fission-track mapping. He identified most of these grains as being either zircons or apatites using a microprobe attachment to a scanning electron microscope and proposed that they could be separated from the matrix material after crushing using heavy liquids and magnetic techniques; he concluded that such grains can be used for thermoluminescence dating (treated farther on in text) and alpha-recoil dating.

A method of geologic age determination using electron spin resonance signals of crystals or mineral powders to measure trapped charges has been developed and has undergone preliminary testing by Zeller, Levy, and Mattern (1967) and Zeller (1968). The method is based on the premise that the trapped charge accumulation in crystals is a function of the total radiation dose (from U, Th, and K impurities and thus a function of the decay constants of these elements) to which the crystal has been subjected. According to this early work, the trapped charge population is a linear function of dose if less than one-fifth of the total number of traps is populated. For age determination, the trapped charge population must be a real measure of the total radiation dose accumulated in specific traps. Electron spin resonance can be a sensitive detector that directly measures the charge contained in certain traps. Only unpaired electrons in traps or otherwise localized in a crystal lattice can be detected directly. The measurement is made by placing the sample in a magnetic field that causes the unpaired electrons to precess as the spin at a precession rate dependent on the strength of the field. There is a simultaneous application of an oscillating electric field perpendicular to the direct current magnetic field. When the precession frequency of the electrons equals the microwave field frequency, energy will be absorbed, resonance will be achieved, and the electrons will flip their spins. At any given temperature, the quantity of energy absorbed is directly proportional to the number of unpaired spins and therefore the number of electrons trapped in a specific defect can be determined directly. It is known that recrystallization, heat, pressure, mechanical shock, and other geological processes can affect both the number of traps and their charge population; in addition, the presence of ESR-active impurity atoms such as Mn and the rare earths may mask any effect from geologically significant defects. In a test of the theory and instrumentation of the electron spin resonance technique, Zeller and his co-workers have dated apatite crystals from Durango, Mexico, at 2 to 4 million years old; the geology of the Durango region puts an upper limit on the age of the apatite crystals at about 30 million years. The age of 2 to 4 million years, then, probably indicates the time elapsed since the traps were last empty. The ESR method cannot distinguish between the time of formation of a crystal and the occurrence of a possible subsequent thermal event high enough to empty and perhaps anneal the traps.

Such a heating event has undoubtedly affected the apatite crystals used in the investigation. No research has yet been done on authigenic crystals formed in the marine environment, but the ESR geologic age determination technique shows promise and is being studied further by Zeller at his University of Kansas laboratory.

As indicated above, certain minerals possess the ability to store energy in the form of electron energy. An electron, displaced by some external form of energy such as natural alpha radiation, can be moved from its thermodynamically most stable lattice position (that with the minimum free energy) and become trapped by some type of imperfection in the crystal lattice structure. Any excess energy associated with the electron becomes temporarily frozen in the trap. When released by the application of energy in the form of heat, the excess energy is dissipated in the form of heat and light. The light thus produced is termed thermoluminescence. Commercial equipment is available for measuring the thermoluminescence glow both as to temperature of occurrence and relative intensity. The possibility of using the thermoluminescence characteristics of sedimentary carbonate rocks as the basis for their absolute age determinations was proposed by Zeller and empirically applied with positive results (Zeller, Wray, and Daniels, 1957). Subsequently, error factors have been found and evaluated, and in 1968 Zeller reviewed the strong and weak points of the technique. Aitken (1968) has summarized the thermoluminescence age determination as it has been rather successfully applied to archaeological specimens such as pottery sherds.

In unaltered minerals and/or rocks, a fundamental relationship exists between the accumulated (or accumulatable) thermoluminescence and the age of the sample being analyzed. The premise is made that the thermoluminescence of rocks and minerals can be used as natural radiation dosimeter measurements in a manner analogous to the use certain alkali halides that serve as artificial radiation dosimeters. Thus, if the total accumulated radiation dose equivalent can be measured and if the rate at which the radiation is applied is known, the length of exposure of the material being studied can be calculated by dividing the total accumulation by the rate of accumulation.

Variations on the age determination using thermoluminescence have been applied. In the first, the natural thermoluminescence of a sample is used as an indicator of the accumulated radiation dose, with the basic assumption being that the number of charged trapping centers is essentially constant throughout the geological history of the sample; this assumption is now known to be incorrect because new traps continually develop from natural radioactive impurities in the crystal. During the age-determination procedure, a sample is drained of its thermoluminescence by heating, after which thermoluminescence is induced by ionizing radiation until the amount induced is equal to the natural accumulated thermoluminescence, and from this, the relative equivalent radiation dose is determined; the assumption is made here that no trap destruction takes place if the sample is properly drained. After determining the natural radioactivity (alpha activity) of the sample, the relative equivalent radiation dose accumulated divided by the natural radioactivity results in a factor that is considered to be a function of the geologic age. During the development stages of this technique, it was recognized that any significant loss of charges from the high-temperature traps (those used to measure accumulated natural radiation dose) would introduce substantial errors in the calculated ages. However, it was assumed that the presence of the middle-temperature traps in the natural samples provides some assurance that no appreciable drainage from the high-temperature traps took place, expecially if the thermoluminescence curve shows the presence of a large natural middle-temperature peak (or a large number of middle-temperature traps).

The use of only artificially-induced thermoluminescence for age determinations is based on the hypothesis that new charge-trapping centers are constantly developed by radiation damage in crystals caused by alpha particles and associated recoil nuclei from natural radioactive impurities. The middle-temperature thermoluminescence glow curve peak representative of middle-temperature energy trapping centers is used and it is assumed that such middle-temperature trapping centers are essentially absent from freshly formed limestones and dolomites. The artificially induced saturation thermoluminescence must be critically determined and the radiation dose equivalent for attaining the saturation is divided by the natural radioactivity of the sample with the resultant figure being considered as a function of geologic age. Both in this technique and the preceding one discussed above, corrections must be made to compensate for many of the potential errors involved not only in the actual measurements, but in the sample collection, treatment, and preparation previous to analysis (Zeller, 1968).

Vaz (1969) has investigated the metamictization–thermoluminescence relations in zircon as a potential age-dating method. He studied the effects of natural radiation damage on thermoluminescence artificially induced by gamma radiation in gem-quality zircons of known age from Ceylon. This was done by analyzing the thermoluminescence response to irradiation of polished pieces of the gemstones with a range of radioactive impurity content (0.004 to 0.6% equivalent U), before and after heating of the samples at 950°C for about 20 hours in an inert nitrogen atmosphere. Vaz demonstrated that natural radiation damage reduces the thermoluminescence output of the material, but that thermal annealing of the samples causes an enhancement of the light output. The enhancement is alpha-dose dependent and is greatest in the samples with relatively low total alpha doses. Age determinations carried out on nine highly metamict zircon concentrates from different localities, using the thermoluminescence–alpha dose relation established from the Ceylon zircons, showed relatively good agreement with their lead-alpha ages.

Calcium phosphate is thermoluminescent, and with existing commercial equipment, the thermoluminescence characteristics of materials as small as individual conodonts can be studied (Siegel and Vaz, 1968). Although charges in trapping centers of calcium phosphate may not be stable for long periods of geologic time, it is probable that they will be stable for 1 to 2 million years. If this is so, thermoluminescence research on authigenic calcium phosphate from Recent marine sediments and sediment cores might reveal relations that can provide a basis for age determinations in such sediments in the time gap between ^{14}C and/or the members of the U and Th natural radioactive series and the K-Ar or paleomagnetic techniques.

The previously cited age determination methods are based on radiation damage in minerals, but work continues on potentially useful isotopes whose half-lives are known to a good degree of accuracy. It was proposed that because of their similar chemical properties, half-lives, and estimated very short residence times (τ) in the oceans (i.e., the average time that a unit of an element remains in seawater before removal by some precipitation or deposition process), the cosmic-ray-produced ^{10}Be (τ = 150 years, half-life = 2.5 $\times$ 10^6 years) and ^{26}Al (τ = 100 years, half-life = 0.74 $\times$ 10^6 years) might be a radionuclide pair that would serve for the determination of absolute ages of marine sediments in the interval between the limits of the ^{14}C technique and the K-Ar technique, and beyond, and might be useful therefore as an excellent check on the age determinations made using the uranium and thorium series radionuclides. However, present indications are that the contribution of ^{26}Al from cosmic dust is significantly greater than the amount of this radionuclide formed by spallation in the atmosphere of the Earth (the *only* known source of the radionuclide ^{10}Be). Because of this observation, it is unlikely that the ratio of $^{10}Be/^{26}Al$ can be used to measure the ages of marine sediments until a confident numerical relation between the ^{26}Al contribution from cosmic dust to the ^{26}Al contribution from spallation is satisfactorily determined. Any age determination from measurements of this radionuclide pair (or other radionuclides produced by cosmic-ray interaction with the atmosphere) is necessarily based on several assumptions; first that the cosmic-ray flux has not changed significantly during the span of time measured; second, that the rate of flux of cosmic dust has remained nearly constant during the same time interval; and third, that the two isotopes have been rapidly incorporated into the sediments being studied so that no chemical process could have effected a fractionation or depletion of one at the expense of the other. With advanced detection capability via instrumentation development, the ^{10}Be can be used alone in the same manner as ^{14}C for the dating of marine sediments, especially if the above-cited relation between the two ^{26}Al contributors cannot be determined so that the nuclide pair dating cannot be achieved.

Also in the marine environment (and lacustrine environments), Koide, Bruland,

and Goldberg (1973) extended an earlier investigation on ^{210}Pb geochronology (Koide, Soutar, and Goldberg 1972) and obtained average rates of sediment accumulation from decreases in ^{210}Pb activity as a function of depth in cores, although surface layers may not have been recovered by the coring operation. The ^{210}Pb derives primarily from the decay of ^{226}Ra in the water column (^{226}Ra develops during the ^{238}U series decay sequence). In the marine coastal environment of Santa Barbara, California, ^{210}Pb concentrations showed an exponential decrease with depth in varved sediments; using a 22.3 year half-life for ^{210}Pb and counting varves, a sedimentation rate of 0.4 cm/year was determined. An unvarved marine coastal sediment from Baja, California (1000 miles away from Santa Barbara), also showed an exponential decrease of ^{210}Pb concentrations with depth, and by using the 22.3 year half-life, dates of deposition could be assigned to various sedimentary levels and a rate of sedimentation of 0.27 cm/year was established. This was in agreement with the observation that at Baja there was less rainfall and a smaller influx of lithogenous material than at Santa Barbara. Similar relations were found for lake sediments, but it was deemed essential that the annual fluxes of ^{210}Pb in rivers and rains be known for proper application of the dating technique. In some cases the authors found that the uppermost few centimeters of marine (or lake) sediments had slightly less ^{210}Pb than the layers immediately below, and this was attributed to the possible mobility of dissolved lead in the surface layers of the coastal sediments with a migration of the isotope in the sediment column as follows: deposition of ^{210}Pb under oxidizing conditions and partitioning between the solid phases and interstitial waters; if the sediment becomes reducing slightly below the surface and H_2S is generated, the lead will precipitate, perhaps as PbS, with a decrease in the activity of the dissolved Pb species in the interstitial waters of the reducing zone, thus establishing a concentration gradient with the waters in the oxidizing environment of the uppermost sediment. If, then, there is a migration downward into the reducing zone or upward into the water, this would result in lower ^{210}Pb concentrations in the upper layers of the sediments. This ^{210}Pb dating technique appears valid for about 100 years.

Koide, Bruland, and Goldberg (1973) also investigated the ^{232}Th natural radioactive series and found that there is an excess (in the surface sediments studied) of ^{228}Th about one order of magnitude over and above that supported by the parents ^{232}Th or ^{228}Ra in such deposits; they also found that the amounts of ^{232}Th supported and ^{228}Ra supported ^{228}Th are similar indicating that both Ra and Th isotopes are removed from solution in coastal zones quite rapidly after introduction or formation. They found that the excess ^{228}Th can be used for dating purposes over time periods of about a decade and for permissive evidence that the uppermost levels of a sediment deposit were obtained during coring operations. The authors' preliminary results from other testing in lakes indicate the possibility that ^{228}Th/^{232}Th geochronologies are held by their sediments. Thus, this technique and that based on ^{210}Pb offer interesting possibilities for dating short-term sequences.

Still considering isotope techniques, Boudin and Deutsch (1970) described a new dating method based on the relation [176]Lu/[176]Hf that requires a 10-gram sample of minerals containing rare earths. This method was made possible by the developments in analytical techniques and in the isotope dilution technique so that precise determinations of Lu and Hf could be made. Equally important to its acceptable application was the agreement between the geologic determination of the half-life of [176]Lu ($3.3 \pm 0.5 \times 10^{10}$ years) and the recent physical measurements of the half-life ($3.6 \pm 0.1 \times 10^{10}$ years, 3.2×10^{10} years, $3.5 \pm 0.14 \times 10^{10}$ years, $3.68 \pm 0.06 \times 10^{10}$ years, and $3.56 \pm 0.05 \times 10^{10}$ years). The [176]Lu/[176]Hf method can be used when the sample is composed of 10 grams of minerals that contain 100 ppm of Lu and has an age of at least 1×10^{10} years.

The possibility of determining the absolute age of sedimentary rock from the measurement of nonradioactivity or isotope-related functions was proposed by Trukhin (1967) who worked with the amount of natural viscous magnetization measured in the rocks. Theoretically, the amount of viscous magnetization increases proportionally to a power of the logarithm of time; some studies indicate that there is an increase proportional to the first power of the logarithm of time, but others indicate that the increase is proportional to the square of the logarithm of time. Obviously, the most important premise that must be made is that the natural viscous magnetization will increase uniformly throughout the time from which the rock formed to the time of measurement, at which point it is still not saturated with the magnetization. Therefore, rocks formed after the last inversion (reversal) of the geomagnetic field (the Matuyama-Brunhes reversal that occurred 700,000 years ago) should be datable from the amount of natural viscous magnetization they contain. Thus far, no date has been obtained using this method. The major difficulty seems to be that no single technique exists for the separation of natural viscous magnetization from other forms of magnetization in a rock. Russian scientists have been working with laboratory models and have had some success in determining the natural viscous magnetization by difference; that is, they have determined total magnetization and nonviscous magnetization and by difference have arrived at a figure for the natural viscous magnetization.

A geomagnetic reversal time scale for the past 3.5 million years has been defined by Cox, Dalrymple, and Doell (1967) on the basis of paleomagnetic data and radiogenic age determinations on young lava flows from many areas of the world. Opdyke et al. (1966) presented confirming data after studying the paleomagnetism of deep-sea sediment cores from Antarctica. From such information, sea floor spreading rates have been calculated and from a premise that sea floor spreading rates have remained essentially constant in given oceanic areas, the reversal time scale has been extrapolated for the last 11 million years. Calculations of the "suggestive" absolute ages of marine samples from paleomagnetic polarity data and proposed relative constancy of spreading rates are consistent with the breakup of a supercontinent and the formation of the present oceanic basins during the past 200 million years. Vine (1969) reviews the sea floor spreading concept initially proposed by Hess (1962) and presents a strong suite of evidence to support the continental drift theory. He states that ages can be assigned to anomalies in magnetic profiles and hence to the underlying oceanic crust and shows that these are consistent with existing knowledge of the age of ocean floors from sediment thicknesses and age determinations of sediment cores and dredged rocks. Similarly, by using paleomagnetic data, Kennett, Watkins, and Vella (1971) used paleomagnetic chronology to set the Pliocene–Pleistocene boundary in New Zealand at 1.79×10^6 years ago, that is, at the base of the Gisla geomagnetic polarity event.

Finally, the fact that organic geochemistry is being evaluated for dating must also be recorded. Bada, Luyendyk, and Maynard (1970) found that isoleucine undergoes a slow racemization reaction with increasing depth in a marine sediment core 5 meters long from a depth of 4014 meters at $30°15.7'$N, $43°18.9'$W. This racemization of isoleucine was used to calculate an age of 1.23×10^6 years for the lower part of the core. The racemization reaction may be written as follows:

$$\text{L-isoleucine} \xrightleftharpoons[k_{rac} \text{ (alloiso)}]{k_{rac} \text{ (iso)}} \text{D-alloisoleucine}$$

There is an interconversion of these chemical compounds and the half-life for the reaction is 4.4×10^{10} years when the environmental pH = 7.6 and the temperature is $0°C$. This testing of the technique was applied in an area where the sedimentation rate is 1 cm/1000 years. There were 0.1 to 2.0 mg of the amino acids for each gram of dry sediment, and of this, 2 to 4% was isoleucine. In a subsequent study of 13 well-dated cores, Wehmiller and Hare (1971) found that isoleucine, one of several amino acids isolated from the > 74 μm fraction (so-called foram fraction selected as the study material because the mineral test is most resistant to contamination from outside sources) shows a progressive increase in the degree of racemization with the age of the sediment. However, the amino acids in sediments show an initial rate of racemization almost an order of magnitude faster than the rate observed for amino acids at a comparable pH and temperature. The observed kinetics depend on a variety of diagenetic processes, but it appears that the ratio of alloisoleucine to isoleucine is a reliable indicator of age for samples less than 400,000 years old; for older samples, the results are more ambiguous. Thus, because of the realization of the nonlinear kinetics of the racemization process and an initial rate of racemization that did not agree with the report by Bada, Luyendyk, and Maynard (1970) based on a single core, it would seem that the dating reported by Bada, Luyendyk, and Maynard is subject to serious question. The empirical approach of comparing sediments of known ages in a core to the alloisoleucine/isoleucine ratio (Wehmiller and Hare, 1971), would appear at this time to be more reliable, although a 400,000-year limit to the confident applicability of the technique was established.

The refinement and extension of age-dating capability is a vibrant field of research in geochemistry and in many cases serves not only to establish a date for a geologic event, but also can often help the geochemist to understand or clear up uncertain aspects of the chemical behavior of the elements. One major gap exists in the geochemist's capability in dating rocks and minerals and that is a technique for dating material greater than 350,000 years old and less than 1 million years old, especially in sedimentary sequences. The potential for establishing this capability is now available and undoubtedly the gap will be bridged in the near future.

REFERENCES

Aitken, M. J., 1968, Thermoluminescence dating in archaeology: Introductory review, in *Thermoluminescence of Geological Materials* (McDougall, ed.), Academic, London, pp. 369–378.

Amin, B. S., Kharkar, D. P., and Lal, D., 1966, Cosmogenic Be-10 and Al-26 in marine sediments, *Deep-Sea Res.,* **13,** 805–824.

Arrhenius, G., 1968, Deep-sea sedimentation; A critical view of U.S. work, Int'l Union of Geodesy and Geophysics Quadrennial Report, Section 2, Radiochronometry, pp. 610–615.

Bada, J. L., Luyendyk, B. P., and Maynard, J. B., 1970, Marine sediments: Dating by the racemization of amino acids, *Science,* **170,** 730–732.

Bender, M. L., 1973, Helium-uranium dating of corals, *Geochim. Cosmochim. Acta,* **37,** 1229–1247.

Boudin, A., and Deutsch, S., 1970, Geochronology: Recent development in the Lutetium-176/Hafnium-176 dating method, *Science,* **168,** 1219–1220.

Cox, A., Dalrymple, G. B., and Doell, R. R., 1967, Reversals of the earth's magnetic field, *Sci. Am.,* **216,** 44–54.

Faul, H., 1954, *Nuclear Geology,* Wiley, New York, 357 pp.

Faul, H., 1966, *Ages of Rocks, Planets, and Stars,* McGraw-Hill, New York, 109 pp.

Fleisher, R. L., Price, P. B., and Walker, R. M., 1965, Tracks of charged particles in solids, *Science,* **149,** 383–393.

Fleisher, R. L., Price, R. L., Walker, R. M., and Leakey L. S. B., 1965, Fission-track dating of Bed I, Olduvai George, *Science,* **148,** 72–74.

Hamilton, E. I., 1965, *Applied Geochronology,* Academic, London, 267 pp.

Harper, C. T., (ed.), 1973, Geochronology: Radiometric Dating of Rocks and Minerals, Dowden, Hutchinson & Ross, Inc., Stroudsburg, Penna., 469 pp.

Hess, H. H., 1962, History of the ocean basins, in *Petrologic Studies,* A Volume to honor A. F. Buddington, Geological Society of America, pp. 599–620.

Huang, W. H. and Walker, R. M., 1967, Fossil-alpha recoil tracks: A new method of age determination, *Science,* 155, 1103–1106.

Katcoff, S., 1969, Alpha-recoil tracks in mica: registration efficiency, *Science,* 166, 382–384.

Kaufhold, J., and Herr, W., 1968, Factors influencing dating CaF_2 by thermoluminescence, in *Thermoluminescence of Geological Materials* (McDougall, ed.), Academic, London, pp. 153–167.

Kennett, J. P., Watkins, N. D., and Vella, P., 1971, Paleomagnetic chronology of Pliocene-Early Pleistocene climates and the Plio-Pleistocene boundary in New Zealand, *Science,* 171, 276–279.

Knutson, D. W., Buddemeier, R. W., and Smith, S. W., 1972, Coral chronometers: seasonal growth bands in reef corals, *Science,* 177, 270–272.

Koide, M., Bruland, K. W., and Goldberg, E., 1973, Th-228/Th-232 and Pb-210 geochronologies in marine and lake sediments, *Geochim. Cosmoschim. Acta,* 37, 1171–1187.

Koide, M., Soutar, A., and Goldberg, E., 1972, Marine geochronology with Pb-210, *Earth Planet. Sci. Letters,* 14, 442–446.

Opdyke, N. D., Glass, B., Hays, J. D. and Foster, J., 1966, Paleomagnetic study of Antarctic deep-sea cores, *Science,* 154, 349–357.

Pasteels, P., 1968, A comparison of methods in geochronology, *Earth Sci. Rev.,* 4, 5–38.

Siegel, F. R., and Vaz, J. E., 1968, Thermoluminescence of conodonts, Abstracts, Annual Meeting Geological Society of America, Mexico City, p. 278.

Tilton, G. R., and Hart, S. R., 1963, Geochronology, *Science,* 140, 357–366.

Trukhin, V. I., 1967, The possibility of determining the absolute age of rocks from the amount of viscous magnetization, *Izv., Acad. Sci., USSR, Physics of the Solid Earth, Engl. Trans.,* 2, 138–141.

U.S. Geological Survey, in preparation, Geochronology, in *Data of Geochemistry* (Fleischer, tech. ed.), Section HH, *U.S. Geol. Surv. Profess. Papers.* 440.

Vaz, J. E., 1969, *Metamictization-Thermoluminescence Relations in Zircon,* Ph.D. dissertation, The George Washington University, Washington, D.C., 54 pp.

Vaz, J. E., and Senftle, F. E., 1971a, Thermoluminescence study of the natural radiation damage in zircon, *J. Geophys. Res.,* 76, 2038–2050.

Vaz, J. E., and Senftle, F. E., 1971b, Geologic age dating of zircon using thermoluminescence, *Mod. Geol.,* 2, 239–245.

Vine, F. J., 1969, Sea-floor spreading – new evidence, *J. Geol. Educ.,* XVII, 6–16.

Wehmiller, J., and Hare, P., 1971, Racemization of amino acids in marine sediments, *Science,* 173, 907–911.

Wetherill, G. W., and Tilton, G. R., 1967, Geochronology, in *Researches in Geochemistry* (Abelson, ed.), Vol. II, Wiley, New York, pp. 1–28.

Zeller, E. J., 1965, Modern methods for measurement of geologic time, *Calif. Div. Mines Geol.* 18, 1–15.

Zeller, E. J., 1968, Geologic age determination by thermoluminescence in *Thermoluminescence of Geological Materials* (McDougall, ed.), Academic, London, pp. 311–326.

Zeller, E. J., Levy, P. W. and Mattern, P. L., 1967, Geologic dating by electron spin resonance, in *Radioactive Dating and Methods of Low Level Counting,* International Atomic Energy Agency, Vienna, pp. 531–540.

Zeller E. J., Wray, J. L., and Daniels, F., 1957, Factors in age determination of carbonate sediments by thermoluminescence, *Bull. Am. Assoc. Petrol. Geologists,* **41,** 121–129.

Zimmerman, D. W., 1971, Uranium distribution in archeologic ceramics: dating or radiactive inclusions, *Science,* **174,** 818–819.

AUTHOR INDEX

SUBJECT INDEX